Environmental Stress
Individual Human Adaptations

ACADEMIC PRESS RAPID MANUSCRIPT REPRODUCTION

Proceedings of a symposium held
at the University of California,
Santa Barbara, August 31–September 3, 1977

Environmental Stress
Individual Human Adaptations

Edited by

Lawrence J. Folinsbee
Jeames A. Wagner
Julian F. Borgia
Barbara L. Drinkwater
Jeffrey A. Gliner
John F. Bedi

Institute of Environmental Stress
University of California, Santa Barbara

ACADEMIC PRESS   NEW YORK   SAN FRANCISCO   LONDON   1978
A Subsidiary of Harcourt Brace Jovanovich, Publishers

ACADEMIC PRESS, INC.
111 Fifth Avenue, New York, New York 10003

United Kingdom Edition published by
ACADEMIC PRESS, INC. (LONDON) LTD.
24/28 Oval Road, London NW1 7DX

Library of Congress Cataloging in Publication Data

Main entry under title:

Environmental stress.

"Proceedings of a symposium held at the University
of California, Santa Barbara, August 31-September 3,
1977."
Includes index.
1. Adaptation (Physiology)—Congresses.
2. Man—Influence of environment—Congresses.
3. Stress (Physiology)—Congresses.   I. Folinsbee,
Lawrence J. [DNLM: 1.   Adaptation, Physiological—
Congresses.   2. Stress—Congresses.   3. Environment
—Congresses.   QT140 E61   1977]
QP82.E63   612'.0144   78-18229
ISBN 0-12-261350-3

Contents

Contents

Contributors

Numbers in parentheses indicate pages on which authors' contributions begin.

Irma Åstrand (149), Work Physiology Unit, National Board of Occupational Safety and Health, Stockholm, Sweden

Per-Olof Åstrand (149), Department of Physiology III, Karolinska Institute, Stockholm, Sweden

M. Barac-Nieto (165), Department of Physiological Sciences, Universidad del Valle, Cali, Columbia

D. V. Bates (85), Faculty of Medicine, University of British Colombia, Vancouver, Canada

J. F. Bedi (125), Institute of Environmental Stress, University of California, Santa Barbara, Santa Barbara, California

Clark M. Blatteis (351), Department of Physiology and Biophysics, University of Tennessee Center for the Health Sciences, Memphis, Tennessee

Robert A. Bruce (205), Department of Medicine, Division of Cardiology, Department of Biostatistics, University of Washington, Seattle, Washington

Ramon D. Buckley (111), Rancho Los Amigos Campus, School of Medicine, University of Southern California, Downey, California

E. R. Buskirk (249), Laboratory for Human Performance Research, Intercollege Research Programs, The Pennsylvania State University, University Park, Pennsylvania

Clarence R. Collier (111), Rancho Los Amigos Campus, School of Medicine, University of Southern California, Downey, California

J. Côté (267), Department of Physiology, School of Medicine, Laval University, Quebec, Canada

D. S. Covert (91), Departments of Environmental Health and Atmospheric Sciences, University of Washington, Seattle, Washington

G. J. A. Cropp (101), Department of Physiology, National Asthma Center, Denver, Colorado

Timothy A. DeRouen (205), Department of Medicine, Division of Cardiology, Department of Biostatistics, University of Washington, Seattle Washington

D. W. Dickey (101), Department of Physiology, National Asthma Center, Denver, Colorado

B. L. Drinkwater (125), Institute of Environmental Stress, University of California, Santa Barbara, Santa Barbara, California

F. N. Dukes-Dobos (71), National Institute for Occupational Safety and Health, Cincinnati, Ohio

S. Dulac (267), Department of Physiology, School of Medicine, Laval University, Quebec, Canada

P. R. Fine (71), Department of Rehabilitation Medicine, University of Alabama in Birmingham, Birmingham, Alabama

L. J. Folinsbee (125), Institute of Environmental Stress, University of California, Santa Barbara, Santa Barbara, California

R. Frank (91), Departments of Environmental Health and Atmospheric Sciences, University of Washington, Seattle, Washington

Ralph F. Goldman (53), U.S. Army Research Institute of Environmental Medicine, Natick, Massachusetts

G. W. Gray (373), Defence and Civil Institute of Environmental Medicine, Toronto, Ontario, Canada

Robert F. Grover (325), Cardiovascular Pulmonary Research Laboratory, Division of Cardiology, Department of Medicine, University of Colorado Medical Center, Denver, Colorado

Jack D. Hackney (111), Rancho Los Amigos Campus, School of Medicine, University of Southern California, Downey, California

John P. Hannon (335), Department of Comparative Medicine, Letterman Army Institute of Research, Presidio of San Francisco, California

Seiki Hori (39), Department of Physiology, Hyogo College of Medicine, Mukogawacho 1-1, Nishinomiya, Hyogo, Japan

Steven M. Horvath (125, 385), Institute of Environmental Stress, University of California, Santa Barbara, Santa Barbara, California

C. S. Houston (373), The University of Vermont, Burlington, Vermont

W. R. Keatinge (299), Department of Physiology, The London Hospital Medical College, University of London, London, England

Ralph H. Kellogg (317), Department of Physiology, University of California, San Francisco, California

Kando Kobayashi (279), Research Center of Health, Physical Fitness and Sports, Nagoya University, Nagoya, Japan

L. Kuehn (303), Defence and Civil Institute of Environmental Medicine, Downsview, Ontario, Canada

K. V. Kuhlemeier (71), Department of Rehabilitation Medicine, University of Alabama in Birmingham, Birmingham, Alabama

T. L. Kurt (101), Department of Physiology, National Asthma Center, Denver, Colorado

T. V. Larson (91), Departments of Environmental Health and Atmospheric Sciences, University of Washington, Seattle, Washington

J. LeBlanc (267), Department of Physiology, School of Medicine, Laval University, Quebec, Canada

R. Limmer (303), Defence and Civil Institute of Environmental Medicine, Downsview, Ontario, Canada

A. R. Lind (195), Department of Physiology, St. Louis University Medical School, St. Louis, Missouri

William S. Linn (111), Rancho Los Amigos Campus, School of Medicine, University of Southern California, Downey, California

S. Livingstone (303), Defence and Civil Institute of Environmental Medicine, Downsview, Ontario, Canada

Jack A. Loeppky (225), Department of Physiology, Lovelace Foundation for Medical Education and Research, Albuquerque, New Mexico

Ulrich C. Luft (225), Department of Physiology, Lovelace Foundation for Medical Education and Research, Albuquerque, New Mexico

M. McFadden (373), University of Toronto, Toronto, Ontario, Canada

M. G. Maksud (165), Department of Physical Education, University of Wisconsin-Milwaukee, Milwaukee, Wisconsin

A. L. Mansell (373), University of Toronto, Toronto, Ontario, Canada

Hideji Matsui (279), Research Center of Health, Physical Fitness, and Sports, Nagoya University, Nagoya, Japan

Makoto Mayuzumi (39), Department of Physiology, Hyogo College of Medicine, Mukogawa-cho 1-1, Nishinomiya, Hyogo, Japan

Mochiyoshi Miura (183), Research Center of Health, Physical Fitness, and Sports, University of Nagoya, Nagoya, Japan

Miharu Miyamura (279), Research Center of Health, Physical Fitness, and Sports, Nagoya University, Nagoya, Japan

Mitsumasa Miyashita (183), Laboratory for Exercise Physiology and Biomechanics, University of Tokyo, Tokyo, Japan

John G. Mohler (111), Rancho Los Amigos Campus, School of Medicine, University of Southern California, Downey, California

C. G. Morrill (101), Department of Physiology, National Asthma Center, Denver, Colorado

Yutaka Murase (183), Department of Physical Education, Nagoyagakuin University, Seto, Japan

Ethan R. Nadel (29), John B. Pierce Foundation Laboratory and Departments of Epidemiology and Public Health, and Physiology, Yale University School of Medicine, New Haven, Connecticut

J. S. Petrofsky (195), Department of Physiology, St. Louis University Medical School, St. Louis, Missouri

A. C. P. Powles (373), Department of Medicine, McMaster University, Hamilton, Ontario, Canada

Michael F. Roberts (29), John B. Pierce Foundation Laboratory and Departments of Epidemiology and Public Health, and Physiology, Yale University School of Medicine, New Haven, Connecticut

Loring B. Rowell (3), Department of Physiology and Biophysics and of Medicine, University of Washington School of Medicine, Seattle, Washington

Kiyoshi Shimaoka (279), Research Center of Health, Physical Fitness, and Sports, Nagoya University, Nagoya, Japan

G. B. Spurr (165), Department of Physiology, The Medical College of Wisconsin, and Research Service, Veterans Administration Center, Wood, Milwaukee, Wisconsin

J. R. Sutton (373), Department of Medicine, McMaster University, Hamilton, Ontario, Canada

Nobuo Tanaka (39), Department of Physiology, Hyogo College of Medicine, Mukogawa-cho 1-1, Nishinomiya, Hyogo, Japan

Junzo Tsujita (39), Department of Physiology, Hyogo College of Medicine, Mukogawa-cho 1-1, Nishinomiya, Hyogo, Japan

F. Turcot (267), Department of Physiology, School of Medicine, Laval University, Quebec, Canada

Michael D. Venters (225), Department of Physiology, Lovelace Foundation for Medical Education and Research, Albuquerque, New Mexico

B. Weatherson (303), Defence and Civil Institute of Environmental Medicine, Downsview, Ontario, Canada

P. C. Weiser (101), Department of Physiology, National Asthma Center, Denver, Colorado

C. Bruce Wenger (29), John B. Pierce Foundation Laboratory and Departments of Epidemiology and Public Health, and Physiology, Yale University School of Medicine, New Haven, Connecticut

Keiji Yamaji (183), Department of Physical Education, University of Toyama, Toyama, Japan

Hisato Yoshimura (293), Department of Physiology, Hyogo College of Medicine, Mukogawa-cho 1-1, Nishinomiya, Hyogo, Japan

Preface

This volume, *Environmental Stress: Individual Human Adaptations,* is the proceedings of a symposium that brought together scientists who have attempted to answer questions about individual variability in response to different environments. Many of us have been interested in the ability of man to cope with and adapt to changes in his environment, but questions have remained unanswered. This symposium arose from a need to identify and create further interest in the role of age, gender, genetic heritage, and other individual differences in the response to environmental stressors.

This meeting was held in honor of Steven M. Horvath, who throughout his long and productive career has made and continues to make extensive contributions to our knowledge of man's adaptability to environmental stress.

The contributors have dealt with four aspects of environmental stress: heat, cold, altitude, and air pollution. Although exercise was inevitably involved as a compounding stress under each environmental condition, one section of the symposium deals specifically with exercise as a stressor.

The section on heat stress includes chapters on circulatory adaptations to heat and exercise, prediction of human heat tolerance, and thermoregulatory and heart rate responses of males and females to exercise and heat stress. The effect of ambient temperature on oxygen uptake is also discussed.

The history of the study of altitude stress provides a unique introduction to the altitude section. The role of circulatory adaptations to altitude, differences between altitude residents and sojourners, and the adaptability of females to altitude are discussed extensively. Studies of sleeping patterns associated with high altitude hypoxia and of alterations in muscular efficiency on initial exposure to altitude offer interesting contributions to the physiology of altitude stress.

The effects of different air pollutants such as ozone, carbon monoxide, and sulfuric acid aerosol are reported. Recent investigations of ways in which man may adapt to air pollution, such as neutralization of acidic aerosols by reaction with respiratory ammonia, and habituation or adaptation to repeated exposure to ozone, provide information on some new avenues of research in air pollution stress. The effects of combined exposure to carbon monoxide and altitude and methods of predicting man's response to ozone exposure are also described.

The cold stress section is opened with an informative review covering adaptation, body fat, thermal protection, cardiovascular responses, and swimming in cold water. In other chapters, cooling rates and pressor responses are discussed with relation to body fat, age, sex, and physical fitness. New evidence is presented that demonstrates the lack of effect of ethanol on heat exchange in men exposed to cold. Japanese studies on the seasonal variations in aerobic power and on cold adaptation are also presented.

Age, gender, biomechanical efficiency, fatness, and gravitational stress are considered among the many factors that affect the physiological response to exercise. The chapter on the role of nutritional status in child and adult work capacity demonstrates the continuing importance of interdisciplinary research. In addition, the role of exercise training in the aging process and in the prevention and treatment of heart disease is considered.

The fulfillment of the aims of this symposium (to discuss the importance of age, gender, and other individual characteristics on the response of man to various environmental stressors) must be credited to the participants, who freely exchanged ideas and expanded our knowledge in this area. The enthusiasm and personal warmth that characterized this symposium were beyond our expectations.

This symposium would not have been possible without the generous support of numerous individuals and institutions. We are grateful for the financial support of the Aluminum Company of America, the Aluminum Company of Canada, Hoffman-La Roche Inc., the United States Air Force Office of Scientific Research (Grant AFOSR 77-3310), and the University of California, Santa Barbara.

Although the contributions of the scientists who served as the chairpersons for each symposium session are not apparent in the proceedings, we very much appreciate their efforts in generating and directing discussion, and also for their participation in the panel discussion that has been summarized by Dr. Horvath.

We are indebted to the fellows, graduate students, and the technical and clerical staff of the Institute of Environmental Stress for their considerable effort in helping to make the symposium run smoothly. The final copy used to reproduce the proceedings was prepared expertly and painstakingly by Ms. Mary Gaines Read, whose efforts are deeply appreciated.

Environmental Stress

Individual Human Adaptations

I

Heat Stress

Chairperson

Austin Henschel

National Institute for Occupational Safety and Health
Cincinnati, Ohio

HUMAN ADJUSTMENTS AND ADAPTATIONS
TO HEAT STRESS–WHERE AND HOW?

Loring B. Rowell

Departments of Physiology and Biophysics and of Medicine
University of Washington School of Medicine
Seattle, Washington

Humans, young and old, male and female, adjust and adapt to heat stress. Large variability can occur from one individual to another and from day to day in a given subject. Physiology seeks to describe the mechanisms responsible for these responses in all individuals. In this paper are defined some basic mechanisms of cardiovascular adjustment to heat stress that are thought to be common to all normal individuals. Sudomotor adjustments are beyond the scope of this paper.

ACTIVE, REFLEX VASODILATION
OF THE CUTANEOUS CIRCULATION

Active vasodilation of cutaneous blood vessels and increased sweating are man's two major adjustments to heat stress. Approximately 90% to 100% of the human cutaneous vascular response is mediated by an active vasodilator mechanism that requires an intact sympathetic nervous system (14,20,43). Only a small portion of the cutaneous vasomotor response can be attributed to *passive* vasodilation, or release of tonic vasoconstrictor activity, and it occurs only when skin temperature (T_{sk}) or core temperature (T_c) is low and vasoconstrictor tone is elevated (43,56). We know this distribution of the response to be true for all portions of the human limbs except the hands and feet (3,21,56). It is assumed to be true for all regions that

3

show thermal sweating, but as yet has not been established for the head and torso because skin blood flow cannot be measured in these regions.

So far, no pharmacological agent has been found that can block the vasodilator mechanism. In 1956, Fox and Hilton (18) proposed that bradykinin released from sweat glands is the mediator of active cutaneous vasodilation. The postulate was based on a finding of bradykinin in the sweat of only one heat-stressed subject and a rather tenuous analogy between the function of salivary glands and that of the embryologically related sweat glands. The postulate appears to be untenable also in light of recent experiments (1,19,65,66). Thus the mechanism underlying the major controller of skin blood flow during heat stress remains to be discovered. It is amazing that the only active vasodilator system with massive, sustained action and established physiological importance has somehow escaped the attention of most cardiovascular physiologists and pharmacologists, since it was documented as early as 1938.

A quest for the mediator of active cutaneous vasodilation will undoubtedly require an appropriate animal model—one that shows this response. Although active vasodilation has been observed in the dog's paw during stimulation of the sympathetic nerves, chemoreceptors, and hypothalamus, no experiments analogous to those conducted on heat-stressed man have been carried out on animals until recently (23). The baboon, a sweating primate that is phylogenetically related to man, has been found to have a humanlike active cutaneous vasodilator mechanism despite previous pharmacological evidence to the contrary (67). Experiments were recently conducted on awake, heat-stressed baboons by Hales, Rowell, and Strandness (23) similar to those conducted on man by Roddie and Shepherd (43) and Edholm et al. (14). Heat stress, subsequent to surgical ablation of the left lumbar sympathetic chain, prevented most of the increase in skin blood flow to the left leg, as determined by the quantity of radioactive microspheres within cutaneous vessels. In other experiments on baboons, procaine injected into chronically implanted catheters to block the left lumbar sympathetic chain at the peak of heating reduced skin blood flow in the left leg from 50 to 60 ml · 100 g^{-1} · min^{-1} down to levels approaching preheating control values (10 to 15 ml · 100 g^{-1} · min^{-1} (Fig. 1). With luck we may be able to identify from the baboon the transmitter eliciting man's primary adjustment to heat stress. Identification of the mechanism of active cutaneous vasodilation is one of the greatest unsolved problems in human temperature regulation.

REFLEX CONTROL OF
VASOCONSTRICTOR OUTFLOW TO SKIN

In contrast to active vasodilation, *passive* vasodilation through release of tonic vasoconstrictor activity appears to play only a minor role over most of the skin surface and to depend on T_c and T_{sk}. The release of vasoconstrictor tone to skin is minimal at normal T_c and T_{sk} and will cause no increase in skin blood flow. It will cause at most a doubling of skin blood flow when T_c and/or T_{sk} are low (43,56). Acral regions (hands, feet) are the exception, as they receive high rates of tonic vasoconstrictor discharge even in normothermic subjects (3,21,56). This condition is especially true in the A-V anastomoses that are present in acral regions (21). When the tone is released by warming, blood flow will increase markedly so that maximal values (i.e., maximal without direct local influences of heat) are achieved by the passive vasodilator mechanism. Taken together, a variety of results suggests that changing T_{sk} acts mainly upon cutaneous vasoconstrictor outflow (46).

Contrary to earlier conclusions (c.f., 56), the skin is on the efferent side of reflexes involved in blood pressure regulation, chemoreceptor stimulation, and in responses to exercise and upright posture (46). An important point is that many reflex drives operate upon the same cutaneous vasoconstrictor outflow. In understanding the human circulatory response to heat stress, it is particularly important to realize that *heat stress is often combined with other stresses that augment vasoconstrictor outflow to skin—and to other regions as well. These drives will compete with thermoregulatory needs of skin blood flow.* In short, the skin cannot serve temperature regulation alone. This is discussed further below.

REFLEX CONTROL OF VASODILATOR OUTFLOW TO SKIN

What reflex drives act upon the cutaneous vasodilator system? If we first focus on thermoregulatory drives, the main question to be answered is the relative importance of reflex drives from cutaneous thermoreceptors as opposed to reflex drives

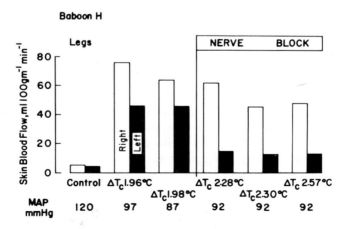

Fig. 1.   Effect of blockade of left lumbar sympathetic chain on left leg skin blood flow (black bars) in an awake baboon. Nerves were blocked after T_c had risen and skin blood flow had reached high levels. Changes in body temperature (ΔT_c) and mean arterial pressure (MAP) are given at the time of each measurement (bottom of graph).

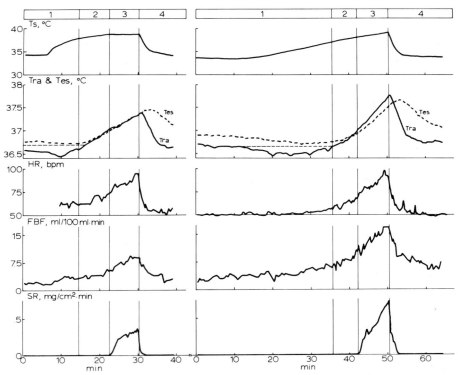

Fig. 2. Results from one subject in whom body skin temperature (T_s) was driven in different temporal patterns (right vs. left upper panels) to achieve periods of separation between changes in T_s and right atrial blood temperature (T_{ra}). Note the close correspondence between rising T_{ra}, heart rate (HR), forearm skin blood flow (FBF), and sweat rate (SR). [From Wyss *et al.* (65), by permission of the American Physiological Society.]

from central thermoreceptors. If it is true that skin temperature affects mainly vasoconstrictor outflow and central temperature affects the cutaneous vasodilator system, then central temperature should dominate in the control of skin blood flow over its full range because most of the increase in skin blood flow during heat stress is due to active vasodilation. To test the hypothesis, one must achieve separate control of skin and central temperatures, a difficult task in man. Wyss *et al.* (65) approached this problem by controlling whole body T_{sk} and driving it in different temporal patterns by means of water-perfused suits (Fig. 2). This design provided some separation of changes in T_{sk} from changes in T_c. The separate protocols precluded protocol dependency in their conclusions. The ramp of increased T_{sk} allowed longer periods for separation of changing T_{sk} from changing T_c. Regression analysis of the results, described in terms of a linear combination of T_{sk} and T_c, yielded coefficients for T_c and T_{sk} with a ratio of 20:1. That is, a $1^{\circ}C$ increase in T_{sk} produces only 1/20th the effect on skin blood flow of a $1^{\circ}C$ increase in T_c. Further, rate of change of T_{sk} was not a significant component. Thus, T_c in resting man is by far the major drive.

The validity of the assumptions made by Wyss *et al.* rests upon the measurement of a "proper" T_c. Haywood and Baker (24) found that blood temperature is the optimal dynamic measure of deep brain temperature in monkeys. Accordingly, they used right atrial temperature (T_{ra}) as the most rapidly responding measure of T_c. Use of a lagging esophageal temperature (T_{es}) resulted in an overestimation of the role of T_{sk} because of the large skin blood flow response that had already occurred when T_{es} rose (the rise of skin blood flow coincided with the earlier rise in T_{ra}).

To further test the validity of their model, Wyss *et al.* (66) applied a different heating protocol as shown in Fig. 3. To eliminate periods wherein T_{sk} and T_c are highly correlated, they raised T_{sk} to $38^{\circ}C$, held it there for 30 min, and then raised it suddenly to $40^{\circ}C$ and then reduced it back to $38^{\circ}C$. This altered T_{sk} in two distinct regions of a T_c-T_{sk} plane. If the previous linear model is correct, then the calculated effects of T_{sk} and T_c on skin blood flow should be the same during periods A, B, and C in Fig. 3. The effect of T_c on skin blood flow was found to be constant and agreed with their previous analysis. However, the influence of T_{sk} depended upon the level of T_c and/or T_{sk}; thus, the system is nonlinear and complex. At elevated T_{sk} and T_c, there was essentially no influence of T_{sk}.

The findings of Wyss *et al.* are summarized in Fig. 4 (65,66). The dashed lines show the calculated influences of T_c and T_{sk} on forearm skin blood flow when a linear model is assumed, i.e, where T_c and T_{sk} have the same relative influence over the full range of T_c. The solid lines reveal the true influence of T_c and T_{sk} and the large error made in the calculation of the influence of T_c by an assumption of linear control of skin blood flow in the face of a nonlinear T_{sk} influence. The middle panel shows the forearm skin blood flow that would be observed from true influences of T_c and T_{sk}; it corresponds to experimentally observed values. These findings are consistent with the observation of Roddie and Shepherd (43) of an initial increase in skin blood flow due to release of vasoconstrictor tone, but of active vasodilation attendant on the rise in T_c. Thus two different vasomotor systems, possibly under different control, are operative. Conclusive proof requires studies in which α-adrenergic activity in the forearm is blocked.

As yet we know nothing about the relative roles of T_{sk} and T_c in the control of skin blood flow during exercise. If changes in T_{sk} do act primarily upon vasoconstrictor activity to skin, then a sudden increase in skin temperature during exercise would be expected to have larger effects than at rest, since vasoconstrictor tone to skin is augmented by exercise (see below). There is some evidence supporting this idea (28).

The findings of Wyss *et al.* (65,66) in man received further support from studies on awake baboons. In these studies, Proppe *et al.* (40) attained separate control of T_{sk} and T_c over a much greater range than was possible in human studies. Control of T_c was achieved by warming or cooling of a chronically implanted femoral A-V heat exchanger. T_{sk} was controlled by regulation of ambient temperature. A typical experiment in which a rise in T_{sk} caused only a small rise in mean right iliac blood flow is shown in Fig. 5. The rise in iliac blood flow is confined to skin [in man (11, 13,29) and baboons, high T_{sk} and T_c do not increase muscle blood flow; see

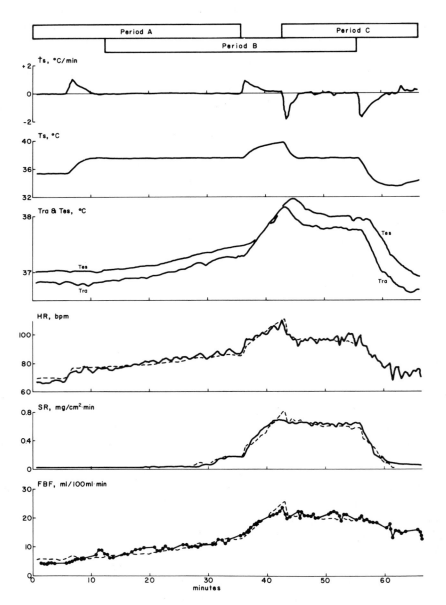

Fig. 3. Results from one subject in whom body skin temperature (T_S, second panel) was held at 38°C and then increased and decreased in two steps to produce high rates of change in T_S ($\dot{T}_S$, top panel). Note how T_{es} lags behind T_{ra} (third panel). See text for explanation. [From Wyss *et al.* (66), by permission of the American Physiological Society.]

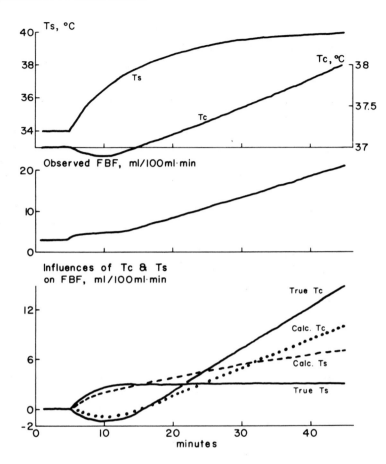

Fig. 4 Results of linear analysis of FBF control when T_S influences are nonlinear. Top panel shows typical changes in T_S and T_C. Supposing that solid lines in bottom panel are true influences of T_C and T_S, then observed FBF will be as shown in the second panel (i.e., assuming that T_C influence is constant and the T_S influence is x times the change in T_S up to a T_S of 37°C; above a T_S of 37°C, T_S influence is nonlinear, having no influence). Dotted and dashed lines in bottom panel show T_C and T_S influences on FBF that would be calculated assuming linear influences of both T_C and T_S. This assumption produces large errors in estimating the influence of T_C on FBF because of nonlinear influence of T_S. [From Wyss *et al.* (66), by permission of the American Physiological Society.]

Fig. 11]. With T_{sk} held constant, raising blood temperature caused a marked increase in iliac blood flow. In short, these experiments reveal that cutaneous blood flow is ten times more sensitive to changes in T_C (measured as arterial blood temperature) than to changes in T_{sk}. A similar relationship was observed in anesthetized cats by Ninomiya and Fujita (39). In general, increased T_C during heat stress appears to be the major reflex drive causing increased cutaneous blood flow. In man and baboons, this drive acts predominantly upon an active vasodilator mechanism.

INTERACTION BETWEEN REFLEX VASOCONSTRICTION
AND VASODILATION DURING HEAT STRESS

Understanding man's circulatory adjustment to heat stress is complicated by the fact that the cutaneous sympathetic vasoconstrictor nerve fibers constitute the efferent arm of so many different reflexes. For example, skin is reflexly vasoconstricted by exercise (6,31), upright posture (2,31,38), blood loss and/or decrements in either arterial pressure (via arterial baroreflex) (4,51) or central venous blood pressure (via low-pressure baroreflexes from cardiopulmonary stretch receptors) (32). Conversely, a rise in arterial blood pressure will lead to release of tonic vasoconstrictor tone to skin (4,48).

A competitive interaction between cutaneous vasoconstrictor and vasodilator reflexes has been demonstrated in man. Some early studies reviewed by Amberson (2) showed that heat was gained more slowly in upright than in supine man during external heating. Johnson $et\,al.$ (31) demonstrated that the slower rise in T_c occurred because skin blood flow was lower at any given T_c in the upright posture. A similar competitive interaction was observed when hemorrhage was simulated by lower body negative pressure in heated man (30). In Fig. 6 is illustrated the magnitude of vasoconstriction (fall in forearm vascular conductance) with each application of lower body negative pressure as T_c and skin blood flow rise progressively during prolonged direct whole-body heating. Thus, the skin can undergo vasoconstriction in response to baroreflexes even when cutaneous vasodilator activity is intense.

Perhaps the most common occurrence of interaction between cutaneous vasoconstrictor and vasodilator activities is observed during upright exercise. In exercise, sympathetic vasoconstrictor outflow is increased to most organs in proportion to the relative severity of exercise (44). A relative vasoconstriction of cutaneous blood vessels has been observed during exercise (31), but it may not be proportional to the severity of exercise (59). Johnson $et\,al.$ (31) showed that at any given T_c, skin blood flow during brief (10-min) periods of upright exercise was much lower than during either upright or supine rest, as illustrated in Fig. 7. Note in Fig. 7 how the skin blood flow-T_{es} relationship is shifted to the right by upright exercise. Brengelmann $et\,al.$ (8), using the same basic experimental design, extended these observations to more prolonged periods of exercise (17 to 30 min). Again, skin blood flow was lower at any given T_{es} during exercise than during rest in these subjects. However, the finding of an apparent upper limit in skin blood flow above a T_{es} of 38°C, as shown in Fig. 8, was unexpected. This response was not the result of reaching a true upper limit of skin blood flow, since the same subjects could reach flows that were twice this during supine rest. Saturation of the response may have been caused by a baroreflex associated with falling arterial blood pressure. In a previous study of similar design, the rate of decrease in arterial blood pressure was reduced at a similar internal temperature and at a time when cardiac output seemed to be leveling off (50).

Cutaneous vasoconstriction is essential in heat-stressed upright man because it is his only defense against rapid pooling of blood in dependent cutaneous veins. Since

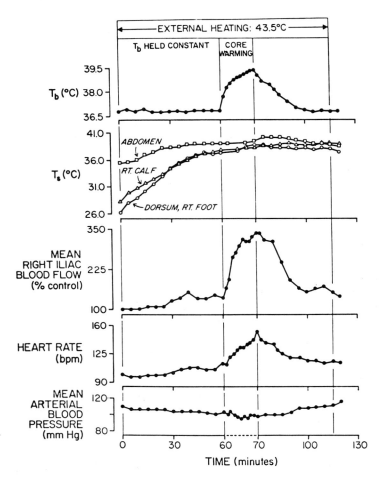

Fig. 5. Effects of separately controlling central or blood temperature (T_b, top panel) and body skin temperature (T_s, second panel) on mean right iliac blood flow in an awake baboon. T_s was increased by raising air temperature in a heated chamber while T_b was held constant by cooling a chronically implanted femoral A-V heat exchanger. Note the small effect of increasing T_s on blood flow with T_b held constant (0 to 60 min). In contrast, increasing T_b (60 to 70 min) had a large effect on iliac blood flow. Note how blood flow tracked T_b during core cooling (70 to 115 min). [From Proppe et al. (40), by permission of the American Physiological Society.]

we lack venoconstrictor reflexes to counteract heat-induced orthostatic shifts in blood volume, our only defense is to vasoconstrict the skin and to reduce the rate at which cutaneous veins fill. The hemodynamic consequences of blood volume displacement into cutaneous veins in upright man have been discussed in detail elsewhere (44,45).

   The main thrust of this discussion of vasomotor interaction is that the *full potential of the cutaneous vasodilator system is not realized during heat stress under a
wide variety of conditions.*

   Although skin is the major target for temperature regulation, it is an important
target for baroreceptor and other reflexes even when skin is maximally vasodilated.
Thermoregulatory control of skin blood flow during heat stress cannot be considered separate from other reflex inputs, as illustrated in Fig. 9. *The temperature
we maintain in conditions other than supine rest is the resultant of competing vasoconstrictor and vasodilator drives to skin.* For example, the so-called change in
thermoregulatory "set point" during exercise may reflect nothing more than the
effects of the background bias of vasoconstrictor activity associated with exercise.
This possibility is undoubtedly true for other situations as well.

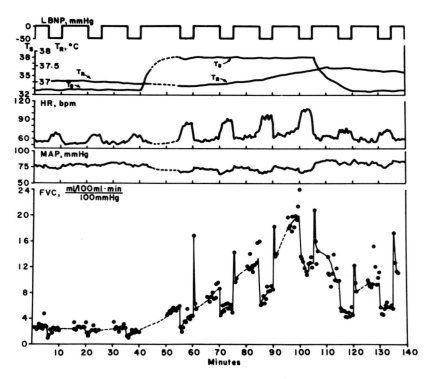

Fig. 6. Cutaneous vasoconstriction during combined lower body negative pressure (LBNP,
top panel) and hyperthermia. The subject was heated by holding body skin temperature (T_s,
second panel) at 38°C. Heart rate (HR, third panel) rose with each application of LBNP; the
increase was accentuated by heat stress. Note that the fall in forearm vascular conductance
(FVC, bottom panel) increased with each application of LBNP during heat stress. Mean arterial
blood pressure (MAP, fourth panel) was well maintained by regional vasoconstriction. [From
Johnson *et al.* (30), by permission of the American Physiological Society.]

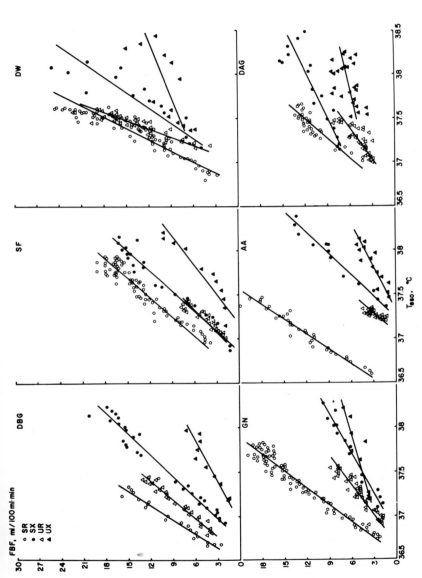

Fig. 7. Regressions of forearm blood flow (FBF) vs. esophageal temperature (T_{eso}) in six normal men during supine rest (SR, open circles), supine exercise (SX, solid circles), upright rest (UR, open triangles), and upright exercise (UX, solid triangles). Under all conditions, skin temperature was held constant at 38°C in order to raise T_{eso} and FBF during upright and supine rest. Note that at any given T_{eso}, upright rest and supine exercise decreased FBF below that at supine rest. Upright exercise caused the greatest decrease. [From Johnson et al. (31), by permission of the American Physiological Society.]

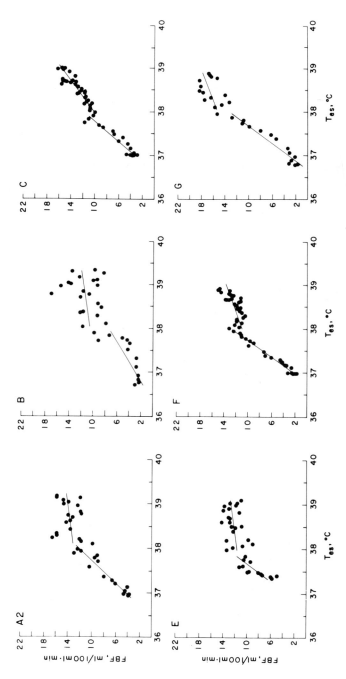

Fig. 8. Forearm blood flow vs. esophageal temperature (T_{es}) in six subjects who exercised for 17 to 30 min at work loads of 86 W to 147 W. During exercise, skin temperature (T_{sk}) was clamped at 38°C by water-perfused suits. Note the leveling off of skin blood flow as T_{es} rose above 38°C. Leveling off occurred at flows far below those reached by these subjects during supine rest. [From Brengelmann et al. (8), by permission of the American Physiological Society.]

Where the integration of vasoconstrictor and vasodilator drives takes place is unknown. Possibly norepinephrine, the transmitter of neurogenic vasoconstriction, and the unknown vasodilator transmitter interact competitively at receptor sites. Conversely, vasodilator outflow may also be centrally modulated by nonthermo-regulatory reflexes, so that reduced vasodilator drive might augment cutaneous vasoconstriction attending baroreflexes, exercise, etc. during heat stress. How this interaction takes place is a major unsolved problem in human temperature regula-tion, and we must find a way to separate the two vasomotor effector mechanisms.

INTERACTION BETWEEN REFLEX AND LOCAL
EFFECTS OF HEAT ON SKIN BLOOD FLOW

In addition to being on the efferent side of thermoregulatory and nonthermo-regulatory reflexes, the skin is also uniquely responsive to local heating (and cooling). Normally, *reflex* control of skin blood flow is studied in skin kept as close to normal T_{sk} as possible in order to avoid the complicating influences of direct local effects. Local T_{sk} will modify in important ways the cutaneous response to increases in both vasoconstrictor and vasodilator outflow, as recently illustrated by Johnson *et al.* (29). Earlier experiments by Roddie and Shepherd (42) suggested that elevated local temperature augmented the reflex effects of increasing T_c on hand blood flow. In the hand, vasodilation occurs only by withdrawal of vasoconstrictor tone (3,21, 43, 56). Johnson *et al.* (29) found that increased local T_{sk} modifies cutaneous

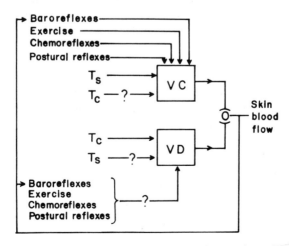

Fig. 9. Diagram of known inputs to cutaneous vasoconstrictor and vasodilator systems. The only factor known to influence the vasodilator system is changing T_c. On the other hand, the vasoconstrictor system receives input from many receptors. The key unanswered question is whether these nonthermoregulatory reflexes also influence vasodilator outflow.

responses to increased vasodilator outflow as well. During whole-body heating, increasing the local T_{sk} of one arm caused skin blood flow to rise earlier than in the arm with temperature maintained at a neutral (32°C) level. In the cool arm, the rise in skin blood flow paralleled the rise in T_{es} and often did not reach values as high as those seen in the heated arm. Because of this interaction of direct and local reflex effects, it is crucial in investigations of *reflex* control of skin blood flow to properly control local T_{sk}.

An important observation made by Mosley (38) and Johnson *et al.* (29) was that skin vasodilated by locally applied heat still maintained the ability to vasoconstrict. Even the most intense local heating of forearm skin (42.5°C for 60 min) did not abolish the skin's ability to vasoconstrict in response to a mild vasoconstrictor stimulus [lower body negative pressure at only -10 Torr (29)]. However, vasoconstriction never reduced skin blood flow to the low levels caused by a given vasoconstrictor stimulus when skin was not directly heated (c.f. Fig. 6). As with competing reflex drives to skin, the mechanism of interaction between local temperature and neurogenic mechanisms is poorly understood.

OVERALL VASOMOTOR OUTFLOW DURING HEAT STRESS

In 1884, Dastre and Morat (10) made the generalization that changes in blood flow to superficial vessels are compensated for by reciprocal changes in blood flow to deep tissues. This generalization appears to be valid in several species including man (44), when skin is vasodilated by heat stress. In an extensive series of experiments, Simon's group selectively heated either the spinal cord or the hypothalamus and produced antagonistic changes in blood flow and sympathetic nervous activity to skin and intestine with little change in skeletal muscle blood flow (27,34,52,58). Although changes in visceral blood flow were quite small, possibly owing to anesthesia, similar patterns of response have been measured in conscious sheep by Hales (22), who used the radioactive microsphere technique. Since changes in blood flow parallel changes in sympathetic nerve activity (27,34,58), the implication is that heat stress causes increased sympathetic vasoconstrictor outflow, which in turn reduces blood flow to visceral organs.

Man (44) and unanesthetized baboons (Hales, Rowell, and King, unpublished observations) appear to show more striking decreases in visceral blood flow as well as far greater increases in cutaneous blood flow during heat stress than other species investigated. The changes observed in man during prolonged, direct, whole-body heating are summarized in Fig. 10 (44). The distribution of blood flow in heat-stressed, conscious baboons whose T_c was increased by 1.5°C to 2.0°C is shown in Fig. 11. The decrements in visceral blood flow are caused primarily by increases in vascular resistance; blood pressure does tend to decrease for a time, but only a little, and then tends to return to control levels toward the end of heating (see Fig. 10). Fig. 11 also shows that blood flow to the muscles was reduced by heating, whereas flows to brain and spinal cord were unaffected.

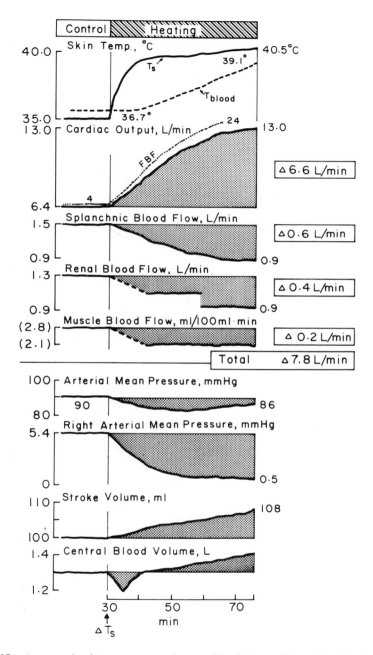

Fig. 10. Average circulatory responses in man directly heated by maintaining body skin temperature (T_{sk}) at 40°C to 41°C. Contributions of increased cardiac output and reduced regional blood flows to total skin blood flow (7.8 ℓ/min) are shown in boxes on the right. All data except splanchnic blood flow (seven men) and renal blood flow [data from one man in this series plus data from the literature (44)] and atrial pressure (four men) are averaged from 12 to 17 men. [From Rowell (44), by permission of the American Physiological Society.]

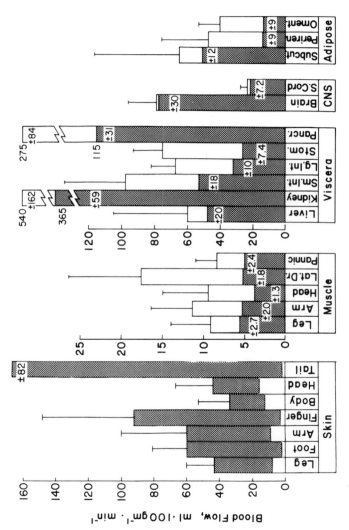

Fig. 11. Redistribution of blood flow in heat-stressed, awake baboons as determined by radioactive microspheres (13 observations from eight animals). Unshaded bars represent control values (these are lower than heat stress values only for cutaneous regions). The shaded bars show the values during heating when T_C was 1.5°C to 2.0°C above control levels (shaded bars show the reductions in flow to most non-cutaneous regions). Increments in skin blood flow and decrements in head and Latissimus Dorsi (Lat. Dr.) muscle and in visceral blood flows (except liver) were significant at $P < 0.05$ (vertical lines show 1 S.D.). (Hales, Rowell and King, unpublished observations.)

The first continuous measurements of renal blood flow during heat stress, made in awake baboons by the use of an electromagnetic flow probe on one renal artery, revealed a progressive rise in renal vascular resistance that paralleled the rise in T_c (15). The surprising finding in these experiments, however, was that most of the renal vasoconstriction attending heat stress was attributable to the renin-angiotensin system. As in man (5,33), plasma renin activity increased with increasing T_c. The β-blocker propranolol prevented increased plasma renin activity and also most of the renal vasoconstriction. In addition, infusion of the angiotensin II analog, saralasin acetate (1-sar-8-ala, angiotensin II), which competitively inhibits angiotensin II at receptor sites, also prevented most of the renal vasoconstriction. Blockade of both plasma renin release and renal vasoconstriction by propranolol indicates several things. First, renin release is caused by a β-adrenergic mechanism that appears to be neurally mediated (68); thus, the response does reflect the increased sympathetic outflow to the kidneys. In addition, Ninomiya and Fujita (39) directly measured increases in sympathetic firing rate in the renal nerve of anesthetized heated cats. Second, the increase in plasma renin activity is not due to decreased hepatic removal of renin, as this would not be affected by β-blockade—nor would the concentration of renin substrate in plasma. This substrate is not increased by heat stress (5). Finally, increased sympathetic vasoconstrictor discharge plays only a minor role in renal vasoconstriction, although the importance of this system may increase at higher levels of heating than tolerated by the awake baboons.

The assumption that heat stress causes visceral vasoconstriction only through increased sympathetic adrenergic activity (44) may require modification. In studies on heat-stressed man, Berlyne et al. (5) found that propranolol had the same action as in baboons; i.e., the rise in plasma renin activity was prevented. As yet, however, the effect of blockade of the renin-angiotensin system on renal blood flow in heat-stressed man is unknown. Taken together, several recent findings raise the question of just how much of the visceral vasconstriction attending heat stress might be attributed to circulating angiotensin II. Because of the kidney's uniquely high sensitivity to angiotensin II (26), it may be that only renal blood flow is affected by this mechanism during heat stress. This remains to be established.

Visceral vasoconstriction attending heat stress has clinical significance (35,62). Severe heat stress is often associated with gastrointestinal and renal symptoms that probably result in part from prolonged periods of reduced blood flow. Metabolic changes have been observed in the human liver suggesting developing hepatic ischemia (44).

CIRCULATORY ADAPTATION TO HEAT STRESS

Acclimatization to work in the heat is a rapid and dramatic process as revealed in the decrements in heart rate, T_c, T_{sk}, and the rise in sweat rate at a given T_c and T_{sk} (35,44,62). The large decrements in T_c and T_{sk} have been attributed primarily

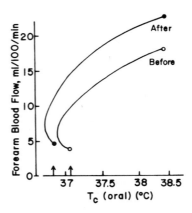

Fig. 12. Shift in forearm blood flow-T_c relationship following repeated daily heatings in water baths to a target oral temperature of 38.5°C. Blood flow (average of 20 subjects) was higher at any given T_c after acclimatization (sweating rate was similarly altered). [Modified from Fox et al. (17).]

to augmentation of sweating and associated increases in evaporative cooling. This appears to be especially true in hot, humid environments (37). One of many unsolved problems is just how much the nature and magnitude of the adaptation are influenced by environmental water vapor pressure. Acclimatization is a complex process. Even with the focus of this paper upon circulatory adaptations, there are still too many problems to deal with in depth.

A pivotal question concerning heat acclimatization has been whether or not cardiac output and skin blood flow are increased or decreased by the adaptation. In contrast to adjustments to chronic heat stress during exercise, adjustments at rest seem more difficult to detect (c.f. 35). In natural environments, cardiac output at rest probably decreases during acclimatization (53) along with decreasing skin blood flow (25). The changes are probably referable to increased evaporative cooling of skin and progressive lowering of T_{sk} and T_c. If attenuation of the cutaneous vasomotor response due to feedback from the sudomotor system is eliminated by direct whole-body heating, then repeated heatings cause skin blood flow to rise at a lower T_{sk} and T_c as acclimatization progresses (Fig. 12) [the sweating response parallels changes in skin blood flow (17)].

During exercise at a given oxygen uptake, cardiac output appears to be unaffected by acclimatization when first and final days of the adaptation are compared (49,63). However, Wyndham et al. (64) have recently observed considerable variation in cardiac output and stroke volume from day to day and from subject to subject during the adaptation. A substantial part of this variation could be due to methodology. The progressive decrease in heart rate during acclimatization is compensated for by a proportional increase in stroke volume (17,49). The mechanism by which stroke volume is increased is unknown. One hypothesis has been that acclimatization might restore cutaneous venous tone during exercise, so that displacement of blood volume into cutaneous veins would be reduced, and cardiac filling pressure and thus stroke volume would be restored. Experimental evidence suggests that venous compliance does not decrease (60,61).

The extent to which plasma volume might contribute to the rise in stroke volume is unknown. Recent literature on adjustments of plasma volume to heat stress and acclimatization is extensive and cannot be covered here. Whether or not plasma volume is increased significantly at the end of acclimatization has been a matter of considerable disagreement (c.f. 44). Traditional methods of labeling plasma proteins have drawn words of caution from Senay (54), as these proteins may exchange between intravascular and extravascular spaces during heat stress. Senay and colleagues' own estimates of changing plasma volume based upon changes in hematocrit and plasma protein suggest that plasma volume is significantly increased by acclimatization (54,55). Senay and coworkers have productively pursued the idea that cutaneous vasodilation produces an increased translocation of protein from cutaneous interstitial space to the intravascular space; this in turn causes an increased plasma volume during heat stress. These investigators conclude that heat acclimatization modifies the protein available within cutaneous interstitial spaces and causes influx to become more rapid. Measurements of plasma volume by Wyndham et al. (63) by contemporary tracer techniques also suggest a rise in plasma volume, but only a small increase—only 5%, or 120 ml, by the 17th day of acclimatization. Bonner et al. (7) observed changes in hematocrit that suggest only a 6.7%, or 200-ml, increase in plasma volume during acclimatization. Of course, in such estimates, a stable red cell volume and mass are assumed.

The question is whether the increases in plasma volume are large enough to compensate for the large volume of blood displaced to skin. If so, then cardiac filling pressure might increase sufficiently to raise stroke volume. Total systemic venous compliance in man averages approximately 2.3 ml $\cdot$ kg^{-1} $\cdot$ Torr^{-1} (12). This would mean that venous blood volume would have to increase by approximately 170 ml to increase right atrial pressure by 1 Torr (in a 75-kg man). Rowell et al. (50) saw right atrial pressure decrease 4 Torr as a result of vasodilation of the skin during exercise. Reducing T_{sk}, which causes rapid venoconstriction (47), suddenly increased right atrial pressure by 4 Torr and restored heart rate and stroke volume to control levels. To achieve the same result by expanding plasma volume would require an increase of nearly 700 ml, an increase that far exceeds most estimates of increased plasma volume during acclimatization. However, in Senay and colleagues' (55) most recent experiments, plasma volume was estimated to increase by approximately 23% (average of four subjects whose average weight was 63 kg). In these subjects, the calculated increase in plasma volume of 630 ml could acutely raise central venous pressure by 4 Torr; but over long periods (hours or days), stress-relaxation or other adjustments might markedly alter the relationship between plasma volume and central venous pressure so that the latter is reduced. Clearly, measurements of central venous pressure before and after acclimatization are crucial.

Senay and colleagues' (55) proposal that a causative relationship between increases in plasma volume and cardiovascular changes (i.e., increased cardiac output and stroke volume) exists during the first five days of acclimatization is difficult to accept. Although their subjects did show simultaneous increases in cardiac output, stroke volume and plasma volume during the middle portion of a 10-day period of acclimatization (hot, humid conditions), cardiac output and stroke volume fell during the final days (64) despite maintained and even further increases in plasma volume. These results suggest no close functional link between plasma volume and the observed cardiovascular events. Apparently, the thinking is that increased plasma volume *permits* the earlier cardiovascular changes that in turn pave the way for adaptation in the sweating mechanism. The sweating adaptation persisted along with the increased plasma volume despite the loss of early cardiovascular adjustments. The authors warned, however, that adjustments to hot, humid environments may differ considerably from those to hot, dry environments; sweat rate does not increase as much in the latter.

Other possibilities for explaining the rise in stroke volume are that the reduced T_{sk} and T_c reflexly lower heart rate so that cardiac output is maintained by a compensatory rise in stroke volume. It can be questioned, however, how stroke volume could increase in the face of continued displacement of blood volume to cutaneous veins (44). An obvious answer would be that skin blood flow decreases (causing a passive decrease in cutaneous venous volume) along with falling T_{sk} and T_c. Again, experiments by a variety of methods have yielded conflicting results (c.f. 41,44). Recently, Roberts *et al.* (41) measured forearm blood flow in subjects before and after 10 days of acclimatization to exercise in humid heat. The authors' plots of forearm blood flow vs. T_{es} resembled that shown for resting subjects in Fig. 12, except that the relationship appeared linear in Roberts and colleagues' study. In general, acclimatization appeared to shift the forearm blood flow:T_c relationship to the left, so that skin blood flow was higher at any given T_c. Unfortunately, Roberts and coworkers made these post-acclimatization measurements during only 15 min of exercise in a $25^\circ C$ environment. It cannot be inferred that this relationship would hold during prolonged work at higher ambient temperatures. If skin blood flow is reduced by acclimatization, this change follows the decreases in T_c and T_{sk} and could contribute to the rise in stroke volume by passively reducing cutaneous venous volume.

Changes in the distribution of cardiac output attending heat acclimatization at rest or during exercise have not been defined. In either case, an increase in visceral blood flow would be expected to accompany the fall in T_c and heart rate during acclimatization. Under a variety of conditions, splanchnic blood flow is reciprocally related to heart rate (44), and heart rate is directly related to increases in plasma norepinephrine concentration (9) and to plasma renin concentration (16) as illustrated in Fig. 13. An important generalization follows from this; namely, under a variety of stresses, increased sympathetic nerve activity to the heart, which raises heart rate, is directly proportional to sympathetic vasoconstrictor outflow to visceral organs. As acclimatization progresses, heart rate and urinary norepinephrine excretion (36) decline, so that sympathetic nervous activity decreases along with a progressive fall in T_c. Thus, visceral blood flow would be expected to increase

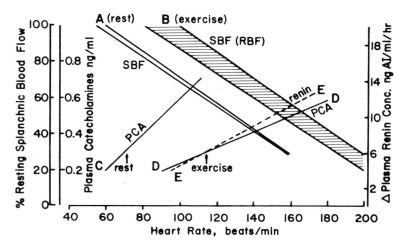

Fig. 13. Reductions in splanchnic blood flow (SBF) in relation to heart rate in supine resting man exposed to heat stress and lower body negative pressure (line A) and in upright man exercising under a variety of conditions (line B) (44). Line B also includes renal blood flow (RBF) responses. The shaded area for line B includes all the regression lines from several investigations from different laboratories (44). Exercise shifts the SBF-heart rate relationship to the left so that decreases in SBF occur at higher heart rates. The stress of exercise causes a similar leftward shift in plasma catecholamine concentration (PCA, line D) (primarily norepinephrine caused by increased neuronal leakage) when values are compared with those obtained in orthostatically stressed resting subjects (line C) (9). The rise in plasma renin activity [expressed as ng of angiotensin I (AI) produced per min] also parallels the rise in heart rate. Resting heart rates from Christensen's and Brandsborg's data (9) were normalized to 60 bpm. Changes in heart rate, SBF, RBF, PCA, and plasma renin concentration all track progressive increases in sympathetic vasoconstrictor outflow.

during acclimatization, and a smaller fraction of a constant cardiac output would be expected to go to the skin. This idea is supported by the gradual diminution of heat-induced gastrointestinal symptoms (57) as acclimatization progresses, together with the widening of the T_c-T_{sk} gradient (in hot, dry environments).

Although some of the pieces appear to fit together, others do not. There are still major unanswered questions which were put into clear perspective in a recent comprehensive review by Wyndham (62). Perhaps the biggest problem is the tendency to mix together findings from humid and dry environments in an effort to explain circulatory changes that appear to be unique to a particular environment.

ACKNOWLEDGMENTS

Studies from this laboratory were supported by National Heart and Lung Institute Grants HL-09773 and HL-16910, and through the Clinical Research Center Facility at the University of Washington, supported by National Institutes of Health Grant RR-37.

REFERENCES

1. Allwood, M.J. and G.P. Lewis. Bradykinin and forearm blood flow. *J. Physiol., London* 170: 571-581, 1964.
2. Amberson, W.R. Physiologic adjustments to the standing posture. *Maryland Univ. Sch. Med. Bull.* 27: 127-145, 1943.
3. Barcroft, H. Sympathetic control of vessels in the hand and forearm skin. *Physiol. Rev.* 40, Supp. 4: 81-92, 1960.
4. Beiser, G.D., R. Zelis, S.E. Epstein, D.T. Mason, and E. Braunwald. The role of skin and muscle resistance vessels in reflexes mediated by the baroreceptor system. *J. Clin. Invest.* 49: 225-231, 1970.
5. Berlyne, G.M., J.P.M. Finberg, and C. Yoran. The effect of β-adrenergic blockade on body temperature and plasma renin activity in heat-exposed man. *Brit. J. Clin. Pharmacol.* 1: 307-312, 1974.
6. Bevegård B.S. and J.T. Shepherd. Regulation of the circulation during exercise in man. *Physiol. Rev.* 47: 178-213, 1967.
7. Bonner, R.M., M.H. Harrison, C.J. Hall, and R.J. Edwards. Effect of heat acclimatization on intravascular responses to acute heat stress in man. *J. Appl. Physiol.* 41: 708-713, 1976.
8. Brengelmann, G.L., J.M. Johnson, L. Hermansen, and L.B. Rowell. Altered control of skin blood flow during exercise at high internal temperatures. *J. Appl. Physiol.* 43:790-794, 1977.
9. Christensen, N.J. and O. Brandsborg. The relationship between plasma catecholamine concentration and pulse rate during exercise and standing. *Europ. J. Clin. Invest.* 3: 299-306, 1973.
10. Dastre, A.F. and J.P. Morat. Influence du sang asphyxique sur l'appareil nerveux de la circulation. *Arch. Physiol. Norm. Pathol.* 3, Ser. 3: 1-45, 1884.
11. Detry, J.-M.R., G.L. Brengelmann, L.B. Rowell, and C. Wyss. Skin and muscle components of forearm blood flow in directly heated resting man. *J. Appl. Physiol.* 32: 506-511, 1972.
12. Echt, M., J. Düweling, O. H. Gauer, and L. Lange. Effective compliance of the total vascular bed and the intrathoracic compartment derived from changes in central venous pressure induced by volume changes in man. *Circulation Res.* 34: 61-68, 1974.
13. Edholm, O.G., R.H. Fox, and R.K. McPherson. The effect of body heating on the circulation in skin and muscle. *J. Physiol., London* 134: 612-619, 1956.
14. Edholm, O.G., R.H. Fox, and R.K. MacPherson. Vasomotor control of the cutaneous blood vessels in the human forearm. *J. Physiol., London* 139: 455-465, 1957.
15. Eisman, M.M. and L.B. Rowell. Renal vascular response to heat stress in baboons–role of renin-angiotensin. *J. Appl. Physiol.* 43: 739-746, 1977.
16. Finberg, J.P.M., M. Katz, H. Gazit, and G.M. Berlyne. Plasma renin activity after acute heat exposure in nonacclimatized and naturally acclimatized man. *J. Appl. Physiol.* 36: 519-523, 1974.
17. Fox, R.H., R. Goldsmith, D.J. Kidd, and H.E. Lewis. Blood flow and other thermoregulatory changes with acclimatization to heat. *J. Physiol., London* 166: 548-562, 1963.
18. Fox, R.H., and S.M. Hilton. Bradykinin formation in human skin as a factor in heat vasodilatation. *J. Physiol., London* 142:219-232, 1958.
19. Freewin, D.B., D.J. McConnell, and J.A. Downey. Is a kininogenase necessary for human sweating? *Lancet* 2: 744, 1973.
20. Grant, R.T., and H.E. Holling. Further observations on the vascular responses of the human limb to body warming: evidence for sympathetic vasodilator nerves in the normal subject. *Clin. Sci.* 3: 273-285, 1938.
21. Greenfield, A.D.M. The circulation through the skin. *Handbook of physiology*. Washington, D.C.: American Physiological Society, Sec. 2, Vol. II, pp. 1325-1351, 1963.

22. Hales, J.R.S. Effects of exposure to hot environments on the regional distribution of blood flow and on cardiorespiratory function in sheep. *Pflügers Arch.* 344:-133-148, 1973.
23. Hales, J.R.S., L.B. Rowell, and D.E. Strandness, Jr. Active cutaneous vasodilatation in the hyperthermic baboon. *Proc. Aust. Physiol. Pharmacol. Soc.* 8: 70P, 1977.
24. Hayward, J.N. and M.A. Baker. Role of cerebral arterial blood in the regulation of brain temperature in the monkey. *Am. J. Physiol.* 215: 389-403, 1968.
25. Hellon, R.F. and A.R. Lind. Circulation in the hand and forearm with repeated daily exposures to humid heat. *J. Physiol., London* 128: 57P-58P, 1955.
26. Hollenberg, N.K., H.S. Solomon, D.F. Adams, H.L. Abrams, and J.P. Merrill. Renal vascular responses to angiotensin and norepinephrine in normal man. *Circulation Res.* 31: 750-757, 1972.
27. Iriki, M., W. Riedel, and E. Simon. Regional differentiation of sympathetic activity during hypothalamic heating and cooling in anesthetized rabbits. *Pflügers Arch.* 328: 320-331, 1971.
28. Johnson, J.M. Regulation of skin circulation during prolonged exercise. *Ann. N. Y. Acad. Sci.* 301: 195-212, 1977.
29. Johnson, J.M., G.L. Brengelmann, and L.B. Rowell. Interaction between local and reflex influences on human forearm skin blood flow. *J. Appl. Physiol.* 41: 826-831, 1976.
30. Johnson, J.M., M. Niederberger, L.B. Rowell, M.M. Eisman, and G.L. Brengelmann. Competition between cutaneous vasodilator and vasoconstrictor reflexes in man. *J. Appl. Physiol.* 35: 798-803, 1973.
31. Johnson, J.M., L.B. Rowell, and G.L. Brengelmann. Modification of the skin blood flow-body temperature relationship by upright exercise. *J. Appl. Physiol.* 37: 880-886, 1974.
32. Johnson, J.M., L.B. Rowell, M. Niederberger, and M.M. Eisman. Human splanchnic and forearm vasoconstrictor responses to reductions of right atrial and aortic pressures. *Circulation Res.* 34: 515-524, 1974.
33. Kosunen, K.J., A.J. Pakarinen, K. Kuoppasalmi, and H. Adlercreutz. Plasma renin activity, angiotensin II and aldosterone during intense heat stress. *J. Appl. Physiol.* 41: 323-327, 1976.
34. Kullman, R., W. Schönung, and E. Simon. Antagonistic changes of blood flow and sympathetic activity in different vascular beds following central thermal stimulation. I. Blood flow in skin, muscle and intestine during spinal cord heating and cooling in anesthetized dogs. *Pflügers Arch.* 319: 146-161, 1970.
35. Leithead, C.S. and A.R. Lind. *Heat stress and heat disorders.* Philadelphia: Davis, 1964.
36. Maher, J.T., D.E. Bass, D.D. Heistad, E.T. Angelakos, and L.H. Hartley. Effect of posture on heat acclimatization in man. *J. Appl. Physiol.* 33: 8-13, 1972.
37. Mitchell, D., L.C. Senay, C.H. Wyndham, A.J. van Rensburg, G.G. Rogers, and N.B. Strydom. Acclimatization in a hot, humid environment: energy exchange, body temperature, and sweating. *J. Appl. Physiol.* 40: 768-778, 1976.
38. Mosley, J.G. A reduction in some vasodilator responses in freestanding man. *Cardiovascular Res.* 3: 14-21, 1969.
39. Ninomiya, L. and S. Fujita. Reflex effects of thermal stimulation on sympathetic nerve activity to skin and kidney. *Am. J. Physiol.* 230: 271-278, 1976.
40. Proppe, D.W., G.L. Brengelmann, and L.B. Rowell. Control of baboon limb blood flow and heart rate—role of skin vs. core temperature. *Am. J. Physiol.* 231: 1457-1465, 1976.
41. Roberts, M.F., C.B. Wenger, J.A.J. Stolwijk, and E.R. Nadel. Skin blood flow and sweating changes following exercise training and heat acclimation. *J. Appl. Physiol.* 43: 133-137, 1977.
42. Roddie, I.C., and J.T. Shepherd. The blood flow through the hand during local heating, release of sympathetic vasomotor tone by indirect heating, and a combination of both. *J. Physiol., London* 131: 657-664, 1956.

43. Roddie, I.C., J.T. Shepherd, and R.F. Whelan. The contribution of constrictor and dilator nerves to the skin vasodilatation during body heating. *J. Physiol., London* 136: 489-497, 1957.

44. Rowell, L.B. Human cardiovascular adjustments to exercise and thermal stress. *Physiol. Rev.* 54: 75-159, 1974.

45. Rowell, L.B. Competition between skin and muscle for blood flow during exercise. In: *Problems with temperature regulation during exercise.* Edited by E.R. Nadel. New York: Academic Press, 1977, pp. 49-76.

46. Rowell, L.B. Reflex control of the cutaneous vasculature. *J. Invest. Dermatol.* 69: 154-166, 1977.

47. Rowell, L.B., G.L. Brengelmann, J.-M.R. Detry, and C. Wyss. Venomotor responses to rapid changes in skin temperature in exercising man. *J. Appl. Physiol.* 30: 64-71, 1971.

48. Rowell, L.B., L. Hermansen, and J.R. Blackmon. Human cardiovascular and respiratory responses to graded muscle ischemia. *J. Appl. Physiol.* 41: 693-701, 1976.

49. Rowell, L.B., K.K. Kraning, II, J.W. Kennedy, and T.O. Evans. Central circulatory responses to work in dry heat before and after acclimatization. *J. Appl.. Physiol.* 22: 509-518, 1967.

50. Rowell, L.B., J.A. Murray, G.L. Brengelmann, and K.K. Kraning, II. Human cardiovascular adjustments to rapid changes in skin temperature during exercise. *Circulation Res.* 24: 711-724, 1969.

51. Rowell, L.B., C.R. Wyss, and G.L. Brengelmann. Sustained human skin and muscle vasoconstriction with reduced baroreceptor activity *J. Appl. Physiol.* 34: 639-643, 1973.

52. Schönung, W., H. Wagner, C. Jessen, and E. Simon. Differentiation of cutaneous and intestinal blood flow during hypothalamic heating and cooling in anesthetized dogs. *Pflügers Arch.* 328: 145-154, 1971.

53. Scott, J.C., H.C. Bazett, and G.C. Mackie. Climatic effects on cardiac output and the circulation in man. *Am. J. Physiol.* 129: 102-122, 1940.

54. Senay, L.C., Jr. Changes in plasma volume and protein content during exposures of working men to various temperatures before and after acclimatization to heat: separation of the roles of cutaneous and skeletal muscle circulation. *J. Physiol., London* 224: 61-81, 1972.

55. Senay, L.C., D. Mitchell, and C.H. Wyndham. Acclimatization in a hot, humid environment: body fluid adjustments. *J. Appl. Physiol.* 40: 786-796, 1976.

56. Shepherd, J.T. *Physiology of the circulation in human limbs in health and disease.* Philadelphia: Saunders, 1963.

57. Taylor, H.L., A.F. Henschel, and A. Keys. Cardiovascular adjustments of man in rest and work during exposure to dry heat. *Am. J. Physiol.* 139: 583-591, 1943.

58. Walther, O.-E., M. Iriki, and E. Simon. Antagonistic changes of blood flow and sympathetic activity in different vascular beds following central thermal stimulation. II. Cutaneous and visceral sympathetic activity during spinal cord heating and cooling in anesthetized rabbits and cats. *Pflügers Arch.* 319: 162-184, 1970.

59. Wenger, C.B., M.F. Roberts, J.A.J. Stolwijk, and E.R. Nadel. Forearm blood flow during body temperature transients produced by leg exercise. *J. Appl. Physiol.* 38: 58-63, 1975.

60. Whitney, R.J. Circulatory changes in the forearm and hand of man with repeated exposure to heat. *J. Physiol., (London)* 125: 1-24, 1954.

61. Wood, J.E. and D.E. Bass. Responses of the veins and arterioles of the forearm to walking during acclimatization to heat in man. *J. Clin. Invest.* 39: 825-833, 1960.

62. Wyndham, C.H. The physiology of exercise under heat stress. *Ann. Rev. Physiol.* 35: 193-220, 1973.

63. Wyndham, C.J., A.J.A. Benade, C.G. Williams, N.B. Strydom, A. Goldin, and A.J.A. Heyns. Changes in central circulation and body fluid spaces during acclimatization to heat. *J. Appl. Physiol.* 25: 586-593, 1968.

64. Wyndham, C.H., G.G. Rogers, L.C. Senay, and D. Mitchell. Acclimatization in a hot, humid environment: cardiovascular adjustments. *J. Appl. Physiol.* 40: 779-785, 1976.

65. Wyss, C.R., G.L. Brengelmann, J.M. Johnson, L.B. Rowell, and M. Niederberger. Control of skin blood flow, sweating and heart rate: role of skin vs. core temperature. *J. Appl. Physiol.* 36: 726-733, 1974.

66. Wyss, C.R., G.L. Brengelmann, J.M. Johnson, L.B. Rowell, and D. Silverstein. Altered control of skin blood flow at high skin and core temperatures. *J. Appl. Physiol.* 38: 839-845, 1975.

67. Wyss, C.R. and L.B. Rowell. Lack of humanlike active vasodilation in skin of heat-stressed baboons. *J. Appl. Physiol.* 41: 528-531, 1976.

68. Zanchetti, A., A. Stella, G. Leonetti, A. Morganti, and L. Terzoli. Control of renin release: a review of experimental evidence and clinical implications. *Am. J. Cardiol.* 37: 675-691, 1976.

THERMOREGULATORY ADAPTATIONS
TO HEAT AND EXERCISE:
COMPARATIVE RESPONSES OF MEN AND WOMEN

Ethan R. Nadel
Michael F. Roberts
C. Bruce Wenger

John B. Pierce Foundation Laboratory
and
Departments of Epidemiology and Public Health,
and Physiology
Yale University School of Medicine
New Haven, Connecticut

In order to resist hyperthermia during heat exposure, the body must have the capability of sensing the excess heat or a correlate thereof, must be able to control the transfer of heat from the body core to the skin surface, and must have the capability to dissipate the heat from the skin surface, even when the skin is cooler than the environment.

It has been well established, almost exclusively from animal studies, that the body is abundantly endowed with temperature-sensitive free nerve endings. These sensors have been isolated nearly everywhere on the skin surface (6). Temperature-sensitive neurons have also been isolated in the preoptic region of the anterior hypothalamus, the region that is characterized as the thermoregulatory center (10). Because of the convincing evidence that discrete changes in skin temperature and/or hypothalamic temperature elicit appropriate thermoregulatory responses (13), these are almost certainly primary loci of sensation of the body's thermal state. From these primary sensory loci, the integrating center in the hypothalamus receives a constant appraisal of the body's thermal state.

It has also been well established that the body has the capability to increase its skin blood flow on the order of 10 to 20 times its basal level (8). This provides for a range of overall skin heat conductances from 5 to 10 $W \cdot m^{-2} \cdot {}^{\circ}C^{-1}$ during intense cutaneous vasoconstriction up to 100 $W \cdot m^{-2} \cdot {}^{\circ}C^{-1}$ during vasodilated conditions. In conditions of thermal loading, either internal loading as during exercise, or external loading as during exposure to a hot environment, there is a controlled increase in skin blood flow, thereby increasing the carriage of heat from core to skin. Once the heat reaches the skin, it can be dissipated to the environment by radiation and convection, if the skin-to-environment temperature gradients are favorable, and by evaporation. Thus, the ability to increase the skin circulation in the presence of a thermal load is a primary defense against hyperthermia.

Dissipation of heat by evaporation depends largely upon the ability of the sweat glands to provide a layer of water onto the skin surface. During conditions of thermal loading, accompanying the controlled increase in skin blood flow, there is an increase in sweating rate, which increases the skin-to-ambient gradient of water vapor pressure, thereby increasing the rate of evaporation in given environmental conditions.

Although there have been a number of studies comparing thermoregulatory responses of men and women during heat exposure, few of these have evaluated the relation of these responses to body temperatures. Over the past few years, we have attempted to characterize the physiological systems which control against hyperthermia as control systems, relating the primary physiological responses, the rates of sweating and skin blood flow, to the primary physical stimuli, internal and skin temperatures. From these studies, using male subjects only, we found that both sweating rate and forearm blood flow were well accounted for by the following linear additive models:

$$E_{sk} = 197 \, (T_{es}) + 23 \, (\overline{T}_{sk}) - 8012$$

$$BF = 8.01 \, (T_{es}) + 0.85 \, (\overline{T}_{sk}) - 321$$

where: E_{sk} is the skin evaporative rate in $W \cdot m^{-2}$,

BF is the forearm blood flow in $ml \cdot (100 \, ml)^{-1} \cdot min^{-1}$, and

T_{es} and $\overline{T}_{sk}$ are internal (esophageal) and mean skin temperatures in ${}^{\circ}C$ (7,14).

The primary purpose of the present study was to evaluate and compare the characteristics of skin blood flow and sweating rate control in men and women. A second purpose of the study was to determine whether the lowered internal temperature which occurs in heat-acclimated individuals during exercise in the heat could be ascribed to an increased sweating and/or skin blood flow sensitivity with respect to changes in internal temperature, or whether the thresholds for these responses were lowered following acclimation. Finally, we were interested in whether there were any differences in the acclimation responses of men and women.

METHODS

Eight healthy, sedentary subjects, four men and four women, volunteered for participation in the study. The men were larger than the women (75 kg vs. 50 kg), and their initial determinations of maximal aerobic power were also greater (45.8 ml • min^{-1} • kg^{-1} vs. 39.6 ml • min^{-1} • kg^{-1}). Maximal aerobic power ($\dot{V}O_2$max) was determined for each subject in a standard incremental test on an upright cycle ergometer, with $\dot{V}O_2$ measured by an open circuit technique. Next, the BF:T_{es} and chest sweating rate:T_{es} relations were determined in duplicate from standardized tests, which consisted of exercise at 60%-70% $\dot{V}O_2$max for 15 min in a 25°C ambient temperature. Subjects exercised on a modified cycle ergometer (2) sitting in a contour chair behind the pedals, with legs nearly horizontal and arms supported at shoulder level. BF was measured every 30 s by venous occlusion plethysmography, using an electro-capacitance gauge (14), and chest sweating rate ($\dot{m}_{sw}$) was recorded continuously by resistance hygrometry (7) from a 13 cm^2 collection capsule. T_{es} and $\overline{T}_{sk}$ were also monitored continuously. Figure 1 illustrates representative data from a standardized test. By plotting physiologic responses (BF and $\dot{m}_{sw}$) against the primary stimulus (T_{es}), we were able to obtain slopes and T_{es} thresholds for each response from each subject.

Following these baseline tests, the subjects underwent 10 days of physical training, which consisted of cycle ergometer exercise for one hour per day (four 15-min bouts separated by 5-min rest periods) at an intensity that maintained heart rate between 160 and 170 beats per min. Then we repeated the standardized test in duplicate and determined $\dot{V}O_2$max . The subjects then underwent 10 days of heat acclimation, which consisted of one hour per day (two 30-min bouts separated by a 15-min rest period) of cycle ergometer exercise at 50% $\dot{V}O_2$max in an ambient temperature of 35°C, with water vapor pressure of 35 torr. Following this, the standardized tests were repeated in duplicate, and $\dot{V}O_2$max was again determined.

Since our standardized tests were done at ambient temperatures of 25°C, all control equations were calculated and corrected (assuming a given change in T_{es} produces 10 times a change in $\dot{m}_{sw}$ or BF as a similar change in $\overline{T}_{sk}$) to T_{sk} = 33.0. All standardized tests were performed at the same time of day for each subject. All subjects were studied in the winter months. Since each subject participated in all phases of the study, each was his or her own control.

RESULTS

Physical training produced significant increases in $\dot{V}O_2$max for both men and women. The men increased $\dot{V}O_2$max from 45.8 to 51.8 ml • min^{-1} • kg^{-1} over the 10 days of training, an increase of 13.1%. The women increased from 39.6 to

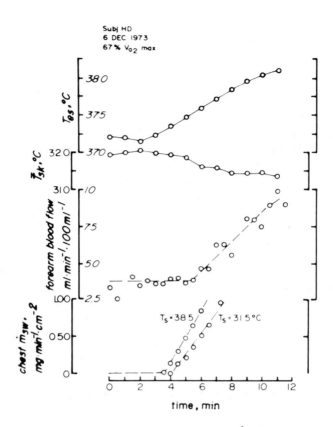

Fig. 1. Time course of a typical standard test to determine $\dot{m}_{sw}$:T_{es} and BF:T_{es} relations.

43.8 ml • min^{-1} • kg^{-1}, an increase of 10.6%. There were no significant changes in $\dot{V}O_2$max following the 10 days of heat acclimation, with the men undergoing a decrease of 4.6% and the women an increase of 1.4%. However, heat acclimation was accompanied by progressively lower heart rates and internal temperatures in all subjects over the 10 days, as we (9,11) and others (1,12) have found previously. There were no differences in the qualitative response of men and women over the 10 days.

Tables 1 and 2 give the mean slopes and T_{es} thresholds, determined from linear regression analysis, for the sweating and vasodilation responses of men and women. Baseline data showed that prior to training, the men had lower T_{es} thresholds than women for both $\dot{m}_{sw}$ and vasodilation and higher $\dot{m}_{sw}$:T_{es} and BF:T_{es} slopes once these responses commenced. Mean sweating thresholds were 37.43°C for the men and 37.58°C for the women. Vasodilator thresholds were 37.30°C and 37.86°C, respectively.   The slopes of the $\dot{m}_{sw}$:T_{es} relations averaged 1.01 and 0.78 mg • min^{-1} • cm^{-2} • °C^{-1} for the men and women, and the slopes of the BF:T_{es}

Table 1. Mean Slopes and T_{es} Thresholds of the $\dot{m}_{sw}:T_{es}$ Relations for Men and Women Subjects Prior to Training, Following Training, and Following Heat Acclimation

	Pretests		Post-training		Post-acclimation	
	Slope[a]	Threshold[b]	Slope	Threshold	Slope	Threshold
Men (N = 4)	1.01	37.43	1.17	37.24	1.66	37.07
Women (N = 4)	0.78	37.58	1.23	37.54	1.08	37.47

[a] $mg \cdot min^{-1} \cdot cm^{-2} \cdot {}^{\circ}C^{-1}$
[b] given in $^{\circ}C$

Table 2. Mean Slopes and T_{es} Thresholds of the $BF:T_{es}$ Relations for Men and Women Subjects Prior to Training, Following Training, and Following Heat Acclimation

	Pretests		Post-training		Post-acclimation	
	Slope[a]	Threshold[b]	Slope	Threshold	Slope	Threshold
Men (N = 4)	14.6	37.30	15.7	37.22	12.8	36.94
Women (N = 4)	10.4	37.86	11.4	37.52	12.7	37.31

[a] $mg \cdot min^{-1} \cdot (100 \ ml)^{-1} \cdot {}^{\circ}C^{-1}$

[b] given in $^{\circ}C$

relations were 14.6 and 10.4 $ml \cdot min^{-1} \cdot (100 \ ml)^{-1} \cdot {}^{\circ}C^{-1}$, respectively. These values are similar to other data recently collected in our laboratory (D. J. Cunningham, *et al.*, manuscript in preparation) which found the average internal temperature sweating thresholds during *resting* heat exposures ($\overline{T}_{sk} \sim 36.0^{\circ}C$) to be $37.20^{\circ}C$ and $37.46^{\circ}C$ for men and women, and the average slopes of the evaporative rate:tympanic temperature relation to be 212 and 169 $W \cdot m^{-2} \cdot {}^{\circ}C^{-1}$, respectively.

Following physical training, female subjects increased the slope of the $\dot{m}_{sw}:T_{es}$ relation from 0.78 to 1.23 $mg \cdot min^{-1} \cdot cm^{-2} \cdot {}^{\circ}C^{-1}$, with practically no change in the T_{es} threshold for sweating. The primary change in BF control was a decrease in the vasodilator threshold from $37.86^{\circ}C$ to $37.52^{\circ}C$, with practically no change in the slope of the $BF:T_{es}$ relation. Following heat acclimation, the women had approximately the same slope for the $\dot{m}_{sw}:T_{es}$ relation and the sweating threshold was lower by around $0.2^{\circ}C$. There was also a decrease of $0.2^{\circ}C$ in the vasodilator threshold, with practically no change in $BF:T_{es}$ slope.

The male subjects responded to physical training and heat acclimation in approximately the same way as did the female subjects. However, the increase in the slope of the $\dot{m}_{sw}$:T_{es} relation with training was somewhat less and the decrease in T_{es} threshold for sweating somewhat more than in the females. Similarly, there was no significant change in the slope of the BF:T_{es} relation with training and there was a decrease of 0.2°C in vasodilator threshold. Following heat acclimation, the men increased $\dot{m}_{sw}$:T_{es} slope by approxmately 25% and the sweating threshold was decreased by nearly 0.2°C. As with the women, the BF:T_{es} slope was essentially unchanged with heat acclimation and there was a decrease in vasodilator threshold of nearly 0.3°C.

An example of the changes in sweating and skin blood flow control from one subject is shown in Figure 2.

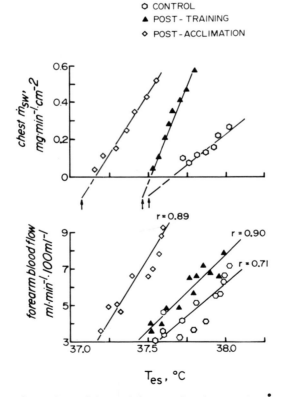

Fig. 2.   Effect of exercise training and heat acclimation on the $\dot{m}_{sw}$:T_{es} and BF:T_{es} relations of one subject. Mean skin temperature, 32°C. Hexagons, control responses. Triangles, responses after exercise training alone. Diamonds, responses after both training and heat acclimation.

DISCUSSION

Physical training and heat acclimation progressively enhance sweating and peripheral vasodilation at a given level of internal temperature, when $\overline{T}_{sk}$ is constant. In both men and women, the increases in $\dot{m}_{sw}$ are brought about by increases in the slope of the $\dot{m}_{sw}:T_{es}$ relation and a reduction in the T_{es} threshold for sweating (Fig. 3). This was more prounced in our male subjects than in our female subjects, primarily because the female subjects showed relatively little change in the sweating response between trained and acclimated conditions. Increased cutaneous circulation at any level of T_{es} following training and acclimation was achieved by a progressive reduction in the T_{es} threshold for vasodilation, without any systematic change in the slope of the $BF:T_{es}$ relation (Fig. 4). The vasodilator threshold was reduced by $0.36^{\circ}C$ in men and by $0.55^{\circ}C$ in women from the untrained to the heat-acclimated condition. Thus, in conditions of thermal load following acclimation, the increased sensitivity of the sweating response would minimize increases in internal body temperature, which in turn would call for a relatively lower blood flow to the periphery. Such a reduction in skin blood flow following heat acclimation has been observed by others (4). Since high skin blood flow is associated with venous pooling, the reduced demand for skin blood flow following acclimation may be a factor in maintaining central blood volume and cardiac filling pressure during exercise in the heat. Other factors which could help acclimated individuals to maintain cardiac filling pressure during exercise in the heat are increased blood volume (1) and a reduction in the loss of plasma from the intravascular space (12).

Others have shown that sedentary male subjects begin sweating at lower temperatures than do sedentary female subjects (3,5) and that males have a higher skin conductance at a given internal temperature than do females (5). The data of this study not only confirm that the internal temperature threshold for these responses is lower in males, but also suggest that the gain of the thermoregulatory responses is higher. However, it is also clear that the differences between sedentary males and females represent quantitative rather than qualitative differences in thermoregulation. These differences may be ascribed in part to differences in normal physical activity levels, as Fox *et al.* have suggested (5). In the preliminary, baseline studies, our male subjects had a higher relative level of physical fitness than the females ($\dot{V}O_2max$ = 45.8 vs. 39.6 ml $\bullet$ min^{-1} $\bullet$ kg^{-1}). However, following 10 days of training and 10 more of heat acclimation, the women (with $\dot{V}O_2max$ = 44.4 ml $\bullet$ min^{-1} $\bullet$ kg^{-1}) were nearly identical in thermoregulatory characteristics to the pre-trained men ($\dot{V}O_2max$ = 45.8 ml $\bullet$ min^{-1} $\bullet$ kg^{-1}). For instance, heat-acclimated women had T_{es} thresholds of $37.47^{\circ}C$ for sweating and $37.31^{\circ}C$ for vasodilation. The thresholds for pre-trained men were $37.43^{\circ}C$ and $37.30^{\circ}C$, respectively. Slopes of the $\dot{m}_{sw}:T_{es}$ and $BF:T_{es}$ relations were 1.08 mg $\bullet$ min^{-1} $\bullet$ cm^{-2} $\bullet$ $^{\circ}C^{-1}$ and 12.7 ml $\bullet$ (min $\bullet$ 100 ml $\bullet$ $^{\circ}C$)$^{-1}$ for acclimated women and 1.01 and 14.6 for the pre-trained men. These results are in general agreement with those of Wyndham *et al.* (15), who reported that women acclimate to heat in the same manner as men while maintaining lower sweating rates at any given internal temperature.

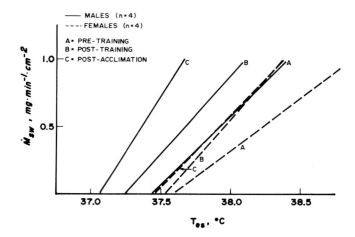

Fig. 3. Progressive alterations in the control of sweating with physical training and heat acclimation.

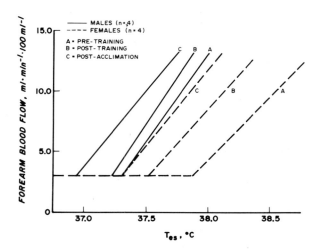

Fig. 4. Progressive alterations in the control of skin blood flow with physical training and heat acclimation.

Like Fox *et al.* (5), we have ignored any possible effects of the menstrual cycle in our analyses, treating our female subjects as if there were no thermoregulatory changes during the cycle. This treatment obscures any effects of the phase of the cycle on observed differences in response between men and women. Bittel and Henane (3) reported that women's body core temperatures in the post-ovulatory phase of the cycle were consistently higher than those of men, and the sweating thresholds were up to $0.4^{\circ}C$ higher. They also found that these differences were much less in the pre-ovulatory phase. These data suggest a role for the sex hormones in thermoregulatory differences between men and women. Although these observations could explain a part of the difference which we have observed between men and women, they do not in themselves account for the thermoregulatory adaptations of our female subjects to training and acclimation.

In conclusion, physical training and heat acclimation lead to reductions in the internal temperature thresholds for sweating and vasodilation and an increase in the sensitivity of the sweating response per unit of internal temperature increase. These changes were seen in both our male and female subjects. Although the men had lower internal temperature thresholds than women for both sweating and vasodilation throughout the entire procedure, the thermoregulatory differences between our untrained male subjects ($\dot{V}O_2max = 45.8$ ml $\cdot$ min^{-1} $\cdot$ kg^{-1}) and heat-acclimated female subjects ($\dot{V}O_2max = 44.4$ ml $\cdot$ min^{-1} $\cdot$ kg^{-1}) were small. These differences may be ascribed at least in part to their relatively higher level of physical condition at any state, owing perhaps to differences in daily living habit. This hypothesis should be followed up with studies of highly trained or heat-acclimated women.

SUMMARY

Eight subjects, four male and four female, underwent a 10-day physical training program followed by a 10-day heat acclimation program. The relations of chest sweating rate and forearm blood flow to internal temperature were determined in standardized exercise tests at $25^{\circ}C$ before training, following training, and following acclimation. Physical training and heat acclimation caused progressive reductions in the internal temperature thresholds for sweating and vasodilation. There was also an increased sensitivity of the sweating response to changes in internal temperature with training and acclimation, but the sensitivity of the forearm vasodilator response was not systematically changed by these procedures. The responses of men and women were similar, except that men maintained lower thresholds throughout the procedure. The observed changes in the blood flow response would tend to produce higher levels of skin blood flow at any given internal temperature following acclimation. Therefore, our data indicate that the reductions in cutaneous circulation which have been reported to accompany acclimation must be the result of the lowered body temperatures.

ACKNOWLEDGMENTS

This study was partially supported by National Institutes of Health Grants ES-00123 and ES-00354 and by National Aeronautics and Space Administration Grant NSG-9023.

All experiments described in this paper were conducted in accordance with the Declaration of Helsinki.

REFERENCES

1. Bass, D.E., E.R. Buskirk, P.F. Iampietro, and M. Mager. Comparison of blood volume during physical conditioning, heat acclimatization and sedentary living. *J. Appl. Physiol.* 12:186-188, 1958.
2. Bigland-Ritchie, B., H. Graichen, and J.J. Woods. A variable speed motorized bicycle ergometer for positive and negative work exercise. *J. Appl. Physiol.* 35:739-740, 1973.
3. Bittel, J. and R. Henane. Comparison of thermal exchanges in men and women under neutral and hot conditions. *J. Physiol., London* 250:475-489, 1975.
4. Eichna, L.W., C.R. Park, N. Nelson, S.M. Horvath, and E.D. Palmes. Thermoregulation during acclimatization in a hot, dry (desert type) environment. *Am. J. Physiol.* 163: 585-597, 1950.
5. Fox, R.H., B.E. Löfstedt, P.M. Woodward, E. Eriksson, and B. Werkstrom. Comparison of thermoregulatory function in men and women. *J. Appl. Physiol.* 26:444-453, 1969.
6. Hensel, H., A. Iggo, and I. Witt. A quantitative study of sensitive cutaneous thermoreceptors with C afferent fibers. *J. Physiol., London* 153:113-126, 1960.
7. Nadel, E.R., R.W. Bullard, and J.A.J. Stolwijk. The importance of skin temperature in the regulation of sweating. *J. Appl. Physiol.* 31:80-87, 1971.
8. Nadel, E.R., I. Holmér, U. Bergh, P.O. Åstrand, and J.A. J. Stolwijk. Energy exchanges of swimming man. *J. Appl. Physiol.* 36:465-471, 1974.
9. Nadel, E.R., K.B. Pandolf, M.F. Roberts, and J.A.J. Stolwijk. Mechanisms of acclimation to heat and exercise in man. *J. Appl. Physiol.* 37:515-520, 1974.
10. Nakayama, T., H.T. Hammel, J.D. Hardy, and J.S. Eisenman. Thermal stimulation of electrical activity of single units of the preoptic region. *Am. J. Physiol.* 204:1122-1126, 1963.
11. Roberts, M.F., C.B. Wenger, J.A.J. Stolwijk, and E.R. Nadel. Skin blood flow and sweating changes following exercise training and heat acclimation. *J. Appl. Physiol.* 43: 133-137, 1977.
12. Senay, L.D., Jr., D. Mitchell, and C.H. Wyndham. Acclimatization in a hot, humid environment: body fluid adjustments. *J. Appl. Physiol.* 40:786-796, 1976.
13. Stitt, J.T., J.D. Hardy, and J.A.J. Stolwijk. PGE_1 fever: its effect on thermoregulation at different low ambient temperatures. *Am. J. Physiol.* 227:622-629, 1974.
14. Wenger, C.B., M.F. Roberts, J.A.J. Stolwijk, and E.R. Nadel. Forearm blood flow during body temperature transients produced by leg exercise. *J. Appl. Physiol.* 38:58-63, 1975.
15. Wyndham, C.H., J.F. Morrison, and C.G. Williams. Heat reactions of male and female Caucasians. *J. Appl. Physiol.* 20:357-364, 1965.

OXYGEN INTAKE OF MEN AND WOMEN
DURING EXERCISE AND RECOVERY
IN A HOT ENVIRONMENT
AND A COMFORTABLE ENVIRONMENT

Seiki Hori
Makoto Mayuzumi
Nobuo Tanaka
Junzo Tsujita

Department of Physiology
Hyogo College of Medicine
Mukogawa-cho 1-1
Nishinomiya, Hyogo
Japan

INTRODUCTION

The effects of different ambient temperatures on physiological reactions to exercise and energy requirements during muscular exercise have been studied extensively (1,14,16,18,22). However, most of the investigations have been performed on male subjects and investigations on female subjects are limited in number. Some differences in thermoregulatory and metabolic responses to heat have been observed between sexes (2,20,21). Women sweat less than men and vasodilation in women is greater than in men when they are exposed to the same environmental heat (2,8,20). According to Hardy and DuBois (9), women showed a marked drop in basal metabolism in an acute reaction to heat exposure, while men showed no such drop. In a hot environment, reduction of metabolic heat is favorable for women because metabolic heat produced in the body must be dissipated into the environment at the rate of its production to maintain a constant body temperature. It was also expected that there are sex differences in thermoregulatory and metabolic adjustment during exercise as the environmental heat stress increases. Thus, it is of interest to compare metabolic rate and physiological reactions of women during exercise and recovery in a hot environment and a comfortable environment with those of men. The main objectives of this study were to investigate sex differences related to changes in energy requirements and physiological reactions during exercise performed under an environmental heat stress, and to discuss the sex differences in physiological reactions during physical exercise in a hot environment from the viewpoint of thermoregulation.

METHODS

Seven young male Japanese and ten young female Japanese in Nishinomiya were selected as subjects. The females participated during the follicular phase of the menstrual cycle as determined by oral temperature measured early in the morning. Table 1 summarizes the physical characteristics of subjects. Physiological reactions and oxygen intake of subjects during exercise and recovery periods were measured under two different environmental conditions: 23°C, 50% R.H. and 35°C, 50% R.H. with a wind velocity of about 17 cm/s. Experiments were randomized to minimize effects of the previous experiment. Experiments were carried out during the spring at around 3 p.m. Subjects were instructed not to eat a meal and to rest at least two hours prior to the experiment in order to minimize effects of specific dynamic action and physical activities. After sitting in a chair at rest in a climatic chamber at least 30 min, male subjects (clad in shorts) and female subjects (clad in swimming suits) rested on the saddle of a Monark bicycle ergometer for 10 min and then pedalled at the constant work load (600 Kg • m/min for male subjects, 300 Kg • m/min and 600 Kg • m/min for female subjects) in time with a metronone at a cycling rate of 50 rpm for 20 min. Following exercise, subjects were seated for 40 min.

Table 1. Physical Characteristics of Subjects

Table 1.Physical characteristics of subjects.

Sex	Number	Age (yr)	Height (cm)	Weight (kg)	B.S.A. (m²)	Skinfold thickness(mm) Upper arm	Subscapular
Male	7	28.1 ±1.2	170.2 ±4.2	62.20 ±10.10	1.730 ±0.140	5.3 ±2.2	11.7 ±4.9
Female	10	21.4 ±0.6	160.0 ±2.6	55.42 ±5.30	1.625 ±0.077	10.0 ±2.8	12.5 ±2.4

B.S.A.: Body surface area.
Mean values are given with their standard deviations.

Body weight was measured prior to and immediately after the experiment with an accuracy of ±5 g and the net body weight was obtained by subtracting the weight of shorts or swimming suits. Sweat volume was obtained from the difference between net body weight before and after the experiment without correction for weight loss through respiration. During periods of rest, exercise and recovery, expired gas samples were collected by Douglas bags. Aliquots of expired gas samples were analyzed for oxygen and carbon dioxide in duplicate by Haldane apparatus. Heart rate was taken by electrocardiogram. Rectal temperature and local skin temperatures were recorded continuously by copper-constantan thermocouples throughout the experiment. The local skin temperatures were measured at the following four sites:

Forehead: central part
Chest: adjacent to the xiphoid process
Forearm: halfway between the elbow and wrist in the inner portion
Calf: halfway down in the frontal portion

The skinfold thicknesses were measured on the back of the upper arm and the skin under the scapula. Blood pressure was measured by brachial auscultation every 5 min. Oxygen requirement was calculated by subtracting the value corresponding to the resting oxygen intake for 60 min from total oxygen intake during exercise and recovery. Oxygen intake during the 40-min recovery period in excess of that required during the resting state was defined as oxygen debt. The ratio of oxygen debt to oxygen requirement was defined as the anaerobic fraction of the oxygen requirement. Net efficiency (μ) was calculated as follows:

$$\mu = [(1 + F) \cdot W]/[427 \cdot M] \times 100\%$$

where: W = work done calculated from brake force and distance moved (Kg·m),
F = factor involved in friction in the transmission of a bicycle ergometer
which increases work load, 0.09 (19), and
M = energy requirement for the exercise corresponding to oxygen
requirement (Kcal).

The weighting factor of Harashima and Kurata (7) was used for calculating the mean skin temperature ($\overline{Ts}$). The following formulas were used to calculate the mean body temperature ($\overline{Tb}$):

$$\overline{Tb} = 0.65 \, Tre + 0.35 \, \overline{Ts} \text{ at } 23^{o}C$$
$$\overline{Tb} = 0.8 \, Tre + 0.2\overline{Ts} \text{ at } 35^{o}C \, (3,8)$$

where: Tre = rectal temperature (^{o}C).

RESULTS

Table 2 presents changes in oxygen intake during exercise and recovery at $23^{o}C$ and $35^{o}C$. Oxygen intake increased sharply with the onset of exercise, then remained at a relatively steady level during the period from 5 min after exercise to the end of the exercise. After cessation of exercise, oxygen intake decreased sharply within 5 min, followed by a slower decrease to pre-exercise values. The values of oxygen intake per body surface area of men at rest and during the late stage of recovery

Table 2. Changes in Oxygen Intake During 20-min Exercise and Recovery Period
of 40 min in 23°C, 50% R.H., and 35°C, 50% R.H.

Sex	Condition (°C)	Load (kp)	at rest	exercise period				recovery period		
				0 - 5	5 - 10	10 - 15	15 - 20	0 - 5	5 - 20	20 - 40
Male	23	2	9.04	41.09	53.02	53.25	54.44	16.41	10.27	9.62
			±1.27	±6.69	±3.95	±5.18	±4.50	±1.95	±1.05	±1.17
	35	2	10.13	41.92	52.33	57.57	57.12	18.89*	11.60*	11.41*
			±1.02	±7.27	±4.80	±6.80	±4.92	±1.29	±0.80	±0.83
			(+12.06)	(+2.02)	(-1.30)	(+8.11)	(+4.92)	(+15.11)	(+12.95)	(+18.61)
Female	23	2	7.63	43.02	52.43	55.86	55.71	17.06	9.13	8.38
			±0.96	±5.06	±4.85	±3.34	±4.79	±4.41	±0.78	±0.76
	35	2	8.24	45.96	56.44	55.67	56.19	19.44	9.67	8.96
			±0.75	±2.07	±2.53	±3.47	±5.91	±3.50	±0.68	±0.75
			(+7.99)	(+6.83)	(+7.65)	(-0.35)	(+0.86)	(+13.95)	(+5.91)	(+6.92)
	23	1	7.87	27.95	32.42	32.49	33.64	12.15	8.98	8.50
			±1.05	±2.18	±1.99	±3.68	±3.53	±1.14	±1.35	±1.40
	35	1	8.26	29.64	32.22	32.56	32.89	13.08	8.86	8.65
			±1.10	±1.51	±2.54	±2.02	±1.51	±1.81	±1.07	±0.78
			(+4.96)	(+6.05)	(-0.62)	(+0.22)	(-2.23)	(+7.65)	(-1.34)	(+1.76)

Mean values in $l/m^2/hr$ are given with their standard deviations. Figures in parentheses indicate changes in the mean values expressed as percentage deviation from the mean values in 23°C. Significant difference between 23°C and 35°C, * at 5% level.

Table 3. Oxygen Intake, Oxygen Debt, Anaerobic Fraction,
Net Efficiency, and Sweat Volume

Sex	Condition (°C)	Load (kp)	T.VO$_2$ (l/60min)	VO$_2$E. (l/20min)	O$_2$D. (l)	An.F. (%)	E.R. (Cal)	N.E. (%)	Sweat V. (kg)
Male	23	2	40.94	28.95	2.18	48.2	125.7	24.5	0.200
			2.05	1.09	0.91	9.3	9.8	1.9	0.043
	35	2	44.89**	30.04	3.20	50.9	133.1	23.2	0.627****
			1.95	2.70	1.63	17.9	12.1	2.2	0.047
			(+9.64)	(+3.77)	(+46.79)	(+5.60)	(+5.89)	(-5.31)	(+213.5)
Female	23	2	38.65	28.10	2.42	43.2	131.3	23.5	0.221
			2.84	1.80	1.30	15.9	11.0	2.1	0.063
	35	2	40.32	28.92	2.55	41.9	133.8	23.0	0.600****
			2.62	1.99	0.71	16.9	8.3	1.5	0.165
			(+4.32)	(+2.92)	(+5.37)	(-3.01)	(+1.90)	(-2.13)	(+171.5)
	23	1	26.89	17.06	1.25	47.9	68.6	22.5	0.129
			3.23	1.58	0.60	17.9	6.2	2.0	0.025
	35	1	27.22	17.16	1.14	41.9	67.9	22.6	0.331****
			2.55	1.36	0.30	16.9	4.0	1.3	0.029
			(+1.23)	(+0.59)	(-8.80)	(-12.53)	(-1.02)	(+0.44)	(+156.6)

Mean values are given with their standard deviations. T.VO$_2$: Total O$_2$ intake. VO$_2$E.: O$_2$ intake during exercise. O$_2$D.: O$_2$ debt. An.F.: Anaerobic fraction. E.R.: Energy requirement. N.E.: Net Efficiency. Sweat V.: Sweat volume. Figures in parenthese indicate changes in the mean values expressed as percentage deviation from the mean values in 23°C. Significant difference between 23°C and 35°C, ** at 2% level, and **** at 0.1% level.

were higher than those of women, and values of oxygen intake per body surface area of women during exercise tended to be slightly higher than those of men. In both sexes, oxygen intake at rest and during exercise had a tendency to increase with rise in ambient temperature. Oxygen intake during recovery increased more as a result of heat exposure in men than in women. Total oxygen intake, oxygen intake during exercise, oxygen debt, anaerobic fraction of oxygen requirement, energy requirement, net efficiency, and sweat volume at 23°C and 35°C are shown in Table 3. In men, the mean value of total oxygen intake (48.89 ℓ) at 35°C was significantly greater than that (40.94 ℓ) at 23°C. The mean value of oxygen debt (3.2 ℓ) at 35°C was considerably greater than that (2.18 ℓ) at 23°C, and the mean value of oxygen intake during exercise (30.04 ℓ) at 35°C was slightly greater than that (28.95 ℓ) at 23°C, although these differences were not statistically significant. In women, total oxygen intake at both work loads had a tendency to rise with increased ambient temperature. However, the percentage increase of oxygen intake for women from 25°C to 35°C was smaller than that for men.

Oxygen debt for women working at a load of 2 kp at 35°C was slightly greater than that at 23°C, while oxygen debt at a work load of 1 kp at 35°C was smaller than at 23°C. Thus, the influence of hot environments on the oxygen intake for women was smaller than that for men. The mean value of anaerobic fraction for men at 35°C was greater than that at 23°C, while the mean values of anaerobic fraction for women at 35°C were smaller than those at 23°C. The effect of hot ambient temperature on energy requirement and net efficiency was greater in men than in women. Increase in energy requirement at the work load of 2 kp in a hot environment was 5.9% for men and 1.9% for women. The mean value of energy requirement at the work load of 1 kp for women was smaller at 35°C than at 23°C. The value of net efficiency is inversely proportional to that of energy requirement, i.e., the mean value of net efficiency for men was smaller at 35°C than at 23°C, while changes in these values for women were smaller. The mean values of sweat volume for men at 23°C and 35°C were 200 g and 627 g, respectively. The mean values of sweat volume at the work load of 2 kp for women at 23°C and 35°C were 221 g and 600 g, respectively; those at the work load of 1 kp at 23°C and 35°C were 129 g and 331 g. Thus, the ratio of sweat volume at 35°C to that at 23°C was greater in men than in women.

The time course responses of heart rate of both sexes are shown in Figs. 1, 2, and 3. The time course of the elevation of heart rate was similar in both sexes. However, absolute values of heart rate during the exercise at the same work load were (on the average) higher for the women than for the men. Immediately after the subject participated in exercise, heart rate increased sharply and continued to increase gradually during exercise. After cessation of exercise, heart rate decreased sharply and was followed by a more gradual decrease. In both sexes, heart rates during the latter periods of exercise and recovery were significantly higher at 35°C than at 23°C. The increase in heart rate due to heat exposure had a tendency to be higher in men than in women. The time course responses of blood pressure of both sexes at 23°C and 35°C are shown in Figs. 4, 5, and 6. In both sexes, systolic

Seiki Hori *et al.*

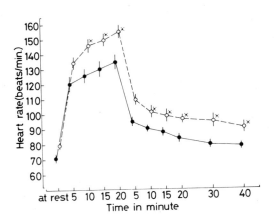

Fig. 1. The time course responses of heart rate of male subjects during 20-min exercise at work load of 2 kp and recovery period of 40 min in 23°C and 35°C. * Significant difference between 23°C and 35°C, * at 5% level. Vertical bars indicate standard errors. — in 23°C, --- in 35°C.

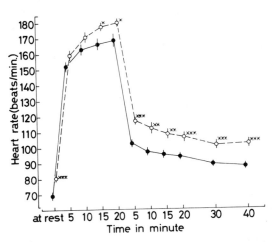

Fig. 2. The time course responses of heart rate of female subjects during 20-min exercise at work load of 2 kp and recovery period of 40 min in 23°C and 35°C. * Significant difference between 23°C and 35°C, * at 5% level, ** at 2% level, *** at 1% level. Vertical bars indicate standard errors. — in 23°C, --- in 35°C.

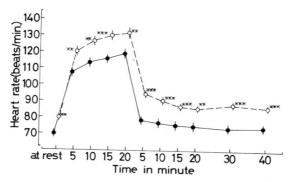

Fig. 3. The time course responses of heart rate of female subjects during 20-min exercise at work load of 1 kp and recovery period of 40 min in 23°C and 35°C. ∗ Significant difference between 23°C and 35°C, ∗∗ at 2% level, ∗∗∗ at 1% level. Vertical bars indicate standard errors. − in 23°C, --- in 35°C.

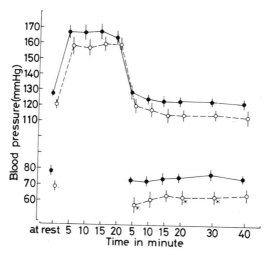

Fig. 4. The time course responses of blood pressure of male subjects during 20-min exercise at work load of 2 kp and recovery period of 40 min in 23°C and 35°C. ∗ Significant difference between 23°C and 35°C, ∗ at 5% level. Vertical bars indicate standard errors. − in 23°C, --- in 35°C.

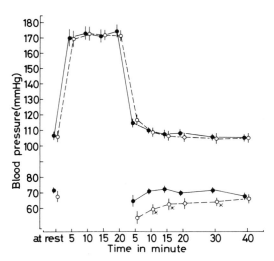

Fig. 5. The time course responses of blood pressure of female subjects during 20-min exercise at work load of 2 kp and recovery period of 40 min in 23°C and 35°C. ✷ Significant
difference between 23°C and 35°C, ✷ at 5% level. Vertical bars indicate standard errors. – in
23°C, - - - in 35°C.

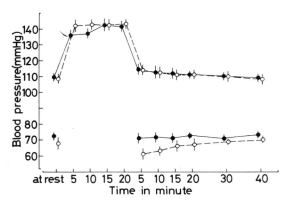

Fig. 6. The time course responses of blood pressure of female subjects during 20-min exercise at work load of 1 kp and recovery period of 40 min in 23°C and 35°C. Vertical bars
indicate standard errors. – in 23°C, - - - in 35°C.

pressure increased sharply as soon as exercise started and reached a sustained level within 5 min. When exercise stopped, systolic pressure decreased sharply, followed by a slower decrease, and fell below the pre-exercise levels. Throughout exercise and recovery, systolic blood pressure for men was lower at 35°C than at 23°C, while that for women was approximately the same under both conditions. Diastolic pressure before exercise and recovery period was lower at 35°C than at 23°C. After cessation of exercise, diastolic pressure was considerably lower than pre-exercise level. It increased gradually thereafter, but continued to remain lower than pre-exercise levels for at least 30 min into the recovery period. The magnitude of fall in diastolic pressure at 35°C and at 23°C tended to be greater in men than in women.

The time course responses of rectal temperature, mean skin temperature, and mean body temperature at 23°C and 35°C are shown in Figs. 7,8,9. The mean values of rectal temperature throughout the experiment at 23°C and 35°C were lower in women than in men. At 23°C, rectal temperature in men fell within 5 min of the onset of exercise, increased gradually during the period from 5 min after exercise to the end of exercise, was unchanged or slightly increased immediately after cessation of exercise, and was then followed by a slow decrease. At 23°C, mean values of rectal temperature for women increased with the onset of exercise and decreased after cessation of exercise. The increase in rectal temperature during

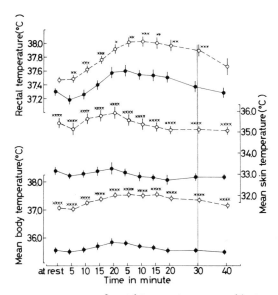

Fig. 7. The time course responses of rectal temperature, mean skin temperature, and mean body tempersature of male subjects during 20-min exercise at work load of 2 kp and recovery period of 40 min in 23°C and 35°C. * Significant difference between 23°C and 35°C, * at 5% level, ** at 2% level, *** at 1% level, and **** at 0.1% level. Vertical bars indicate standard errors. − in 23°C, --- in 35°C.

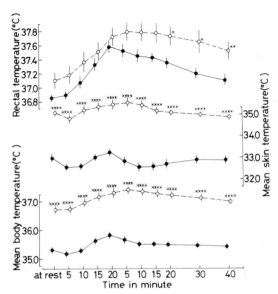

Fig. 8. The time course responses of rectal temperature, mean skin temperature, and mean body temperature of female subjects during 20-min exercise at work load of 2 kp and recovery period of 40 min in 23°C and 35°C. ✳ Significant difference between 23°C and 35°C, ✳ at 5% level, ✳✳ at 2% level, and ✳✳✳✳ at 0.1% level. Vertical bars indicate standard errors. — in 23°C, --- in 35°C.

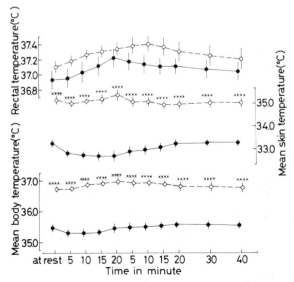

Fig. 9. The time course responses of rectal temperature, mean skin temperature, and mean body temperature of female subjects during 20-min exercise at work load of 1 kp and recovery period of 40 min in 23°C and 35°C. ✳ Significant difference between 23°C and 35°C, ✳✳✳✳ at 0.1% level. Vertical bars indicate standard errors. — in 23°C, --- in 35°C.

exercise was considerably greater in women than in men. The slope of the decrease in rectal temperature during recovery was steeper in women than in men. The mean values of rectal temperatures in both sexes increased during exercise at 35°C and continued to increase within 5 to 10 min after cessation of exercise, then decreased slowly. The magnitude of decrease in rectal temperature during recovery at 35°C was greater in women than in men. Increased ambient temperature caused a greater rise in rectal temperature in men than in women.

The mean skin temperature fell during the initial 5 min of exercise, then increased gradually to the end of exercise, except in the case of exercise at the work load of 1 kp for women at 23°C. The mean skin temperatures for men and women decreased at 23°C and 35°C when exercise ceased; those for women decreased at 23°C, then increased slightly, followed by a decrease. When women exercised at a work load of 1 kp, their mean skin temperatures increased at 23°C and decreased at 35°C after cessation of exercise. The mean skin temperatures throughout the experiment were significantly lower at 23°C than at 35°C.

DISCUSSION

It is well known that metabolic heat produced in the body must be dissipated into the environment at the rate of its production in order to maintain thermal equilibrium. When muscular exercise is performed, the large quantities of metabolic heat produced tend to accumulate in the body. With an increase in ambient temperature, the extra heat produced during exercise accumulated more rapidly in the body as a consequence of insufficient heat dissipation. Thus, the mean values of rectal temperature increased during exercise in both sexes. The increases observed at 35°C were significantly greater than those at 23°C (Figs. 7, 8, 9). At 23°C, a higher rise in rectal temperature for women at the end of exercise of 2 kp could be attributed to a greater work rate per body surface area for women. The lesser rise in rectal temperature in women due to a rise in ambient temperature probably reflects a lesser increase in the O_2 intake during exercise and consequently a lesser increase in heat production (Table 3). As shown in Figs 7, 8, and 9, fall in mean skin temperature and mean body temperature was observed at the beginning of exercise. The fall in skin temperature might be brought about by the increased convective heat loss due to exercise and cutaneous vasoconstriction caused by increased sympathetic activity. The fall in mean body temperature probably indicates that the formulas used to calculate mean skin temperature and/or mean body temperature in this experiment were inadequate during the early period of exercise. As shown in Tables 2 and 3, total oxygen intake for men was significantly greater at 35°C than at 23°C; consequently, the energy requirement was considerably greater. The increased oxygen intake at 35°C might reflect an increased metabolism due to higher body temperature (4) and an increased energy requirement for heat dissipation as indicated by a higher heart rate, increased ventilatory volume, and more profuse sweating. Brouha et al. (2) reported that oxygen intake during exercise was less in a hot environment than at room temperature. It is known that physiological reactions to heat are affected markedly by humidity (5,6). The lower humidity in the hot environment used by Brouha *et al.* (2) probably accounts for their observation of a decrease in oxygen consumption during work in a hot environment. Table 3 shows net efficiency for men was less at 35°C than at 23°C. Since net efficiency is inversely proportional to energy requirements when the subjects work at a fixed rate, a smaller net efficiency at 35°C reflects the increased energy requirements at 35°C. Changes in oxygen intake and energy requirement occurring with a rise in ambient temperature were similar but smaller in women. According to Hardy and DuBois (4), women are able to reduce their heat production in the basal state when they are exposed to environmental heat. Our results suggest that (compared to men) women could exercise at a given workload without a large increase in energy requirement resulting from a rise in environmental temperature. This relatively moderate increase in metabolism in a hot environment is obviously favorable for temperature regulation in women and probably accounts for the smaller rise in rectal temperature in spite of a smaller increase in sweat volume. As shown in Table 3, the mean values of oxygen debt for men and women were greater at 35°C than at 23°C. However, the magnitude of the increase in oxygen debt at 35°C was markedly greater in men than in women. Gupta *et al.* (13) also reported that the anaerobic fraction of work increased with an increase in environmental temperature. Table 2 shows oxygen intake increased at 35°C during both the early and late periods of recovery. The early phase of oxygen debt is attributable to the alactic phase, while the late phase of oxygen debt represents the lactic phase (10). The increase in oxygen intake during the early period of recovery at 35°C probably reflects the increase in metabolism caused by a higher body temperature, as well as an increased energy requirement for heat dissipation as described above. In a hot environment, increased blood flow through the skin is brought about by cutaneous vasodilation, and blood flow through the working muscle might tend to be reduced (11,15,17,18). Consequently, an increase in oxygen intake during the late period of recovery at 35°C might reflect the increased production of lactic acid in working muscles (10) as a result of an insufficient blood supply to the working muscle, as well as an increase in metabolism due to a higher body temperature during the recovery period. As shown in Figs. 4, 5, and 6, there was a decrease in diastolic blood pressure during rest and recovery as a result of an increase in ambient temperature, and diastolic pressure after exercise was lower than the pre-exercise level (12). The trend of lower diastolic blood pressure during rest and recovery at 35°C might be due to the marked dilation of cutaneous vessels caused by a higher body temperature. The fall in diastolic pressure, during the recovery period, below pre-exercise levels could be induced in part by a higher body temperature and in part by hyperventilation after exercise. In conclusion, changes in physiological responses during exercise, occurring with an increase in ambient temperature, were less in women than in men, and their smaller increase in energy requirement due to heat exposure was favorable for temperature regulation in women during exercise in a hot environment.

SUMMARY

Physiological responses and oxygen intake during exercise and recovery were measured on ten young female Japanese during the follicular phase of their menstrual cycle and seven young male Japanese in a climatic chamber at 35°C, 50% R.H., and at 23°C, 50% R.H., respectively. After sitting on a chair for 30 min in a climatic chamber, subjects rested on the saddle of a Monark bicycle ergometer for 10 min and then pedalled at constant work load (600 kg • m/min for both groups and 300 kg • m/min for women) at a cycling rate of 50 rpm for 20 min. After exercise, subjects recovered while seated for 40 min. Body weights were measured prior to and immediately after the experiment. Oxygen intake, blood pressure, heart rate, rectal temperature, and skin temperature were measured during periods of rest, exercise, and recovery. Higher ambient temperature had remarkable influences on sweat rate and cardiac function as indicated by a higher heart rate at 35°C than at 23°C. The mean values of oxygen intake during exercise and oxygen debt and body (rectal) temperature during exercise and recovery were greater at 35°C than at 23°C. The mean value of diastolic blood pressure during the periods of rest and recovery was lower at 35°C than at 23°C. Body weight loss, oxygen intake during exercise, oxygen debt, and fall in diastolic blood pressure during recovery in the male group at 35°C were considerably greater than for women. Increase of rectal temperature was less in women than in men. Higher heart rates at 35°C than at 23°C might have resulted from the increase in blood flow through the skin for heat dissipation at 35°C. The increase in oxygen intake in a hot environment might reflect an increased metabolism caused by higher body temperature and the increased energy requirement for heat dissipation as indicated by more profuse sweating, higher heart rate, and increased ventilatory volume. When work is done in a hot environment, skin blood flow increases and blood flow through the working muscle might be reduced. Therefore, the marked increase in oxygen debt and oxygen intake during the late period of recovery at 35°C probably reflects the increased production of lactic acid in the working muscle as a result of an insufficient blood supply to the muscle.

The most distinct sex differences of physiological responses during exercise in a hot environment were more profuse sweating and more marked increase in oxygen intake in men. A smaller rise in rectal temperature for women in spite of a smaller increase in sweat volume in women in a hot environment could reflect the smaller increase in oxygen intake in women than in men in a hot environment.

We have complied with the Recommendations of the Declaration of Helsinki in this experiment.

REFERENCES

1. Bean, W.B. and L.W. Eichna. Performance in relation to environmental temperature: reactions of normal young men to simulated desert environment. *Fed. Proc.* 2:144-158, 1943.

2. Brouha, L., P.E. Smith, R. De Lanne, and M. E. Maxfield. Physiological reactions of men and women during muscular activity and recovery in various environments. *J. Appl. Physiol.* 16:133-140, 1961.

3. Burton, A.C. The average temperature of the tissues of the body. *J. Nutr.* 9:261-280, 1935.

4. DuBois, E.F. The basal metabolism in fever. *J. Am. Med. Assoc.* 77:352-355, 1921.

5. Fox, R.H., R. Goldsmith, D.J. Kidd, and H.E. Lewis. Acclimatization to heat in man by controlled elevation of body temperature. *J. Physiol. (London)* 166:530-547, 1963.

6. Fox, R.H., R. Goldsmith, I.F.G. Hampton, and T.J. Hunt. Heat acclimatization by controlled hyperthermia in hot-dry and hot-wet climates. *J. Appl. Physiol.* 22:39-46, 1967.

7. Harashima, S. and S. Kurata. Computation of mean skin temperature. *Proc. Research Committee on Physiological Reaction to Climatic Seasonal Change* 21:7-9, 1952 (in Japanese).

8. Hardy, J.D. and E.F. DuBois. Basal metabolism radiation, convection, and vaporization at temperature of 22 to 35°C. *J. Nutr.* 15:477-497, 1938.

9. Hardy, J.D. and E.F. DuBois. Difference between men and women in their response to heat and cold. *Proc. Nat. Acad. Sci.* 26:389-398, 1940.

10. Hill, A.V., C.N. Long, and H. Lupton. Muscular exercise, lactic acid, and the supply and utilization of oxygen. *Proc. Roy. Soc. (London), Series B* 97:84-138, 1924.

11. Klausen, K., D.B. Dill, E. Phillips, and D. McGregor. Metabolic reaction to work in the desert. *J. Appl. Physiol.* 22:292-296, 1967.

12. Lowsley, O.S. The effects of various forms of exercise on systolic, diastolic, and pulse pressures and pulse rate. *Am. J. Physiol.* 27:446-466, 1911.

13. Sen Gupta, J., G.P. Dimri, and M.S. Malhotra. Metabolic responses of Indians during submaximal and maximal work in dry and humid heat. *Ergonomics* 20:33-40, 1977.

14. Radigan, L.R. and S. Robinson. Effects of environmental heat stress and exercise on renal blood flow and filtration rate. *J. Appl. Physiol.* 2:185-191, 1949.

15. Robinson, S. Physiological adjustments to heat. In: *Physiology of temperature regulation and the science of clothing.* Edited by L.H. Newburgh. Philadelphia: Saunders, 1949, pp. 193-231.

16. Robinson, S., E.S. Turrell, H.S. Belding, and S.M. Horvath. Rapid acclimatization to work in hot climates. *Am. J. Physiol.* 140:168-176, 1943.

17. Rowell, L.B., J.R. Blackman, R.H. Martin, J.A. Mazzarella, and R.A. Bruce. Hepatic clearance of indocyanine green in man under thermal and exercise stresses. *J. Appl. Physiol.* 20:384-394, 1965.

18. Taylor, H.L., A.F. Henschel, and A. Keys. Cardiovascular adjustments of man at rest and work during exposure to dry heat. *Am. J. Physiol.* 139:583-591, 1943.

19. Von Döbeln, W. A simple bicycle ergometer. *J. Appl. Physiol.* 7:222-224, 1954.

20. Weiman, K.P., Z. Slabochova, E.M. Bernauer, T. Morimoto, and F. Sargent II. Reactions of men and women to repeated exposure to humid heat. *J. Appl. Physiol.* 22:533-538, 1967.

21. Wyndham, C.H., J.F. Morrison, and C.G. Williams. Heat reactions of male and female Caucasians. *J. Appl. Physiol.* 20:357-364, 1965.

22. Wyndham, C.H., N.B. Strydom, J.F. Morrison, C.G. Williams, G.A.G. Bredell, J.S. Maritz, and A. Munro. Criteria for physiological limits for work in heat. *J. Appl. Physiol.* 20:37-45, 1965.

PREDICTION OF HUMAN HEAT TOLERANCE

Ralph F. Goldman

U.S. Army Research Institute of Environmental Medicine
Natick, Massachusetts

ASSESSMENT OF A "HOT" ENVIRONMENT

Man interacts with his thermal environment by four different avenues of heat exchange: conduction, convection, evaporation, and radiation. Assessment of heat stress clearly requires measurement of those environmental factors that control the heat transfer between man and his environment by these four avenues.

Conduction of heat occurs at the interface between the body's skin surface and the surface of any contacting substance, be it solid, liquid, or gas. The transfer of energy occurs as the result of a microcosmic billiard game, with direct contact of the molecules at the interface transferring thermal energy in proportion to the temperature difference of the two objects until the two surfaces come to equal temperatures; at that point, although the surface electrons continue to "bang" each other, both surfaces have essentially equal energy. The heat exchange between them is balanced, so no net heat flow occurs between the surfaces. When the human body is in contact with a finite body, the amount of heat exchanged until this equilibrium is established depends on the relative masses (kg) of the two bodies, their specific heats (kcal $\cdot$ kg^{-1} $\cdot$ $^{\circ}$C^{-1}), and the initial temperature difference.

The specific heat of the human body tissues averages 0.83 kcal $\cdot$ kg^{-1} $\cdot$ $^{\circ}$C^{-1}). We can calculate an average "comfortable" body temperature ($\overline{T}_b \simeq 35^{\circ}$C) by considering the body as a two-compartment model with one-third of its mass at skin temperature ($\overline{T}_s = 33^{\circ}$C) and the remaining "core" at rectal temperature ($T_{re} = 37^{\circ}$C).

Thus:
$$\overline{T}_b = 1/3 \, \overline{T}_s + 2/3 \, T_{re}. \qquad Eq \; 1$$

Assume an average man [defined as a man of 70-kg weight (m), 18.5% body fat, 173 cm tall, with 1.8 m² of body surface area (A)] ingests 1 ℓ of 60°C water — carefully, since this is near the maximum tolerable temperature for a hot beverage. Then, since the mass of 1 ℓ of water is 1 kg and the specific heat of water is 1 kcal $\cdot$ kg^{-1} $\cdot$ °C^{-1}, his mean body temperature must rise as a result of the extra 25 kcal of heat energy [i.e. for the water, c_p x m x (60 - 35) = 1 kcal $\cdot$ kg^{-1} $\cdot$ °C^{-1} x 1 kg x 25°C). Assuming none of these 25 kcal is transferred from the body's skin to the environment, then mean body temperature will rise by ~0.4°C [i.e. by 25 ÷ (0.83 x 70)]. Note that this relationship, i.e. that an average man (m = 70 kg) changes his body temperature ($\overline{T}_b$) by 1°C when there is a 58.1 kcal (i.e. 0.83 x 70) change in his heat content, is the same whether heat is gained or lost from the body. Thus, a measured change in mean body temperature can be calculated to represent a change in body heat storage (ΔS) of a given number of kcal over a period of time, i.e.

$$\Delta S = 0.83 \times m \times (\overline{T}_{b_1} - \overline{T}_{b_0}) \qquad\qquad Eq\ 2$$

where $\overline{T}_{b_0}$ represents the initial mean body temperature at some initial reference time, and $\overline{T}_{b_1}$ represents the temperature at the end of the time interval for which the change in heat storage is being calculated.

Conduction of heat is generally of limited interest in thermal environmental heat physiology, since man is seldom immersed in hot water and he rarely lies down on hot desert sand. Thus, the surface area across which heat conduction occurs is limited, and since all heat transfer is a linear function of the surface area involved, the magnitude of human heat transfer by conduction is generally small enough to be ignored. Nevertheless, the equation for conductive heat exchange (H_{con}):

$$H_{con} = k\ A\ (\overline{T}_s - T_{obj}) \qquad\qquad Eq\ 3$$

where: k is the coefficient of conduction
           A is the contact area
           $\overline{T}_s$ is the body skin temperature, and
           T_{obj} is the object surface temperature

is of significance in calculating time/temperature relationships for contact burns, with long (3- to 5-hr) contacts with 42°C wires producing pinpoint blisters even though the pain threshold for skin contact is closer to 45°C (14).

Heat transfer by convection occurs only as a sequel to conductive heat transfer; the initial transfer of heat energy is by "billiard ball" conduction at the interface of any solid, liquid, or gas. However, with liquids or gases, the initial energy received by conduction can set up a flow of energy away from the surface by "convection" currents. This close association of convection with conduction is reflected in the terminology of earlier thermal physiologists, who occasionally used the terms interchangeably. Heat transfer by convection in air is, again, a linear function of surface area (A); generally, the entire 1.8 m² surface area can be considered as available for convective transfer for the average man, with no allowance for dead air space between the extremities and the torso. Any imprecision introduced is probably no worse than that induced by estimating another value for the available skin surface area, and by the effects of local turbulence around the various body cylinders, since the convective heat transfer coefficient (h_c) is a function of the ambient air motion (V) taken to the 0.6 power ($V^{0.6}$). The amount of heat lost by convection depends upon the difference between the skin surface temperature ($\overline{T}_s$) and the air temperature (T_a).

There is an insulating film of still air (I_a) surrounding every physical object; its thickness depends on the relative motion of the air layer. To avoid errors in measuring air temperature that might result from heat stored in the thermometer being retained by this insulating film, the usual measurement of air temperature is with a dry thermometer bulb exposed to an air motion > 3 m/s. This movement is produced by using a thermometer slung at the end of a chain (a "sling" psychrometer), or an "aspirating" psychrometer with air drawn over the thermometer by a fan. The "dry bulb temperature" (T_{db}) measured is taken as the air temperature.

The equation for convective heat loss (H_c) can therefore be written:

$$H_c = kV^{0.6} \, A \, (\overline{T}_s - T_{db}). \qquad\qquad Eq\ 4$$

The heat transfer coefficient (h_c) equals $kV^{0.6}$, and k depends on such properties of the surrounding medium as its density, viscosity, etc., and a dimension/shape factor for the body, etc. While these details can be calculated fairly rigorously from physical principles for the nude man, the confounding effects of clothing generally defy rigorous physical analysis, and less rigorously derived approximations are used. Belding (1) suggests a value of 12 kcal • hr⁻¹ • °C⁻¹ for an average "nude" laboratory subject (1.8 m²) with a 35°C skin temperature in a hot environment (i.e. convective heat gain) wearing shorts and tennis shoes at a given air motion (V in m/s), e.g.

$$H_c = 12\ V^{0.6}\ (T_{db} - 35), \qquad\qquad Eq\ 5$$

and suggests that adding a light, long sleeved shirt and trousers reduces h_c, and thus H_c, by 40%. An approach involving direct measurement of clothing insulation (clo) can also be used to determine h_c.

Two factors from the thermal environment have thus far been identified if one is to assess a hot environment: air temperature (T_{db}) and air velocity (V). A third factor is required in environments where the temperature of walls or other objects in the surround differs from air temperature. Indeed heat transfer by radiation not only is independent of air temperature, but occurs whether air is present or not. Even in a vacuum, heat is transferred by radiant energy exchange; the wave lengths of the radiant beams exchanged between any two objects are related to the temperature of their surfaces, and the net heat transfer by radiation is proportional to the difference between their absolute temperatures to the fourth power and to the relative reflective and absorptive properties of the two surfaces. Generally, the temperatures of objects which surround the human body, except for the sun, are still low enough that the wave length of the heat radiation allows almost total absorption of the energy at the skin or clothing surface; i.e. the body surface behaves like a black body and its temperature is raised above the ambient air temperature to an extent determined by the rate at which convection prevents the still air layer at the surface from entrapping all the radiant heat arriving at the surface. Mean radiant temperature, as it affects the human body, is most readily assessed using an instrument where the relationship between its convection (h_c) and its radiation coefficient (h_r) is approximately the same as the relationship for a human body. A 15.4-cm globe, painted flat black, has a value of 0.178 for the ratio h_c/h_r, which closely approximates the corresponding ratio for an average human body (16). The globe temperature, measured by a thermometer whose bulb is at the center of the hollow, thin-walled, blackened 15.4-cm sphere, is used directly in several environmental heat stress indices. However, the net radiant heat exchange between a man and his radiant surroundings requires estimation of an average, or mean radiant temperature (MRT); MRT can be calculated from the black globe temperature (T_g) as:

$$MRT = (1 + 0.222\ V^{0.5})(T_g - T_{db}) + T_{db}. \qquad\qquad Eq\ 6$$

Such an approach is, again, of greatest use for assessing the net radiant heat gain (or loss if MRT $< T_s$) for a nude body. Belding suggests estimation of radiant heat exchange for a subject wearing shorts and tennis shoes by the relationship:

$$H_R = 11\ (MRT - \overline{T}_s) \qquad\qquad Eq\ 7$$

with the same reduction to 60% of this level as for convection when a long sleeved shirt and trousers are worn. The area of an average man for radiant exchange is some 20% smaller than the total 1.8 m² of skin surface, because some of the skin surfaces (e.g. legs, arms) face each other rather than the ambient radiant environment. Nevertheless, few workers make any distinction and the entire 1.8 m² is used in the calculations. Direct measurement of clothing insulation with a heated copper manikin avoids this problem, and provides a combined measurement of $H_c + H_R$.

Solar heat load poses a different set of problems. We have suggested (2) that the three components of a solar heat load[a] received by the body can be handled by three pairs of equations, one of each pair treating the fraction of the total solar load absorbed at the body surface, the other that fraction transmitted through the clothing to the skin surface. The total solar heat load to the surface of a clothed man can, we suggest (12), be estimated from the solar constant, zenith angle, cloud cover, humidity and dust content of the air, and the like. Although complex, this approach seems preferable to the oversimplification of simply increasing the T_{db} by 7°C to compensate for a maximum solar load in the desert, as suggested by Lee (9).

The fourth and final environmental factor that must be assessed to describe a thermally hot environment is relevant only for physical objects with water available at their surface, for evaporation. The capacity of air to take up water is limited by the temperature of the air; a thermometer with a wick saturated with water surrounding its bulb is used to measure the ability of air to take up additional moisture. Air holding all the water it can is said to be "saturated," at 100% relative humidity.

---

[a] (1) Incident direct radiation; (2) Diffuse indirect solar radiation; and (3) Albedo terrain reflected solar radiation.

In such an environment, no water will be evaporated from the wetted wick, and thus the measured "wet bulb temperature" (T_{wb}) will be the same as the air temperature (T_{db}). When the environmental air is at less than 100% relative humidity, evaporation can take place, but may be limited by the still air layer surrounding the wet bulb as it becomes saturated. In the measurement of T_{db}, any radiant heating was minimized by using a sling psychrometer or aspirated measurement of dry bulb temperature; the same approach (inducing air movement) is used to maximize evaporative cooling. The psychrometric wet bulb temperature (T_{wb}) so measured is converted to an equivalent relative humidity with a psychrometric chart. Indeed, one can plot tolerance time for heat (and cold) on such a chart (Fig. 1); note that the T_{db} are represented as perpendicular lines (and cold tolerance follows these lines), while the T_{wb} are diagonal lines, and heat tolerance falls along these diagonals. The point of intersection of the perpendicular T_{db} and diagonal T_{wb} lines representing a pair of psychrometric measures for a given environment, gives the relative humidity for that combination of T_{db} and T_{wb}. More importantly, it also establishes the vapor pressure of the water contained in the air; the ambient vapor pressure is given on the y axis, at the level of the intersection of $T_{db} + T_{wb}$. The saturated vapor pressure (P_a) for a given T_{db} is only a function of the T_{db}; P_a is obtained from the chart at a level parallel to the point on the 100% RH line where T_{db} equals T_{wb}. The vapor pressure at different relative humidities (ϕ_a) can be calculated as $\phi_a P_a$. Note that the slope of the T_{wb} lines is constant across all temperatures and approaches 2.2°C/mmHg; this 2.2 value can be derived from the physical relationship between the evaporative heat transfer coefficient (h_e) and the convective heat transfer coefficient (h_c) and is called the "Lewis number" or psychrometric constant. The actual slope of the T_{wb} lines is closer to 2°C, because of radiant and convective heat gained by the cooler, wet surface even with a psychrometric wet bulb measurement. However, the Lewis relationship implies that the evaporative heat transfer coefficient should be directly related to the convective heat transfer coefficient, with the relationship $h_e = 2.2 h_c$.

Water will actually condense out on any surface which is below air temperature in a saturated environment, since vapor pressure in the air is greater than the vapor pressure at the surface. Water will accumulate on the surface, imposing a heat load of 0.58 kcal/cc of water so condensed; i.e. the specific heat of condensation of water is 0.58 kcal/g. Thus, a substantial heat load can be incurred in a typical 49°C "Turkish" steam bath just from the heat of condensation. Conversely, evaporative cooling can occur from a wetted surface even into a saturated environment as long as the surface temperature is above air temperature. The controlling difference for evaporative cooling is not, therefore, a relative humidity gradient, but the gradient between the vapor pressure of water at the surface temperature (P_s) and the water vapor pressure of the air ($\phi_a P_a$). One can consider that the water evaporates from the skin surface into the adjacent, still air layer, which is unsaturated because its temperature is above the air temperature, and then moves away by convection to condense at the interface between the still air layer and the ambient air. Such a reaction produces a readily observed mist when one removes a sweaty glove or sock from the skin in a cold environment. Evaporative cooling removes 0.58 kcal/cc of water evaporated; i.e. the latent heat of evaporation is 0.58 kcal/g.

Evaporative heat transfer (H_E) for a human body is almost always a heat loss (except in a steam room as noted above) and can be quantified as:

$$H_E = 2.2\, h_c\, i_m\, A\, (P_s - \phi_a P_a) \qquad\qquad Eq\ 8$$

where the area for evaporation is generally considered to be the entire surface area (1.8 m^2 for an average man), P_s is taken as the vapor pressure of water at skin temperature (without adjustment for any solute content), h_c is the convective coefficient, as before, and i_m, a dimensionless constant ranging from zero to one, represents the permeability of any clothing and/or the still air layer around the body to evaporative transport of water. This physical relationship is, as for convection and radiation, rigorously definable only for a nude man. Clothing impermeability (i_m) to water vapor transfer, and also the insulation of clothing (clo) involved in the h_c term confounds its application to the clothed man. Belding (1) suggests estimation of evaporative heat loss for a subject wearing shorts and tennis shoes as:

$$H_E = 23\, V^{0.6}\, (42 - \phi_a P_a) \qquad\qquad Eq\ 9$$

under the assumption that the skin temperature of a "nude," sweating man is fixed at 35°C (hence, $P_s = 42$ mmHg); for a man wearing a long sleeved shirt and trousers, he estimates H_E as, again, 60% of the "nude" value. Note that, although Belding's work was carried out before physiologists generally became aware of the Lewis relationship, his empirically derived h_e is not too remote from 2.2 times his h_c (cf. *Eq 5* and *9*; $h_e = (23/12)\, h_c$, or 1.92). Note also that the H_E defined above represents the maximum evaporative cooling (E_{max}) allowed by the environment, and is in no way related to the level of required evaporative cooling (E_{req}) needed by the man. The calculated E_{max} assumes a 100% sweat-wetted surface area, while if the man does not need any evaporative cooling his skin will be dry.

The four thermal environmental characteristics that must be assessed to characterize any hot environment have now been detailed: T_{db}, T_{wb} and its associated RH and $\phi_a P_a$; V; and T_g and its associated MRT. Solar heat load, although beyond the scope of this paper, has also been referenced. Whether or not a given hot environment represents a heat stress requires consideration of two additional factors. One has already been invoked (clothing) in terms of the extent to which its insulation (clo) alters convective and radiative transfer and its permeability (i_m/clo) alters evaporative transfer between the body surface and the environment (cf. *Eq 8*). The difficulties of dealing rigorously with the effects of clothing have been detailed above; the direct measurement of insulation (clo; 1 in. of clothing thickness provides ~ 4 clo of insulation) and permeability (i_m; about 0.45 for conventional clothing) is a simpler approach. The second "human" factor that determines whether a given hot environment is stressful is the rate of metabolic heat production by the individual. As shown in Table 1, this depends on any physical work required of the body,

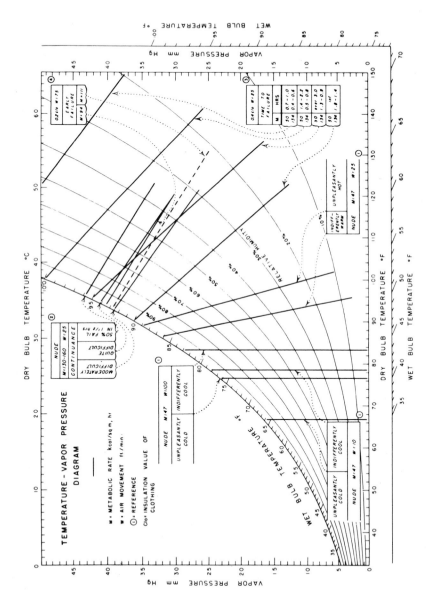

Fig. 1. Psychrometric chart, with observed responses on tolerance limits indicated for nude and partially clothed men.

from the 0.8 MET^a required for the central nervous system, for circulation of blood, for respiration and digestion at rest, to 12.5 MET for exhausting physical work; e.g. 70 kg lifted a distance of 1 m 9 times per min represents 630 kg • m of physical work which, assuming a gross efficiency of 20% for moderate work by the whole body, requires ~ 5 MET of heat production (cf. Table 1 for Physical Work of 640 kg • m • min^{-1}).

The physiologic response of the body to these physical stresses of the work load and thermal load of the environment can be assessed using the classic (15) heat balance equation for the human body-

$$M \pm H_R \pm H_C - E = \Delta S \qquad\qquad Eq\ 10$$

where: M is the metabolic heat production demanded by the work (cf. Table 1)
H_R is the radiant heat exchange (cf. *Eq 7*)
H_C is the convective heat exchange (cf. *Eq 4* and *5*)
E is the actual evaporative heat loss
ΔS is the heat storage (or debt) in the body (cf. *Eq 2*).

This simplified form of the human heat balance equation ignores respiratory heat and water loss and also any evaporation of the moisture continuously diffusing through the semipermeable human skin. These respiratory and moisture diffusion avenues contribute equally to a heat loss totaling about 25% of M at rest in a comfortable environment. The respiratory portion increases with increasing work, but decreases with increasing ambient vapor pressure, while the relatively small diffusional evaporative loss (~ 20 ml/hr or about 12 kcal/hr) becomes an insignificant portion of the overall evaporative cooling when man is actively sweating.

Ideally, the left hand side of the heat balance equation equals zero; i.e. heat production plus heat load equals heat loss, and there is no need for heat storage or debt by the body. Such a balance may be equated to an absence of heat stress, but in fact may not be achieved without considerable heat strain. The sum of the first three terms (heat production ±heat exchange by radiation and convection) provides a useful estimate of the evaporative cooling required (E_{req}), if any; i.e.:

$$E_{req} = M \pm (H_R) \pm H_C. \qquad\qquad Eq\ 11$$

The maximum obtainable evaporative cooling (E_{max}) is constrained by three factors: the ability of the body to produce sweat, the ($\overline{P}_s - \phi_a P_a$) difference discussed above and the extent to which the clothing and still air permeability and insulation, acting in combination (i_m/clo), allow evaporation between the skin surface and the ambient air.

Body temperature is regulated by a variety of behavioral and physiological mechanisms. Homeothermic animals, including man, tend to become less active in the heat, thus reducing M, and to increase their insulation in the cold, fluffing feathers or raising fur (which in man produces "goose flesh" but little benefit). Man can also

---

^{a} The MET is a basic unit of heat production found useful in thermal environmental physiology; 1 MET, by definition = 50 kcal• m^{-2} •hr^{-1}.

Table 1. Relation Between Physical Work and Physiological Cost (Assuming 20% Efficiency), With Some Representative Observations on Fit Young Men[a]

PHYSICAL WORK			ENERGY COST (η = 20%)			ACTIVITY	"MET"	MEASURED "NORMAL" MEN			H.P. (η = 20%)	Efficiency (%)
kg·m·min⁻¹	ft·lb·min⁻¹	watt	kcal/min	watt	$\dot{V}O_2$ ℓ/min			RMV ℓ/min	$\dot{V}O_2$ ℓ/min	H.R b/min		
13	93	2	0.15	10	0.03	Circ. + resp. rest	0.1					
26	185	4	.30	21	.06	C.N.S.	.2					
38	278	6	.45	31	.09	Circ. + resp. work	.3					
64	463	10	.75	52	.15	Gut at rest	.5					
102	741	17	1.2	84	.24	Basal (sleep)	.8					
128	926	21	1.5	105	.30	Sit at rest	1.0					
192	1,389	31	2.25	157	.45		1.5					
224	1,621	37	2.63	183	.53	Very light	1.75	10	0.5	<75	0.05	
256	1,852	42	3.00	209	.60		2.0					
320	2,316	52	3.75	262	.75	Walk 2.75 MPH	2.5					
384	2,779	63	4.50	314	.90		3.0					
416	3,010	68	4.88	340	.98	Light	3.25	20	1.0	75-100	.1	15.7
448	3,242	73	5.25	366	1.05		3.5					
512	3,705	84	6.00	419	1.20		4.0					
640	4,632	105	7.50	523	1.50	Moderate	5.0	35	1.5	100-125		
864	6,253	141	10.13	707	2.03	Heavy 1 hr	6.75	50	2.0	125-150	.2	20
						$\dot{V}O_2$ max						
1,056	7,642	173	12.38	864	2.48	Very heavy	8.25	65	2.5	150-175	.3	20
1,280	9,263	209	15.00	1,047	3.00	10 min $\dot{V}O_2$ max	10.00	85	3.0	>175	.35	20
1,600	11,579	262	18.75	1,308	3.75	Exhausting	12.50				.5	
2,202	15,933	360	25.8	1,800	5.16	2 mile record anaerobic (10 min)	17.2					

[a] RMV = respiratory minute ventilation. $\dot{V}O_2$ = oxygen consumption. H.R. = heart rate. H.P. = horsepower. η = efficiency. 1 MET = 50 kcal·m⁻²·hr⁻¹.

adjust his clothing, adding in the cold and opening closures or removing layers (at least to the social limits of an increasingly permissive society) in the heat; however, the effect of solar heat load is much greater on bare skin than with clothing. These behavioral mechanisms are supplemented by physiological mechanisms, with adjustment of skin temperature by vasomotor regulation serving as the first line of defense in both heat and cold. As the skin receives an increasing heat load from the thermal environment, vasoconstrictor tone is reduced, thus increasing skin blood flow from the core and raising skin temperature. As evident from *Eq 4, 5,* and *7,* raising $\overline{T}_s$ alters convective and radiant heat exchange, either increasing losses or reducing gains. An increase in core temperature may also trigger active vasodilation over most of the body skin surface (but not hands and feet?), resulting in a further increase in $\overline{T}_s$ (13).

As this first line of defense proves inadequate to balance heat losses by radiation and convection against heat production, a second, more powerful defense against heat storage, sweating, is initiated. Sweating seldom begins simultaneously over the entire skin surface, but is progressively recruited as the various skin surface areas increase from a "comfortable" 33°C level to a level where sweating begins, generally at a skin temperature of about 35°C. Sweat production is closely linked with obtaining the required evaporative cooling. Indeed, the "Predicted 4-Hour Sweat Rate" (P4SR) Nomogram (Fig. 2) has been used as a measure of the stress of work in a given environment (10).

As seen in Table 1, a heat production of 700 W represents "Heavy Work," sustainable for only about one hour for a man of average fitness as judged by his maximum oxygen uptake capacity ($\dot{V}O_2$ max). The sustainable level of sweat production is about 1 ℓ per hour, which at ~ 0.58 kcal of potential cooling per ml of sweat evaporated approaches 700 W (580 kcal/hr) of skin surface evaporative cooling. Sweat production can be at higher rates for shorter periods, with short-term maximum rates exceeding 3 ℓ per hour, providing a potential maximum cooling rate of about 2,000 W.

By comparison with the maximum short-term heat production rates in Table 1, it is obvious that sweat production is not usually limiting; the body can produce enough sweat to compensate for most working heat productions with a reasonable surplus of sweat production capacity to offset most convective and radiative heat loads, unless the man is seriously dehydrated, has a heat rash covering a significant portion of his skin surface, or has congenital absence of sweat glands. Otherwise, capacity is clearly adequate to produce enough sweat to meet E_{req} (cf. *Eq 11*). The problem for most heat stress situations is the limitation on E_{max} (cf. *Eq 8* and *9*) imposed by elevated ambient vapor pressures at high temperatures (i.e. $\overline{P}_s \rightarrow \phi_a P_a$) or by clothing impermeability (i_m) and thickness (clo) which limit evaporation by their combined effect (i_m/clo).

Regardless of whether the limiting problem is with sweat production or is with E_{max}, the degree of heat stress placed upon the body can be predicted by comparing the required (E_{req}) and maximum (E_{max}) evaporative losses. Gagge *et al.* (3) have considered the simple ratio E_{req}/E_{max} as a measure of the percent of the skin

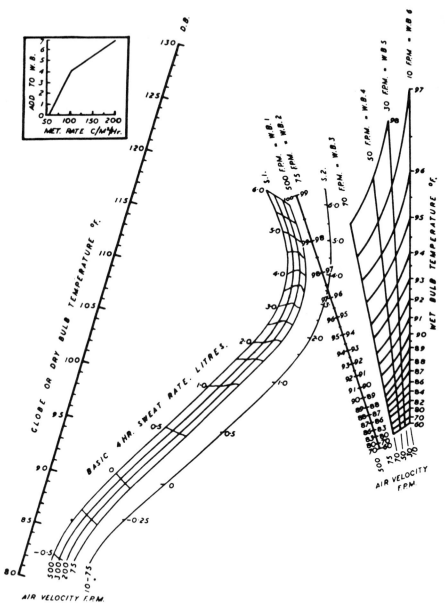

Fig. 2. McArdle's Predicted 4-Hour Sweat Rate (P4SR) nomogram, where values above 3 or 4 ℓ in 4 hr represent severe strain. (From MRC Special Report Series No. 298, *Physiological response to hot environment*, and ref. 10, by permission.)

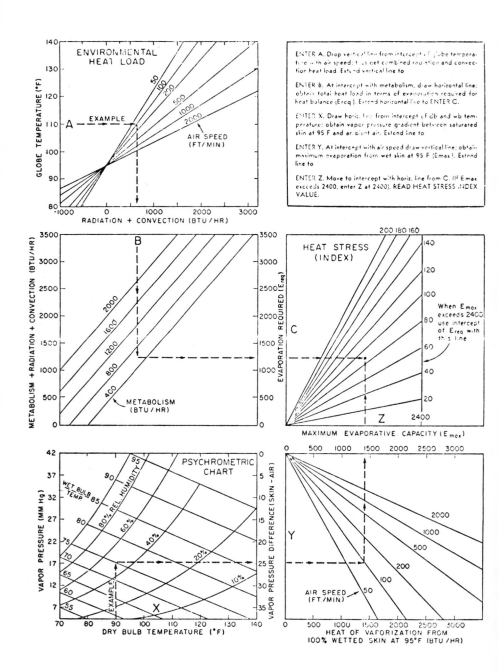

Fig. 3. Belding's Heat Stress Index (HSI) nomogram.

surface that is sweat-wetted (% SA_w) or as the effective relative humidity of the total skin surface; they suggest that values of E_{req}/E_{max} < 20% are compatible with comfort, while increasing percentages indicate tolerance limits. Belding and Hatch (1) use E_{req}/E_{max} as their Heat Stress Index (HSI) (Fig. 3). HSI values > 30% are uncomfortable and may depress mental and fine motor performance, but are tolerable; values of 40% to 60% reduce performance and may result in finite tolerance times; and values of 70% to 100% represent severe, tolerance-limiting situations.

One can also use the classic Effective Temperature Nomogram (ET) (17) as a predictor of heat intolerance, with or without the T_g substituted for the T_{db} to correct for a radiant heat load (CET). Although primarily used for delineating conditions where deficits are likely for psychomotor, cognitive, or light office work performance, the ET nomogram can be modified to reflect the maximum sustainable heat production at a given ET or CET (Fig. 4).

These four charts of environmental effects (Figs. 1,2,3,4; Psychrometric Chart, P4SR, HSI, and ET) can all be used to delineate potential heat tolerance problems. Another environmental index, the Wet Bulb Globe Temperature (WBGT), uses the evaporative cooling of a non-psychrometric, naturally convected wet bulb ($T_{wb_{np}}$). $T_{wb_{np}}$ is obviously a better characterization of the evaporative cooling available to a man who is not being slung by the heels or ventilated by a fan than is the psychrometric T_{wb}. WBGT is calculated as:

$$WBGT = 0.7\,T_{wb_{np}} + 0.2\,T_g + 0.1\,T_{db} \qquad\qquad Eq\ 11$$

and values < 80°F should produce no concern. Values of 80 to 85°F indicate potential problems during work by men not acclimatized (5 to 7 days of prior exposure at work for ~ 2 hr/day at that heat level), while above 85°F, even acclimatized men may have problems, and above 88°F work by fit, acclimatized men will be limited to about 8 to 10 hours (11).

The heat balance equation (*Eq 10*) can also be used to indicate heat tolerance problems and to predict tolerance times when the left hand side of the equation cannot be forced to equal zero; i.e. actual evaporative heat loss E < E_{req}. The heat that cannot be eliminated and must be stored by the body (ΔS) is calculated using the estimates for the individual terms in the heat balance equation. Heat storage of 0 to 25 kcal may not be sensed if incurred at a slow rate, while ΔS of + 80 kcal represents the usual voluntary tolerance limit for continued exposure, ΔS of + 160 kcal represents a 50% risk of heat exhaustion collapse (T_{re} ~ 39.5°C and HR ~ 180 b/min), and $\Delta S \geqslant 240$ kcal cannot be tolerated by fit young men. The time required to incur these totals may be used as estimates of the corresponding tolerance limit; voluntary, 50% or 100% risk of collapse.

Givoni and I, although relying on the above concepts, have taken a different approach (4). We postulate that there exists a level at which the body could achieve a stable state, with no further heat storage, for any combination of metabolic and environmental heat stress. Although it may not be possible to reach such a level before heat exhaustion or death intervenes, nevertheless the body will be driven, inexorably, toward this "equilibrium" level. There are three factors that drive rectal temperature to this equilibrium ($T_{re_{eq}}$). The first two, metabolic and convective plus radiative heat loads, are treated essentially as described in the heat balance equation. First, rectal temperature is elevated by 0.4°C per 100 W of heat production, above a baseline 36.7°C, independently of ambient temperature (as suggested by Nielsen in 1938). Second, it is adjusted by the combined radiative and convective heat transfer predicted by the clothing insulation (clo units),[a] by 0.0128°C/clo · ($\overline{T}_s$ - T_{db}); $\overline{T}_s$ is considered clamped at 35°C for the nude man and at 36°C for the clothed man. The third term, driving T_{re}, is (as might be anticipated from the heat balance equation) a function of the E_{req} in relation to the E_{max}. However, rather than use the ratio E_{req}/E_{max} which suggests that the stress is the same whether the body requires 50 W and can get a maximum of 100, or requires 500 and can get a maximum of 1000, we use the difference between the E_{req} and E_{max} as an exponential forcing function. T_{re} at equilibrium then is predicted as:

$$T_{re_{eq}} = 36.75 + 0.004 \ M + (0.0128/clo) \ (T_{db} - 36) + 0.8e^{0.0047 \ (E_{req} - E_{max})}$$

Eq 12

where: $E_{req} = M + (11.6/clo) \ (T_{db} - 36)$, and
            $E_{max} = 25.5 \ (i_m/clo) \ (44 - \phi_a P_a)$.

The similarity with the terms of the heat balance equation (*Eq 10*) of the body is obvious, although the coefficients are empirically derived.

The time constant for rectal temperature response is also variable, with equilibrium achieved earlier (1 to 2 hr) under low stress conditions; under severe stress, although collapse may occur in 15 to 30 min, the projected time to reach the unattainable balance point may be 5 hr or more. The time constant (k) for the rectal temperature response is a function of the difference (ΔT_{re}) between initial (T_{re_0}) and equilibrium rectal temperature ($T_{re_{eq}}$):

$$k = 0.5 + 1.5e^{-0.3 \ \Delta T_{re}}$$

Eq 13

so that the formula for rectal temperature, at any time t in hours (T_{re_t}), becomes:

$$T_{re_t} = T_{re_0} + \Delta T_{re} \ [1 - e^{-k \ (t - 58/M)}]$$

Eq 14

where (t - 58/M) reflects the time delay of rectal temperature response. This time-based relationship allows one to predict not only whether there are tolerance problems, but also the time to incur them. Whether the problem is an excessively high work load, convective and/or radiative heat transfer, or difficulty in obtaining the required evaporative cooling is delineated, respectively, by the last three terms in *Eq 12*. A level of 38°C has been selected as an upper deep body temperature, as a guide for setting industrial heat stress limits under OSHA. With rising $\overline{T}_s$, a T_{re} of 39.2°C can induce about 10% frank heat exhaustion collapse in fit young men (risk

---

[a] One clo allows a combined total radiation and convective exchange of 6.45 W per °C difference between $\overline{T}_s$ and T_o, where T_o is the operative temperature defined as $T_o = [(h_c \ T_{db} + h_r \ MRT)/(h_r + h_c)]$. With MRT ≈ T_{db}, T_{db} *per se* is used instead of T_o.

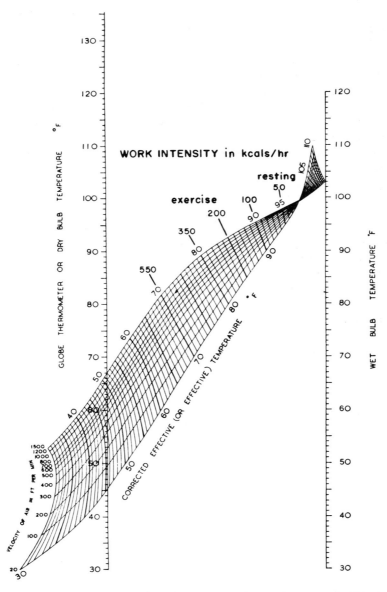

Fig. 4. Effective temperature (ET) nomogram, indicating maximum sustainable metabolic rates at various ET levels.

of heat exhaustion estimated as 25%), while T_{re} levels of 39.5°C, with rising $\overline{T}_s$, represent about a 50% risk of heat exhaustion collapse; T_{re} levels rising above 40°C suggest a risk of heat stroke (when $T_{re} > \sim 42$°C) for the very few, extremely fit individuals whose cardiovascular system allows them to continue to work at $T_{re} > 40$°C with elevated T_s.

Elevated heart rates are associated with these elevated T_{re}; prediction of equilibrium heart rate and time patterns of heart rate response allows prediction of time to heat exhaustion collapse based on heart rates reaching > 190 b/min (5). Factors for heat acclimatization adjustments to T_{re} and HR have been published (6), and factors for dehydration are being developed; both alter the equilibrium levels and time responses of T_{re} and HR, with T_{re_0} reduced by acclimatization.

These predictions have been validated by a number of subsequent experiments and also by reported exposures, with fit young men wearing a variety of clothing (shorts, work clothing, reduced permeability coveralls, armor, etc.) across a range of T_{db} (20 to 50°C), humidity (10 to 100%), and work rate, at low to high wind speeds. Adjustments for fitness (perhaps as a function of $\dot{V}O_2$ demand as a percentage of $\dot{V}O_2$max), for age (perhaps based on the relation $HR_{max} = 220$ - age) and for females (perhaps only as a function of $\dot{V}O_2$ demand/$\dot{V}O_2$max) are being sought.

In earlier work (7,8), concern has been expressed for the contribution of hyperventilation, and the resultant hypocapnia, in heat exhaustion collapse. However, preliminary results from current studies (unpublished) on patients with core temperatures elevated to the 42°C level as part of a cancer therapy regime have failed to show the severe hyperventilation anticipated from our earlier studies. Thus, it appears that the hyperventilation may be more a part of the experimental situational stress and may not need to be factored into the prediction of heat intolerance.

Prediction of $\overline{T}_s$ and core-skin conductance is also available (unpublished) and we are currently addressing convergence of $\overline{T}_s$ toward T_{re} as a heat tolerance limit in cases where E_{max} is severely limited by very high $\phi_a P_a$ or very low i_m/clo. While more work remains to be done, the three approaches given above (charts, heat balance equation, or prediction of T_{re}) should serve as a useful basis for prediction of heat tolerance problems.

SUMMARY

The physical and psysiologic bases for heat tolerance have been delineated. The psychrometric chart, P4SR, ET, HSI, and/or WBGT can all be used as indices of heat intolerance, but not, usually, as predictors of tolerance time *per se*. Use of the human heat balance equation and use of *validated* prediction models of heart rate and rectal temperature can be used to predict tolerance times. Using the latter approach, adjustments can be made for acclimatization and dehydration, and adjustments are being developed to account for age, sex, and physical condition differences.

The views of the author do not purport to reflect the positions of the Department of the Army or the Department of Defense.

Human subjects participated in these studies after giving their free and informed voluntary consent. Investigators adhered to AR 70-25 and USAMRDC Regulation 70-25 on Use of Volunteers in Research.

REFERENCES

1. Belding, H.S. and T.F. Hatch. Index for evaluating heat stress in terms of resulting physiological strain. *Heating, Piping and Air Cond.* 27: 129-136, 1955.
2. Breckenridge, J.R. and R. F. Goldman. Solar heat load in man. *J. Appl. Physiol.* 31: 659-663, 1971.
3. Gagge, A.P. A new physiological variable associated with sensible and insensible perspiration. *Am. J. Physiol.* 120: 277-287, 1937.
4. Givoni, B. and R.F. Goldman. Predicting rectal temperature response to work, environment and clothing. *J. Appl. Physiol.* 32: 812-822, 1972.
5. Givoni, B. and R.F. Goldman. Predicting heart rate response to work, environment and clothing. *J. Appl. Physiol.* 34:201-204, 1973.
6. Givoni, B. and R.F. Goldman. Predicting effects of heat acclimatization on heart rate and rectal temperature. *J. Appl. Physiol.* 35: 875-879, 1973.
7. Goldman, R.F., E.B. Green, and P.F. Iampietro. Tolerance of hot, wet environments by resting men. *J. Appl. Physiol.* 20: 271-277, 1965.
8. Iampietro, P.F., M. Mager, and E.B. Green. Some physiological changes accompanying tetany induced by exposure to hot, wet conditions. *J. Appl. Physiol.* 16: 409-412, 1961.
9. Lee, D.H.K. and J.A. Vaughan. Studies on thermal effects of solar radiation in transportable solar chamber. Am. Soc. Mech. Eng. Paper No. 60-WA-251, 1960.
10. McArdle, B., W. Dunham, H.E. Holling, W.S.S. Ladell, J.W. Scott, M.L. Thomson, and J.S. Weiner. Prediction of physiological effects of hot environments. Med. Res. Counc. London, R.N.P.R.C. 47/39, 1947.
11. Minard, D.H., H.S. Belding, and J.R. Kingston. Prevention of heat casualties (description of Yaglou wet bulb: globe temperature index). *J. Am. Med. Assoc.* 165: 1813-1818, 1957.
12. Roller, W.L. and R.F. Goldman. Estimation of solar radiation environment. *Int. J. Biometeorol.* 11: 329-336, 1967.
13. Rowell, L.B. Human cardiovascular adjustments to exercise and thermal stress. *Physiol. Rev.* 54: 75-159, 1974.
14. Webb, P. Pain-limited heat exposures. In: *Temperature – its measurement and control in science and industry.* Vol. 3, Part 3. New York: Reinhold Publishing Corp., 1963.
15. Winslow, C-E.A., L.P. Herington, and A.P. Gagge. A new method of partitional calorimetry. *Am. J. Physiol.* 116: 641-655, 1936.
16. Woodcock, A.H., R.L. Pratt, and J.R. Breckenridge. Theory of the globe thermometer. Res. Study Rpt. BP-7, Quartermaster R & E Command. U.S. Army Natick Labs, Natick, MA, 1960.
17. Yaglou, C.P. Temperature, humidity and air movement in industries: the effective temperature index. *J. Ind. Hyg.* 9:297-303, 1927.

HEART RATE–RECTAL TEMPERATURE RELATIONSHIPS DURING PROLONGED WORK: MALES AND FEMALES COMPARED

K. V. Kuhlemeier
P. R. Fine

Department of Rehabilitation Medicine
University of Alabama in Birmingham
Birmingham, Alabama

F. N. Dukes-Dobos

National Institute for Occupational Safety and Health
Cincinnati, Ohio

INTRODUCTION

Much effort has been expended by many groups of health workers and scientists to determine what constitutes acceptable limits of physiological strain during work in hot environments. There is, however, no clear consensus on what constitutes "acceptable" limits of strain, nor for that matter which physiologic parameter(s) constitutes the best indicator of strain. The most commonly suggested parameters include deep body temperature (T_{re}) and/or heart rate (HR). Several lines of evidence (9,10,11) suggest that $38.0^{\circ}C$ is a defensible upper limit of T_{re} that can be maintained for long periods without deleterious effects on the health of workers. This principle was also the basis of the Heat Stress Index of Belding and Hatch (1) inasmuch as they calculated the maximum allowable exposure time by determining how long it takes for the T_{re} to increase by 2°F, i.e. from 98.6°F (37.0°C) to 100.6°F (38.1°C), at a given environmental and metabolic load. It has also been suggested by the World Health Organization (WHO) Study Group (16) that HR is not a good indicator for maximum allowable heat exposure in seated or otherwise resting men; nevertheless, they stipulated that 110 bpm might be considered an upper limit that can be tolerated without development of cumulative fatigue. These conclusions were based on data from male subjects and may or may not be valid for females.

In the literature there is much evidence from male subjects (2,5,11,15,17) suggesting that during prolonged (1 to 2 hr) work in heat, subjects with rectal temperatures of 38.0°C will have heart rates substantially over the 110 bpm average, particularly during heavy work. We have also found this to be true in industrial workers during laboratory experiments (8). Furthermore, in male subjects the $HR-T_{re}$ relationship is affected by work rate and the degree of heat acclimatization (unpublished observations). The fact that HR is generally higher than 110 bpm when rectal temperatures average 38.0°C suggests that the strain reflected by a HR of 110 bpm might be less than the strain represented by a T_{re} of 38.0°C, since the cardiovascular and thermoregulatory systems should be under roughly equivalent levels of strain at a given combination of prolonged heat and work.

The present study was conducted to determine heart rate—rectal temperature relationships during prolonged work at different rates over a wide range of environmental temperatures and to determine whether gender and job- and season-acquired heat acclimatization had any influence on the $HR-T_{re}$ relationship.

METHODS

Subjects   Subjects for this study were healthy adult individuals recruited from the Birmingham, Alabama, industrial worker population. All males over 35 years of age and all females were given a graded exercise test by cardiologists prior to acceptance into the study to insure their safety during the testing procedures. Males in the study were recruited from the steel industry (including both open hearth and rolling mill workers), cement manufacture, aluminum reduction, and meat packing plants, as well as from our hospital staff (including students, janitors and orderlies). Female subjects included women from a laundry, a local foundry (which hires primarily women), housewives, students, hospital housekeeping staff, and secretaries. The "cold-neutral" workers referred to in this report include meat packers, students, housewives, hospital orderlies, housekeeping staff, and secretaries. The remaining subjects, i.e. those from the laundry, foundry, steel, aluminum, or cement industries, are considered to be from "hot" industries.

Insofar as possible, subjects were selected so as to obtain a representative sample with regard to age and race. The physical characteristics of the study population appear in Table 1. A total of 46 men and 22 women participated in this study and performed 947 work bouts. Participants ranged in age from 18 to 48 years. Females were tested without regard to the phase of their menstrual cycle. Three females were post-menopausal.

The studies were conducted during the winter (early November to mid-March) and summer (mid-May to early September) seasons. Subjects were not tested during their vacation nor during the week following their return from vacation to insure that they were in their usual state of acclimatization.

Test Protocol Subjects came to the laboratory dressed in their regular work clothing after eating their usual breakfast. The insulative value of their clothing averaged about 0.7 clo. ECG electrodes were attached and the rectal thermocouple inserted. Each subject then walked for one hour on a motor-driven treadmill in environments ranging from 13 to 35°C Corrected Effective Temperature (CET) at a low (22% $\dot{V}O_2$ max) or moderate (29% $\dot{V}O_2$ max) metabolic rate (MR). The specific environmental heat loads used for each work rate in this study are given in Table 2. Subjects generally performed three or four of these walks per day with a rest period of at least one hour between walks. Test days were at least one week apart. During rest periods, the weight loss from the preceding walk was replaced with salted water. Air velocity was measured at chest level and was usually below 0.3 m · s⁻¹. Dry bulb and globe temperature measurements were virtually identical in all experiments.

Measurements HR and T_{re} were determined just prior to cessation of walking. HR was taken from standard ECG tracing and was averaged over 10 beats. T_{re} was measured 9 to 11 cm past the anal sphincter with a copper-constantan thermocouple and a temperature recorder (Honeywell Electronik Model 112), which was calibrated at the beginning of each work day.

Tests were terminated if any of the following criteria was met: (1) HR exceeded

Table 1. Physical Characteristics of Male (M)
and Female (F) Subjects

	Mean		Range	
	M	F	M	F
Age, years	33.3	29.6	20—48	19—47
Height, cm	178	164	165—188	152—173
Weight, kg	82.1	61.4	59.1—118.2	50.9—77.3

Table 2. External Heat Loads Expressed in °C CET
Used to Determine HR—T_{re} Relationship in Males and Females

	Males	Females
Low MR	13,21,24,29,30,31,32,34	22,24,26,27,28,30,32
Moderate MR	13,21,24,28,29,30,31,32,34	22,24,26,27,28,29,30,31

90% of the age-adjusted predicted maximal HR; (2) T_{re} exceeded 39.2°C; (3) blood pressure exceeded 200 mmHg systolic or 110 mmHg diastolic; (4) the subject became faint; or (5) the subject asked that the test be terminated. Data from terminated tests are not included in this study.

Statistical Analysis   A simple linear regression line was fitted to the heart rate (dependent variable) and rectal temperature (independent variable) of each individual in each season at each work rate. The resulting lines were then analyzed to determine (1) the change in heart rate per degree rise in rectal temperature and (2) the predicted heart rate for a rectal temperature of 38.0°C (PHR_{38}) for each subject. The effects of industry classification, season, work rate, and gender on the slope of the heart rate—rectal temperature regression line and the PHR_{38} were then determined using a split plot analysis of variance with industry as the whole unit and season, work rate, and gender as the sub-units (13). Complete models, including all main effects, two-way, three-way, and four-way interactions, were first tested. None of the interaction terms was significant in either the slope ($p > 0.4$) or PHR_{38} ($p > 0.12$) analysis; consequently, the models were reduced to retain only the main effects. The significance levels of the main effects were then retested using a similar split plot analysis of variance.

RESULTS

The magnitude and significance level of work rate, season, industry class, and gender on the slope of the heart rate—rectal temperature regression line and the PHR_{38} are given in Table 3. The slope of the regression line was not significantly affected by any of the two-, three-, or four-way interaction terms ($p > 0.39$), metabolic rate ($p > 0.8$), season ($p > 0.7$), or industry class ($p > 0.6$); it was significantly influenced by the gender of the subjects ($p < 0.02$).   The mean slope was 30.0 bpm/°C for females and 42.6 bpm/°C for the males. Since none of the interactions or main effects, other than gender, was significant, there were no slope adjustments

Table 3. Adjustment Factors for PHR_{38} and Regression
Line Slopes With Their S.E. and Significance Levels

Factor	PHR_{38} Adjustment ± S. E.	P	Slope Adjustment ± S. E.	P
Summer season	-2.37 ± 1.62	0.15	- 1.67 ± 4.53	0.71
Moderate MR	6.37 ± 1.37	0.0001	- 0.78 ± 3.83	0.83
Hot industry	-6.86 ± 3.05	0.03	- 2.53 ± 5.54	0.65
Female gender	0.32 ± 2.96	0.92	-12.59 ± 5.10	0.02

The base values to which these adjustments are made are based on data from males working at the low MR in the winter season. The mean PHR_{38} and slope in these circumstances are 123.85 bpm and 42.63 bpm/°C, respectively.

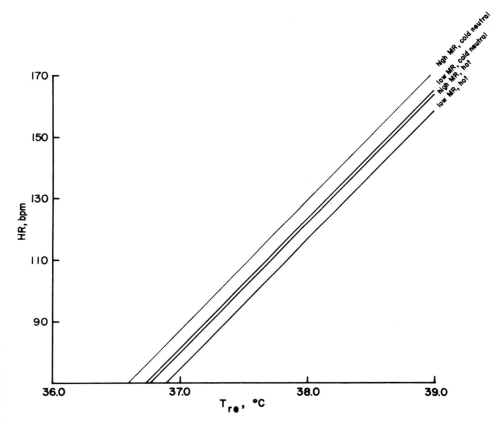

Fig. 1. Heart rate—rectal temperature relationships in males showing effects of work rate and industry group. Regression analysis used is described in text.

for MR, season, or industry class. At a specific rectal temperature of 38.0°C, the heart rate after one hour of work was significantly influenced by work rate ($p < 0.0001$) and industry class ($p < 0.03$), but was not influenced by season ($p > 0.15$), gender ($p > 0.9$), or any of the interaction terms ($p > 0.12$).

The adjusted regression lines are depicted in Fig. 1 for males and Fig. 2 for females. Within a gender there was no statistically significant adjustment to slope, although the slope for males was steeper than the slope for females. The regression line based on data from males from cold-neutral industries working at the lower work rate passed through 124 bpm at a rectal temperature of 38.0°C. The regression line based on the same group of males working at the moderate rate was displaced upward by 6 bpm. Similarly, males from the hot industries were displaced by 7 bpm downward from the regression lines based on men from cold neutral industries. The intercept displacements were the same for women as for the men; only slopes were

Table 4. PHR_{38} of Each Industrial
Group at Each Work Rate

	PHR_{38}, bpm
Low MR, cold-neutral industries	124
Moderate MR, cold-neutral industries	130
Low MR, hot industries	117
Moderate MR, hot industries	123

different. The PHR_{38} is highest for the least acclimatized subjects at the moderate work rate and lowest for the most acclimatized subjects working at the low work rate. The PHR_{38} for each combination of work rate and industry group is given in Table 4. Since gender and season effects were not statistically significant, they were not included in this table.

At a T_{re} of 38.0°C, the one-hour heart rates were quite similar for the genders since the overall regression lines crossed near this point. Since the slope of the regression line based on data from males was significantly steeper than that of the regression line based on data from females, the PHR at rectal temperatures substantially above 38.0°C was higher for men than for women, but the reverse was true at rectal temperatures below 38.0°C.

DISCUSSION

The results of this study reveal a clear difference between males and females in their respective $HR-T_{re}$ relationships. Males show a substantially greater increase in HR per °C change in rectal temperature than do females, but have a markedly lower intercept of the regression line relating HR to T_{re}. The lines cross at approximately 38.0°C, the maximum average rectal temperature recommended during prolonged heavy work. Therefore, the PHR_{38} was not influenced by gender, but a PHR for rectal temperatures above or below 38.0°C could be substantially different for males and females.

On the other hand, the slope of the $HR-T_{re}$ regression line was independent of season, work rate, or industry classification. This implies that neither work rate nor job- or season-acquired heat acclimatization has any influence on the slope of the regression line. The intercept of the regression line moves upward with a higher work rate and downward when the subjects are exposed to an increase in job-related heat stress. Season-acquired heat acclimatization is not sufficient to alter the $HR-T_{re}$ relationship, since neither slope nor intercept is significantly influenced by season.

The fact that above 38.0°C T_{re} men tended to have a higher HR than women at an identical T_{re} seems to contradict the results of several investigators (3,6,17) who found that women had a higher HR than men when exercising in heat at comparable

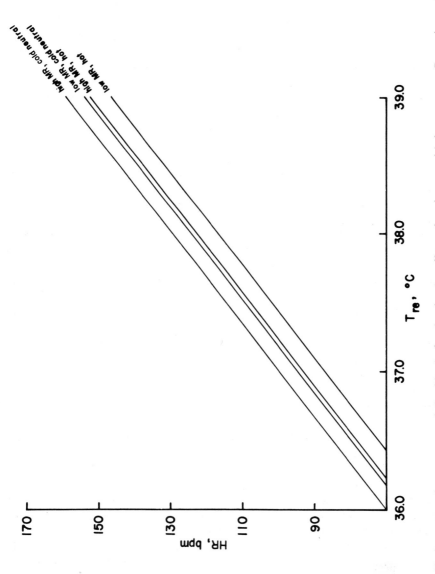

Fig. 2. Heart rate–rectal temperature relationships in females showing effects of work rate and industry group. Regression analysis used is described in text.

rates. However, the present study was conducted under much different conditions than these earlier studies. The subjects of Fox *et al.* (3) wore vapor barrier suits and were resting, while ours wore their usual work clothing and were exercising. Hertig and Sargent's (6) subjects and those of Wyndham *et al.* (17) were nearly nude and were tested only in hot environments. The South African women did the same amount of work (1,560 ft-lb/min) as the men and therefore presumably worked at a higher fraction of their maximum aerobic capacity; in these circumstances, one would expect the women to react more strongly to heat than the men. The subjects of Wyndham *et al.* worked only in hot, very humid environments, while ours worked in a wide range of environments that were drier and had lower CET levels. Air velocities in all three of these studies were higher than in our study. Moreover, the seeming contradiction between the present and earlier studies can be resolved if we consider HR as the independent variable that reflects the combined strain due to exercise plus heat, and T_{re} as the dependent variable. In this case, our results can be interpreted as showing that at a given heart rate, women have a higher rectal temperature than men when the T_{re} exceeds 38.0°C, as shown in Fig. 3. This figure gives the overall regression line for males and for females. It can be seen that on the average, the male with a HR of 150 bpm has a T_{re} of slightly over 38.8°C, while a female with that HR has a T_{re} of 39.0°C. This corresponds with the findings of Löfstedt (10) who showed that when T_{re} exceeds 38.1°C in men or 37.9°C in women, the rate of increase in T_{re} suddenly becomes faster, and in women this phenomenon is more pronounced than in men. Although Löfstedt's subjects were resting during his experiments and ours were exercising, our results also suggest that in the neighborhood of 38.0°C T_{re}, a significant change occurs in thermoregulatory response during exercise as well.

The fact that acclimatized workers have lower heart rates and rectal temperatures than unacclimatized workers when exposed to identical hot environments was described by Taylor *et al.* as early as 1943 (14). Our data suggest a slight trend for the PHR_{38} to be higher in winter than summer, and clearly higher among cold-neutral industry subjects when compared with hot-industry subjects. Thus, acclimatization also affects the relationship between HR and T_{re}.

If one accepts 38.0°C as the maximum permissible T_{re}, it is clear that no single HR can be identified as the maximum allowable. In these experiments none of the groups tested had PHR_{38} of less than 110 bpm. It was found that the job- and season-acclimatized subjects had PHR_{38} of 117 bpm when working, even at the low work rate. The least acclimatized subjects had a PHR_{38} of 130 bpm when working at the moderate work rate. It also must be recognized that the PHR_{38} value is based on the regression line that represents the average response of the subjects. The magnitude of the individual variability around the regression lines is approximately ± 35 bpm; thus, some subjects have a PHR_{38} of over 150 bpm in both the hot and the cold-neutral groups, both in winter and summer at moderate and low work rates. This finding fully supports the statement of the WHO Study Group concerning the inapplicability of HR as the single index for permissible exposure to heat. It is also apparent that the mathematical models developed for predicting T_{re} from HR (4,7,

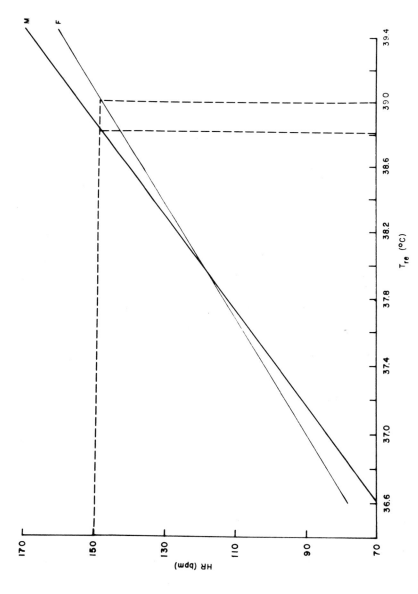

Fig. 3. Heart rate–rectal temperature relationships for men and women. Lines are overall regression lines based on all data, using each simultaneous HR and T_{re} measurement and ignoring effects of work rate, industry group, and season.

12) are valid only for individuals who are acclimatized to heat exactly to the same extent as the subjects who participated in the experiments for developing these models.

SUMMARY

Forty-six men and 22 women performed a total of 947 one-hour walks at a low or moderate work rate (22% or 29% of maximal aerobic capacity, respectively) in environments ranging from cool (13°C CET) to hot (35°C CET) in either the summer or winter seasons. Subjects were recruited from industries requiring their employees to spend considerable time in the heat or from industries whose employees normally worked in neutral or cold ambient temperatures all year. Rectal temperature (T_{re}) and heart rate (HR) were measured immediately prior to cessation of walking. Regression equations were developed to determine the relationship between T_{re} and HR. Males had a significantly (p < 0.02) steeper regression line than females, but the slope of the $HR-T_{re}$ regression line was not influenced by the season in which the testing took place, the degree of job heat exposure of the subjects, or the work rate. At a T_{re} of 38.0°C, the predicted heart rate (PHR_{38}) was significantly influenced by work rate (p < 0.0001) and the industry in which the subjects were employed (p < 0.03), but not by season (p > 0.15) or gender (p > 0.9) of the subjects. The PHR_{38} was higher for subjects in cold or neutral industries than for subjects in hot industries and higher at the moderate work rate than at the lower work rate. The mean predicted HR at a T_{re} of 38.0°C (maximum T_{re} recommended by a WHO Study Group) exceeded the Study Group's recommended HR limit of 110 bpm in all experimental situations.

ACKNOWLEDGMENTS

We thank Erdine Walker, Shirley Bell, and Patricia Saunders for technical assistance and Drs. E. Bradley and D. Naftel for statistical analysis.

This work was supported by NIOSH Contracts No. CDC-210-75-0021 and HSM-00-72-45 and was done in compliance with the Recommendations of the Declaration of Helsinki.

REFERENCES

1. Belding, H.S. and R.F. Hatch. Index for evaluating heat stress in terms of resulting physiological strains. *Heat./Piping/Air Cond.* 27: 129-136, 1955.
2. Ekblom, B., C.J. Greenleaf, J.E. Greenleaf, and L. Hermansen. Temperature regulation during continuous and intermittent exercise in man. *Acta Physiol. Scand.* 81: 1-10, 1971.
3. Fox, R.H., B.E. Löfstedt, P.M. Woodward, E. Eriksson, and B. Werkstrom. Comparison of thermoregulatory function in men and women. *J. Appl. Physiol.* 26: 444-453, 1969.

4. Givoni, B. and R.F. Goldman. Predicting heart rate response to work, environment, and clothing. *J. Appl. Physiol.* 34:201-204, 1973.

5. Greenleaf, J.E. and B.L. Castle. Exercise temperature regulation in man during hypohydration and hyperhydration. *J. Appl. Physiol.* 30: 847-853, 1971.

6. Hertig, B.A. and F. Sargent II. Acclimatization of women during work in hot environments. *Fed. Proc.* 22:810-813, 1963.

7. Kamon, E. Relationship of physiological strain to change in heart rate during work in the heat. *Am. Ind. Hyg. Assoc. J.* 33:701-708, 1972.

8. Kuhlemeier, K.V., J.M. Miller, F.N. Dukes-Dobos, and R. Jensen. Assessment of deep body temperatures of wokers in hot jobs. DHEW (NIOSH) Publ. 77-110, 1976.

9. Lind, A.R. A physiological criterion for setting thermal environmental limits for everyday work. *J. Appl. Physiol.* 18: 51-56, 1963.

10. Löfstedt, B. Human heat tolerance. Department of Hygiene, University of Lund, Lund, 1966.

11. Michael, E.D., Jr., K.E. Hutton, and S.M. Horvath. Cardiorespiratory responses during prolonged exercise. *J. Appl. Physiol.* 16: 997-1000, 1961.

12. Pirnay, F., J.M. Petit, and R. Deroanne. Evolution comparée de la fréquence cardiaque et de la température corporelle pendent l'exercise musculaire à haute température. *In. Z. Angew. Physiol. Einschl. Arbeitsphysiol.* 28: 23-30, 1969.

13. Steel, R.G.D. and J.H. Torrie. *Principles and procedures of statistics.* New York: McGraw-Hill, 1960, pp. 232-251.

14. Taylor, H.L., A.F. Henschel, and A. Keys. Cardiovascular adjustments of man in rest and work during exposure to dry heat. *Am. J. Physiol.* 139: 583-591, 1943.

15. Williams, C.G. and C.H. Wyndham. The problems of "optimum" acclimatisation. *In. Z. Angew. Physiol. Einschl. Arbeitsphysiol.* 26: 298-308, 1968.

16. World Health Organization. Health factors involved in working under conditions of heat stress. Report No. 412, Geneva, 1969.

17. Wyndham, C.H., J.F. Morrison, and C.G. Williams. Heat reactions of male and female Caucasians. *J. Appl. Physiol.* 20: 357-364, 1965.

II

Air Pollution
Chairperson

Mary O. Amdur
Harvard University
Boston, Massachusetts

REVIEW OF AIR POLLUTION

D. V. Bates

Faculty of Medicine
University of British Columbia
Vancouver, Canada

Anyone asked to review our developing knowledge on air pollution may tackle the assignment from a number of vantage points. He may trace the history of recognition of air pollution, customarily beginning with Maimonides, who had some observations to make on this matter and some advice to give; continuing with the early recognition of the hazard of open coal burning; noting the concern of Francis Galton, the man who began the study of biometrics, that if the growth of horse-drawn traffic continued to increase in London, the most serious pollution problem would be that arising from horse manure; and finally bringing the story up to date with the modern history of air pollution control, which dates from the period since World War II. Alternatively, one can review the processes of public decision-making in respect to air pollution, essentially completing a history of the role of industry, the public, and government in the continuing skirmishes that characterize this general battle — and Lord Ashby has recently published a fascinating account of the political science aspects of decision-making in Britain in relation to air pollution, ascribing most of the effective steps that have been taken to public pressure and not to government initiative; or one can review the progress in scientific terms, concentrating on the evolution of knowledge and the developing sophistication of understanding in respect of the environments we are making for ourselves.

Were I to develop any of these approaches in sufficient detail to make them interesting, we would run out of time before I get to the main point I would like to make — and indeed perhaps it would be safest if I began straight away with what that is. Having in mind that the way in which we approach the problem of air pollution or the perspective in which we place it is inevitably closely related to our attitudes as a society to dozens of other questions, we clearly must expect an evolution in public attitudes to air pollution, and acknowledge that this would have occurred even if our understanding of the biological effects of air pollutant had remained stationary. The point I wish to leave you with is that we have now reached a position in which our concern to contain the adverse effects of pollutants is forcing us to look very closely at such things as the definition of a "disease" and leading us to become conscious of new difficulties in respect of what is, after all, a very old problem. I therefore have to abandon rather reluctantly the interesting early history of concern about air pollution and compress into a short space what seems to me to have been occurring in the past 30 years.

We can begin with those air pollutants in the reducing atmospheres caused by coal burning more or less uncontrolled, and leading to an air pollution characterized by relatively large droplets of coal tar and ash, together with relatively high concentrations of sulfur dioxide. This classical urban air pollution existed for about 100 years and first became famous in the literature of London. In that particular climate the problem was compounded by high humidity and temperatures at about freezing level during the winter, and possibly also by the concomitant presence of salt water nuclei in the air as a result of the close proximity of major industrial centers to the sea. This acid type of air pollution wrought havoc on buildings constructed of limestone and was steadily eating into the masonry of the House of Commons in London at a rate that exceeded the growth of concern in that Chamber for what was happening. It all came to a head in the infamous episode of December 1952, and I should perhaps remind you that in that episode there was a high mortality among prize cattle at the Smithfield Cattle Fair, which was going on at the same time. It was later discovered that the prize cattle were more affected than the cattle that hadn't won prizes, because the attendants were quick to remove the excreta from the former animals — for purposes of showing them better — and hence these animals were not protected by the ammonia that would otherwise have surrounded them. (I hope the authors of the paper we are to hear this afternoon, which suggests a protective role for ammonia for humans, have taken note of this early precedent.) It was not difficult at that time to prove an enhanced mortality, and it has become quite clear and fully substantiated that particulate levels above a certain density, when combined with sulfur dioxide above a certain concentration, lead to chronic respiratory morbidity, an increase of respiratory symptoms in children and adults, a considerable intensification of the adverse effects of cigarette smoking in terms of chronic bronchitis and emphysema, and a measurable correlation with the overall mortality from respiratory disease. The decision to attempt to control this state of affairs in Britain by prohibiting the open buring of coal was an important step forward, and a review of subsequent data tends to suggest that the major adverse constituent in such episodes was the very high particulate pollution together with high humidity and cold; and that sulfur dioxide alone, if unaccompanied by those other factors, might well not have been so harmful. We might term this total phenomenon a characteristic pollution of the industrial revolution. You are all aware, of course, of the second type of air pollution, that is a consequence of massive automobile density combined with hydrocarbon release and sunlight — the commonly referred-to type of photochemical air pollution. I suppose we might refer to this as commuter pollution, but we also have to remember that there is a substantial contribution of static oxides of nitrogen sources to it. When I am in California, I always quote Canadian data if I can, and the contribution in terms of oxides of nitrogen to the atmosphere of metropolitan Toronto is that static sources contribute 43,000 tons of oxides of nitrogen per year, which is almost as much as automobiles contribute. It is not generally known that the sunlight intensity is sufficiently great to carry forward the next photochemical steps not only in southern Canada, but during six or seven months of the year, even as far north as Edmonton.

This second form of air pollution is chemically much more complex than the first. As I am sure is true of many of you, I have been fascinated by the extraordinary complexity of the many chemical reactions that may follow the exposure to sunlight of the products of automobile exhaust. I have long since given up any attempt to keep track of the kinetic constants in the couple of dozen reactions that are known to occur, and I am sure you recognize that this type of chemistry has become an important field of study in its own right. It has also given us entirely new problems, since the chemical reactions may continue over the space of several hours and the main burden of the pollutants so produced may fall on regions many miles distant from the heart of the city in which the primary constituents originated. In Canada we are now beginning to measure ozone and oxidants coming at us when the wind blows from the south, and those of you who have followed the long-range transmission information in respect of photochemical air pollutants will not be at all surprised at this. But it is not only the meteorological and chemical factors that have proved to be so very complex in the case of this class of pollutants, but also the biological effects.

In order to make explicit my point concerning the complexity of the questions we are now facing, let me remind you of the present status of our knowledge of the biological effects of ozone. If we consider animal experiments, we find that changes are caused in animal exposures to as little as 0.2 ppm given for some hours each day for periods of 30 days or so. These effects include not only some histological changes visible by electron microscopy in the lung, but changes in the function of alveolar macrophages and alterations in the ability of animals to resist bacterial aerosols. There are also disquieting reports of changes in circulating blood lymphocytes with an increase in chromosome breaks up to two weeks after an ozone exposure to 0.2 ppm for as short as five hours. Many of these experiments cannot be duplicated in man, but the human effects indicate reversible changes in pulmonary function with concentrations of about 0.3 ppm breathed for two hours during intermittent exercise. We have many unanswered questions in respect of ozone, and I shall not go deeply into those controversies or the evolving data. But it brings me to the central question, which is how we are to decide on those effects to which we are going to pay a lot of attention and those which we can afford to dismiss in our minds as probably unimportant. In coming to this conclusion, you quickly find that you are discussing whether or not the probability of repetitive adverse effects on pulmonary function in exercising children really matters. Does it lead to an increased incidence of respiratory disease in adult life? What levels of ozone could the public be exposed to without any potentiation of the virulence of ordinary respiratory infections? How seriously are we to take the evidence that ozone is a mutagen, and what concentration levels are we to set for protection against these effects in ourselves? Here we come face to face with the problem of how many of those effects could be classified as "diseases" in the ordinary sense of the term. We are well aware of this problem and are not helped by increasing efforts to refine the semantic definition. I don't know whether you would call seasickness a disease, or the headache that accompanies the first few days at altitude a disease; but in actual fact, I don't mind much whether you call them diseases or not, because we are now in a position to evaluate environmental effects and to state whether or not they represent an inconvenience or a hazard. Thus, we rightly bypass the question of definition. It always appears odd to me that those who have been stressing the importance of proper standards for protection against the effect of oxidants, and who have defended those that are now in place as representing a minimal safety factor in terms of the known biological effects of the pollutants themselves, are often thought to be "radicals." Personally, I feel that those of us with a knowledge of the lung and its development, and of the complexity of chronic lung disease, and who are conscious of the considerable gaps in our understanding, are in fact being conservative in urging the enforcement of such standards. To me the radicals are those who propose that we can ignore the evidence we have and dismiss the public exposure to photochemical air pollution as necessarily without consequence because of our meager present epidemiological data.

The pioneering leadership given by California in the effort to contain oxidant air pollution has been recognized by us all, but what we now confront is a remarkable situation in which, by combinations of circumstances, we are likely to find that the older reducing form of air pollution and newer oxidant air pollution may combine or co-exist. This may come about because of three quite distinct factors. In countries with a relatively high level of sulfur dioxide pollution but with increasing improvement in particulate pollution, the amount of sunlight may increase and so may accelerate the formation of photochemical air pollutants as a consequence of emitted oxides of nitrogen and hydrocarbons, essentially adding the newer type of air pollution to the old. Second, in cities with major oxidant air pollution, there may be an increasing emission of oxides of sulfur due to the lack of natural gas as a sulfur-free energy source and the necessity to burn higher sulfur fuel oil or even coal. In these environments, the older type of reducing air pollution may be added to the newer. Third, and perhaps most important, the photochemical air pollutants may be transported long distances from one region to another and may suddenly be added to a region that up to that time had not had a sufficient automobile density to generate its own photochemical air pollutants, but that has static sources of sulfur dioxide pollution perhaps coming from a smelter or other industry. In all these three ways, the older and newer forms of pollution are tending to occur together.

I need not remind you that we do not know what the sum total of all these phenomena may be in terms of human morbidity. Some of the papers in this symposium will be given by those whose work has advanced our knowledge considerably in respect of these matters, and we may be sure that there is much work ahead of us.

However, we can have no doubt but that most of the stresses are additive. We do not have a precise way of summating the stress imposed on us by the environments we have made for ourselves. I have always suspected that the products of photochemical air pollution may have behavioral effects far more serious than headaches or irritation of the conjunctivii. Indeed I have wondered whether some of the outbreaks of violence in some cities during hot summer nights may not have been attributable to some product of photochemical air pollution with direct effects on behavior of which we are still in ignorance. I have no evidence to justify such a suggestion, but I think we would all agree that our efforts to deal with air pollution cannot be rigidly confined to morbidity and mortality data. We have to take a much broader view of our responsibilities than to agree to be limited only by those statistics of conditions that other people will accept as being due to specific disease — as defined in the international classification of diseases, very likely. We must push on with the studies of the interaction of air pollutants and remain highly conservative in our approach to the definition of the permitted maxima of exposure for children and for the population as a whole. It may be the hallmark of the successful industrial entrepreneur that he is sometimes prepared to take risks even when he knows he is ignorant, but we cannot accept that it is a responsible attitude to take such a position in respect of the future planning of the environment in which our children will grow up.

RESPIRATORY NH₃ : A POSSIBLE DEFENSE
AGAINST INHALED ACID SULFATE COMPOUNDS

T. V. Larson
D. S. Covert
R. Frank[a]

Departments of Environmental Health
and Atmospheric Sciences
University of Washington
Seattle, Washington

Sulfate compounds in aerosols of the lower troposphere that are attributable to coal combustion consist primarily of sulfuric acid (H_2SO_4), ammonium bisulfate (NH_4HSO_4), and ammonium sulfate [$(NH_4)_2SO_4$] (2,4). NH_4HSO_4 and $(NH_4)_2SO_4$ are products of the reaction of H_2SO_4 with gaseous ammonia (NH_3). Recently, a team of atmospheric scientists reported that the total mass concentration of submicrometric sulfate in the aerosol in and around St. Louis ranged up to about 20 $\mu g/m^3$ (4). The concentration of NH_3 needed to neutralize such concentrations of H_2SO_4 are on the order of parts per billion by volume (ppbv). Professor R. J. Charlson, one of the investigators in the St. Louis study, has suggested that when the prevailing winds are out of the south from the Gulf of Mexico and blow over acid soils that are low in NH_3, the predominant sulfate around St. Louis is likely to be H_2SO_4 ; conversely, when the winds are out of the north and northwest and pass over fertile farm and dairy lands, the products of neutralization will dominate.

[a] This paper was presented by R. Frank.

Table 1 lists the approximate pH for each sulfate, assuming a relative humidity (RH) in the range of 20% to 80%. H_2SO_4 is the strongest acid; $(NH_4)_2SO_4$ is virtually a neutral salt.

On the basis of such observations, we turned our attention to what might be termed the "atmospheric chemistry" of the respiratory system and to the possible role of respiratory NH_3 in protecting the organism against acid sulfate air pollution.

We now hypothesize that acid aerosols undergo at least partial neutralization once they are inhaled, while they are still airborne in the respiratory passageways. In this instance, the NH_3 responsible for the neutralization is released from the tissue surfaces.

If this hypothesis is correct, that is if the hydrogen ion concentration of an inhaled particle is reduced before that particle strikes the mucosal surfaces, the effect should be protective. For there is experimental evidence (not proof) to suggest that the acidity of an inhaled particle is of importance in determining how irritating that particle may be to the mucosa. H_2SO_4 provokes more functional changes in the lung than does $(NH_4)_2SO_4$ or other neutral sulfate compounds (1), and it is reasonable to suppose that the basis for this heightened and adverse effect is related to hydrogen ion.

What may happen to the pH of an inhaled H_2SO_4 particle before impaction occurs? The proposed changes are shown in Table 2. For a wide range of ambient RH, pH before inhalation will be below 0.1. If we also assume that the particle grows threefold in diameter in the high RH (> 99%) of the upper airways owing to hydration, that is if its water content increases 27-fold, pH will rise to a level slightly over 1. If, in addition, the particle were to be completely neutralized by respiratory NH_3, the pH would rise above 5. Thus partial neutralization would result in some intermediate value of pH.

How rapidly these events occurred would largely depend on the original size of the particle: there is a direct relation between the original particle size and the time it takes the particle to reach chemical equilibrium with surrounding gases. Obviously, the slower and less complete the chemical transformation, the less protection likely to be afforded the respiratory system. We shall return to a consideration of the factors governing this process later.

Table 1. Ambient Sulfate Aerosols St. Louis, Missouri (1974)	pH
H_2SO_4	<0.1
NH_4HSO_4	1 − 2
$(NH_4)_2SO_4$	5 − 6

Table 2. Hypothetical Changes in pH of Inhaled H_2SO_4 Particle	pH
Before inhalation	<0.1
After inhalation	
(a) due to hygroscopic growth	∼ 1
(b) due to complete neutralization	5 − 6

EXPERIMENTAL STUDIES

First, we set out to measure how much NH_3 is given off by the human respiratory system and where the release occurs. A more explicit report of this work has been published (8).

We were not the first to measure NH_3 levels in expired air. However, we did refine the technique of measurement and have contributed some insight into the sources of NH_3 within the respiratory system.

For example, it has been customary (among the relatively few investigators concerned with the subject) to consider that the tracheo-bronchial tree is essentially a conducting pathway for NH_3 as it is for oxygen and carbon dioxide, the pulmonary source of NH_3 being blood ammonium ion (NH_4^+). [a] This concept of an NH_3 dead space (VD_{NH_3}) has generally been accepted (5,7,9). At least one investigator (5) has also concluded that the lung is the primary source of NH_3 in air expired from the mouth.

Our initial results did not support either concept. Instead they suggest that the primary source of NH_3 in expired air is in the upper airways, and that there is virtually no VD_{NH_3}, owing to the high diffusivity and solubility of the gas. Of course, in so far as the upper airways are the primary source of NH_3, they constitute a most appropriate site for intercepting and neutralizing acid aerosols.

TECHNIQUE OF MEASUREMENT

We used a chemluminescent instrument that oxidizes NH_3 to nitric oxide (NO), using a high temperaturure (850°C) stainless steel converter. The principle is shown schematically in Fig. 1. We calibrated this technique by means of the wet-chemical indophenol blue analysis, which can be used to rule out interference from oxides of nitrogen or other interfering compounds.

A major problem in the method is to avoid water condensation in the sampling line. Water condensate will absorb NH_3 and cause a falsely low reading. For this purpose, we heated the exhaled air above body temperature and subsequently diluted the sample with dry NH_3-free air to avoid condensation. (The air supplied to the laboratory contained about 5 ppbv NH_3, which is close to the resolution of the technique.)

Our method of sampling gas was designed to identify the sources and sinks of respiratory NH_3. The last structure through which the gas moved before sampling was either the nose, mouth, or upper trachea. The subject breathed either spontaneously by nose, mouth, or through an endotracheal tube, or held his breath while air

---

[a] Blood NH_4^+ may increase significantly in hepatic or renal disease.

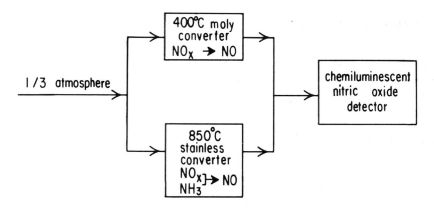

Fig. 1. Schematic of chemiluminescent NH_3 detector.

Table 3. NH_3, $\mu g/m^3$ (25°C) in Respiratory
Gas (Quiet Breathing /Constant Flow)
16 Subjects

	Mean ± SE	Range
Nose	29 ± 4	13 – 46
Mouth	157 ± 27	29 – 520
Trachea[a]	32	–

[a] Breathing by tracheal cannula, 1 subject.

Table 4. Estimated NH_3 Concentrations
over Blood and Saliva

	Reported NH_4^+ 10^{-4} M	Estimated NH_3 $\mu g/m^3$ [a]
Whole Blood (pH = 7.4)	0.15 – 0.4	10 – 25
Saliva (pH = 6.7 – 7.0)	6 – 70	60 – 1750

[a] 1 $\mu g/m^3$ = 1.4 ppbv (25°C)

was pulled at a constant rate: (a) into the mouth and out of the nose, (b) into the nose and out of the mouth, and (c) into one nostril and out of the other.

No attempt was made to control the subject's diet or the interval of time between the last meal and the measurements. Sixteen healthy adults ranging from 23 to 63 years of age were studied.

RESULTS

We found that the concentration of NH_3 in expired air is governed by the last segment of airway traversed. There were distinct differences in the range of values for NH_3 between the nose and mouth. NH_3 concentrations were about five times higher when the mouth was the last segment traversed (Table 3). The one endo-tracheal sample had a value similar to those typical of nasal samples.

Table 4 lists the concentrations of NH_4^+ reported to be found in blood and saliva (10) and our estimates of the equilibrium gas-phase NH_3 concentrations expected over these liquids. The estimated range of NH_3 concentrations over blood is slightly less than those found in the gas samples from the nose and trachea (Table 3).

Actually, the concentration of NH_3 at any particular site will change as a function of flow rate. Fig. 2 illustrates this phenomenon for mouth samples obtained from three subjects. Each subject held his breath, and air was pulled into the nose and out the mouth at several constant flows. The peak flow, 0.6 ℓ/s, is approximately the highest rate achieved during quiet breathing. NH_3 concentrations were highly sensitive to flow in two of the three subjects.

We are currently testing an improved sampling method for NH_3 that relies on having a critical orifice near the tip of the sampling probe. The reduced pressure in

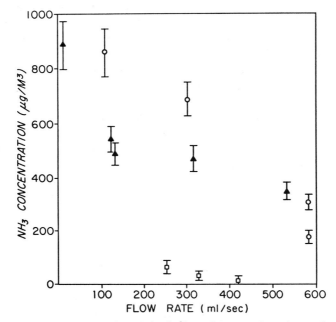

Fig. 2. Dependence of expired NH_3 ($\mu g/m^3$) on flow rate through mouth. Reproduced from *Science* 197:163, 8 July 1977. Copyright 1977, American Association for the Advancement of Science.

the sampling line (about 1/3 atmosphere) prevents condensation. With this method we shall be able to measure the cyclic fluctuations in NH_3 that occur during breathing and the variations that may be found between anatomic sites. Initial NH_3 measurements with this improved method are consistently higher than those reported in Table 3.

Knowing the NH_3 profile in the upper airway, a crucial experiment would have the subject inhale a known concentration of H_2SO_4 aerosol while this aerosol was sampled at different levels of the airway and analyzed chemically. To accomplish this would require an analytical technique that: (a) stopped the chemical reaction at the point of sampling, (b) assessed the degree of chemical transformation, and (c) determined the size-distribution of the aerosol. We are developing such a technique.

The technique relies on the fact that when H_2SO_4 particles ranging in diameter from 0.1 to 1 μm (i.e. accumulation mode aerosol) course a tube heated to $\geqslant 140^{\circ}C$, the particles vaporize. Heating and vaporization require less than 0.1 s. Upon leaving the tube and cooling, the vapor recondenses, forming smaller, nuclei-mode particles (< 0.1 μm MMD). Neither NH_4HSO_4 nor $(NH_4)_2SO_4$ particles vaporize and recondense in this way. The change in size-distribution of the aerosol that is measured at the end of the heated tube can therefore be used to assess the extent of neutralization. The technique cannot distinguish between NH_4HSO_4 and $(NH_4)_2SO_4$. The principle is shown schematically in Fig. 3.

While this analytical development has proceeded, we did make a relatively simple, preliminary observation on one subject that indicated that chemical transformation of the inhaled acid aerosol does occur. To assess gross changes in the aerosol size-distribution due to heating, we used an integrating nephelometer (6).

The subject breathed quietly by mouth. He was exposed to each of two concentrations of H_2SO_4 aerosol for several minutes: ∿ 600 μg/m³ and ∿ 1,200 μg/m³. A large fraction of the inhaled particles were subsequently exhaled. The molar ratio of NH_4^+/SO_4^{2-} in the exhaled particles was determined 0.5 s after they emerged from the respiratory system. (Each respiratory cycle took slightly over 2 s.)

The results of the analysis are shown in Table 5. A significant fraction of the H_2SO_4 was converted to ammonium salts. How much of this conversion occurred during inspiration, expiration, or during the 0.5 s following expiration is unknown. The concentrations of H_2SO_4 used were considerably in excess of any reported community levels (about two orders of magnitude).

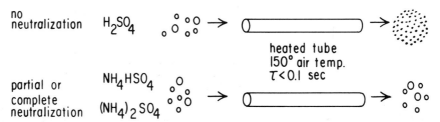

Fig. 3. Principle behind the rapid analysis of the extent of neutralization of H_2SO_4 aerosol.

Table 5. Extent of Neutralization
of *Expired* Aerosol

Expired Aerosol Concentration, $\mu g/m^3$ [a]	% Neutralized
600 ± 100	⩾ 50
1100 ± 200	25 ± 5
1100 ± 200	30 ± 5

Expired NH_3 = 100 $\mu g/m^3$. [a] Equals $\sim$ 90% of inspired H_2SO_4 concentration (0.5 μm MMD).

DISCUSSION

Earlier we referred to the importance of particle size in determining the rate of neutralization by NH_3. It is one of four critical factors, the others being the relative concentrations of the aerosol and NH_3, the residence time of the particle in the NH_3-rich atmosphere, and the relative humidity. Relative humidity is important because it affects the size of the hygroscopic H_2SO_4 particles.

As we have seen, the concentration of NH_3 is highest in the oral cavity and perhaps the oropharynx. The residence time of a particle in these regions is of the order of 0.1 s during quiet breathing and less during vigorous exercise.

In the elevated relative humidity of the upper airways, the particle size increases due to hygroscopic growth so that its surface area for gaseous diffusion increases. The amount and rate of hygroscopic growth for a specific relative humidity is a function of the original size and chemical composition of the particle; the smaller the size, the faster its growth. The relative humidity of the upper airways has not been charted with sufficient precision to model the rate of particle growth in this region, nor has particle growth been measured empirically here or elsewhere within the respiratory system.

A theoretical treatment of the importance of particle size in determining the rate of neutralization is illustrated in Fig. 4. The analysis applies to the accumulation mode of H_2SO_4. The area under each curve represents a theoretical estimate of the mass concentration of H_2SO_4 remaining after the indicated period of time, assuming that the rate of reaction is limited by gaseous diffusion (3). The decrease in mass of H_2SO_4 is due to complete neutralization. The mass distribution per log interval of diameter is given by the shape of each curve.

For example, 0.02 s after exposure of the aerosol to 200 ppbv NH_3, the smallest particles have been neutralized with only a slight reduction in total mass of H_2SO_4. Within 0.2 s, a significant reduction in H_2SO_4 mass is evident. After 1.0 s, H_2SO_4 persists in only the largest particles.

On the basis of such calculations, it might be speculated that the nuclei mode of H_2SO_4 (diam. < 0.1 μm) generated by catalytic converters on cars, is potentially

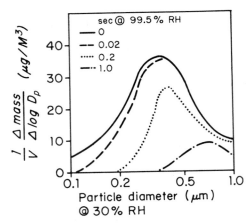

Fig. 4.   Theoretical H_2SO_4 aerosol mass distribution at various times during the aerosol's neutralization by NH_3. The reaction was assumed to occur at 99.5% relative humidity.

less hazardous than is the accumulation mode of H_2SO_4 (1.0 μm > diam. > 0.1 μm). The latter, more typical of ambient air, is largely attributed to coal combustion.

When Dr. Wagner invited us to participate in this program, he asked whether we could present work bearing on the relation between age and susceptibility to air pollution. We thought, mistakenly, that we could collect data on respiratory NH_3 in the newborn in time for the Symposium. Babies might be expected to enjoy less protection against acid aerosols for several reasons: (a) they are obligatory nose-breathers for some time, measured in weeks or longer, following birth; (b) they are likely to have less plentiful and different oral bacterial flora than do adults for several reasons, including their unique diet and lack of teeth. Unfortunately, we were unable to collect this information in time. Certainly, we are intrigued by the possibility that variations in respiratory NH_3 may exist among species and within species and that such variations may correlate with the functional response to acid aerosols.

Supported in part by Environmental Protection Agency, Contract No. DU-76-B156, and by National Institute of Environmental Health Sciences Grant No. PO1 E501478-0.

REFERENCES

1. Amdur, M.O. *Proc. Am. Philos. Soc.* 114:3, 1970.
2. Brosset, C., K. Andreasson, and M. Ferm. *Atmospheric Environment* 9:631, 1974.
3. Cadle, R.D. and R.C. Robbins. *Disc. Faraday Soc.* 30:155, 1961.
4. Charlson, R.J., A.H. Vanderpol, D.S. Covert, A.P. Waggoner, and N.C. Ahlquist. *Atmospheric Environment* 8:1257, 1974.
5. Hunt, R.D. and D.T. Williams. *American Laboratory* June 1977, p. 10.

6. Husar, R.B. In: *Proceedings of the symposium on radiation in the atmosphere.* Edited by H.J. Bolle. Science Press, 1977, pp. 37-40.
7. Jacquez, J., J. Poppell, and R. Jeltsch. *Science* 129:269, 1959.
8. Larson, T.V., D.S. Covert, R. Frank, and R.J. Charlson. *Science* 197:163, Fig. 1, 8 July 1977.
9. Robin, E.D., D.M. Travis, P.A. Bromberg, C.E. Forkner, and J.M. Tyler. *Science* 129: 270, 1959.
10. Searcy, R.L. *Diagnostic biochemistry.* New York: McGraw-Hill, 1969, pp. 60-66.

EFFECTS OF LOW-LEVEL CARBON MONOXIDE EXPOSURE ON THE ADAPTATION OF HEALTHY YOUNG MEN TO AEROBIC WORK AT AN ALTITUDE OF 1,610 METERS

P. C. Weiser

C. G. Morrill

D. W. Dickey

T. L. Kurt

G.J.A. Cropp

Department of Physiology
National Asthma Center
Denver, Colorado

INTRODUCTION

Ambient Air Quality Standards (AAQS) have been established to protect, with a reasonable margin of safety, a susceptible segment of the population as well as healthy individuals. The present AAQS for carbon monoxide (CO) is 9 ppm averaged over eight hours and 35 ppm over one hour, which is sufficient CO to produce a blood carboxyhemoglobin concentration of about 1.5% during these time periods. This AAQS has been established, however, from data acquired from studies conducted at or near sea level. Information necessary to evaluate any additional risk for sojourners or residents at altitude does not presently exist. The Environmental Protection Agency is not now in a position to recommend whether modification of the current AAQS for CO is necessary to protect the health of several million U. S. residents at altitudes above sea level.

Acute exposure to CO impairs exercise performance as measured by maximal aerobic power ($\dot{V}O_2$max) at sea level (2,7,11,13,17,18). The threshold for the decrease in $\dot{V}O_2$max appears to be a blood carboxyhemoglobin (HbCO) level of about 4.0% (2,11). Above this level, the decrease in $\dot{V}O_2$max is proportional to the HbCO level (2,16).

Hypobaric hypoxia that occurs during exposure to altitude also impairs maximal aerobic power. The threshold for an altitude-induced decrease in $\dot{V}O_2$max is about an elevation of 1,600 m. Above this altitude, the loss of $\dot{V}O_2$max is proportional to the decrease in inspired oxygen tension (4,9).

This report offers a preliminary data analysis of a project designed to determine if a 5% HbCO exposure level in healthy young men residing at Denver's altitude (1,610 m) would reduce $\dot{V}O_2$max, alter cardiopulmonary adaptation to submaximal work, and impair post-exercise systolic time intervals to the same extent as that reported for sea level CO exposure.

METHODS

We studied nine healthy male volunteers who had a mean ($\pm$ SEM) age of 24.7 $\pm$ 1.4 years, height of 178 $\pm$ 2 cm, and weight of 76.6 $\pm$ 3.1 kg. Analysis of variance showed no significant change in the subjects' weights during the study. All subjects were non-smokers. After giving informed consent, each subject received a preliminary physical examination which included a resting 12-lead electrocardiogram (ECG), pulmonary function evaluation, and an exercise ECG.

Each subject exercised four times, twice breathing filtered air (FA) and twice breathing filtered air containing CO. These four sessions were divided into two pairs, with each pair consisting of one FA and one CO experiment. The order of FA or CO exposure in each of the two pairs was randomly assigned in a double-blind fashion.

In each experimental session, the same general protocol was used. First, a forearm venous blood sample was taken and analyzed for HbCO. The initial HbCO estimate was always made on a CO-Oximeter with the more accurate measurement made later by modified gas chromotographic method (5) modified for use with a flame-ionization detector. The desired blood HbCO concentration was obtained quickly by adding a bolus of 100% CO to a closed-circuit system (6) from which the subject rebreathed. The volume of the initial CO bolus that increased the subject's Hb CO level to 5% from his baseline value was calculated from a preliminary estimate of total body hemoglobin. About 90% of the calculated CO volume was added over a period of several minutes to the 4.2-ℓ rebreathing circuit. After 8 minutes of rebreathing, another venous blood sample was taken, and the HbCO level measured. Additional CO was administered at that time, if needed to achieve the 5% HbCO level. The total volume of 100% CO administered ranged from 37 to 53 ml (STPD). A third blood sample was taken at the end of the 20-min rebreathing period to determine the final HbCO level. For the remainder of the experiment, the subject breathed from an open-circuit system (6). Throughout the experiment, the inspired and expired CO level was monitored alternately every 1 to 3 min by a long-path infrared analyzer.

Maximal oxygen intake ($\dot{V}O_2$max) was determined by a modified incremental treadmill test (3). The work test was preceded by a 9-min resting period in the supine position. The test began by walking for 4 min at 2.0 mph on a grade; the speed increased at 2-min intervals to 3.0 and 3.75 mph. At 3.75 mph, the grade was increased 2% every 2 min until the subject could no longer continue or until the treadmill had reached its maximum grade of 20%. If the subject could continue at a 20% grade, then the speed was increased every 2 min to 4.5, 5.0, and 5.5 mph or until the subject could not continue to run. Following exercise, the subject walked slowly in place for 1 to 2 min and then rested in the supine position for the remainder of the 60-min recovery period.

During the exercise tests, minute ventilation ($\dot{V}E$,BTPS), expired air fraction of oxygen and carbon dioxide, and heart rate (HR) were measured. The data were acquired by an on-line digital computer (Health Garde), and values were calculated for oxygen intake. During the 9-min rest period, computer data sampling was done over the last 2 min of each 3-min interval. For work rates producing a HR of less than 180 bpm, the second minute of each 2-min period was sampled. During heavy work above a HR of 180 bpm, sampling was done for the last 20 s of each minute of work. Following exercise, sampling was done over the last 20 s of each minute for the first 6 min and then over progressively longer intervals for the rest of the 60-min recovery period. A post-exercise blood sample was taken at the fourth minute of recovery.

Minute ventilation was measured in a 120-ℓ gasometer, and respiratory rate was obtained by monitoring inspiratory flow rate with a Fleisch pneumotachygraph connected to a differential pressure transducer. Oxygen and carbon dioxide contents of mixed expired air were monitored by a mass spectrometer from the outlet of a 9-ℓ mixing chamber in the expiratory side of the breathing circuit. Heart rate was obtained during the sampling periods from the electrocardiogram. Venous blood samples were analyzed for pH (Radiometer), lactate (Sigma), hemoglobin by the cyanmethemoglobin method, and hematocrit by the microhematocrit method.

Systolic time intervals were recorded in the supine position at rest before exercise, 3 to 4 min after exercise, and 19 to 22 min after exercise. Measurements were made by simultaneously recording the ECG, phonocardiogram, and carotid arterial pulse at a paper speed of 100 mm/s with time lines every 20 ms.

The data were analyzed using a one-way analysis of variance with repeated measures across the four exposure conditions. When the F-value was significant at the 5% level, an orthogonal comparison between the individual means was made (20). For these analyses, the selected variables included all physiological variables measured at $\dot{V}O_2$max and the venous blood variables measured immediately following the exercise tests.

RESULTS AND DISCUSSION

Changes in HbCO during the experiments are shown in Table 1. The CO bolus increased HbCO from about 1.0% to about 5.1% HbCO. Maximal exercise during FA breathing significantly decreased %HbCO by approximately 0.1% and the %HbCO exposure by about 0.4%.

Exercise performance was decreased by CO exposure as shown in Table 2. Total exercise time was significantly decreased by 3.8% from an average of 27 min for the two FA experiments to an average of 26 min for the CO experiments. Total physical work done during uphill walking or running (i.e. at a work rate of 3.75 mph, 2% grade or greater) was significantly decreased 10.0% from an average of 1.40 x 10^6 kpm for the FA experiments to 1.26 x 10^6 kpm for the CO experiments (p < 0.05). These changes in work done were also associated with a significant decrease in $\dot{V}O_2$ max expressed as ℓ/min (2.8%) or as ml $\cdot$ kg^{-1} $\cdot$ min^{-1} (3.5%).

Thus, at an altitude of 1,600 m (Denver, Colorado), a significant reduction in exercise performance occurred when the HbCO level was increased to 5.1% from a resting level of 1.0%. CO exposure during a progressive work capacity test resulted in a 3.8% decrease in total work time, a 4.8% decrease in final work rate, and a 10.0% decrease in physical work done. At sea level, Aronow and Cassidy (2) found that raising the HbCO level from 1.07 to 3.95% in 45- to 55-yr-old asymptomatic subjects reduced the total exercise time by about the same amount (5.0%) in a similar stress test. Horvath *et al.* (11) studied the effects of raising HbCO from 0.4 to 4.3% and found a decrease of 8.1% in the total exercise time for a graded stress test. Therefore, maximal exercise at Denver altitude resulted in a decrement in exercise performance similar to but no greater than that found at sea level.

Table 1. Low-Level Blood Carboxyhemoglobin levels (% Sat)

Condition	Baseline	After Bolus	End Exercise
Air	1.07 ± 0.05	1.05 ± 0.04	0.94 ± 0.05
CO	0.99 ± 0.05	5.09 ± 0.11	4.69 ± 0.16

Table 2. Low-Level Carbon Monoxide Exposure Decreases Exercise Performance

Condition	Total Time (min)	Total Work Done (kpm x 10^6)	$\dot{V}O_2$ max (ml$\cdot$kg^{-1}$\cdot$min^{-1})
Air	27.0 ± 1.1	1.40 ± 0.14	44.3 ± 2.2
CO	26.0 ± 1.0	1.26 ± 0.12	42.8 ± 2.0

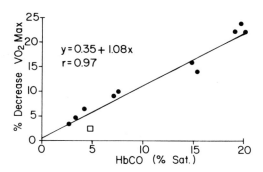

Fig. 1. Impairing effect of elevated blood carboxyhemoglobin levels (HbCO) that decreased physical work capacity ($\Delta \dot{V}_{O_2}$ max) in this study (□) and in others reported in the literature (7,11,13,16,17,18).

Maximal aerobic power was significantly decreased about 3% with low-level CO exposure. Fig. 1 summarizes the decrement in $\dot{V}_{O_2}$ max found in our study and reported by others who used similar exercise testing during CO exposure from 3 to 21% HbCO (7,11,13,16,17,18). The decrement for $\dot{V}_{O_2}$ max is highly correlated (P < 0.01) to HbCO level ($\Delta \dot{V}_{O_2}$ max = 1.08 x HbCO + 0.35) with an intercept that is close to zero HbCO. The decrement in $\dot{V}_{O_2}$ max found at Denver altitude is similar to that found at sea level and is clearly not greater than that observed at sea level. Hypobaric hypoxia produced by elevations above Denver altitude does produce a significantly greater reduction in $\dot{V}_{O_2}$ max (4). Therefore, this observation does not exclude the possibility that at higher altitude the same low-level CO exposure may have detrimental effects on maximal aerobic power compared to those found at Denver altitude or below.

Although work capacity was reduced, certain physiological variables observed during $\dot{V}_{O_2}$ max work during the FA and CO experiments were not significantly different. Maximal heart rate was unaffected by CO in this study as found in other studies done at sea level using low-level CO exposure (2,7,11). With higher HbCO levels at sea level maximal heart rate was decreased (7,13,16,17,18). Maximal minute ventilation (BTPS) was significantly greater at Denver altitude than at sea level; however, $\dot{V}_{Emax}$ during CO exposure in Denver was slightly but not significantly lower than the FA value. This small change was also found at sea level (7,11). Maximal ventilatory equivalent was unchanged as previously observed (7,11), and oxygen debt was unchanged, confirming an earlier report (11). Peak post-exercise venous lactates were the same in both conditions, which is in agreement with the findings of Ekblom and Huot (7). Decreased post-exercise lactates have been reported by Horvath et al. (11), but this may be explained by differences in test procedures, subjects, or altitude conditions. These physiological adaptations to maximal exercise are similar regardless of the gas breathed; however, during CO breathing, these adaptations to maximal work occurred at a significantly lower work rate. This

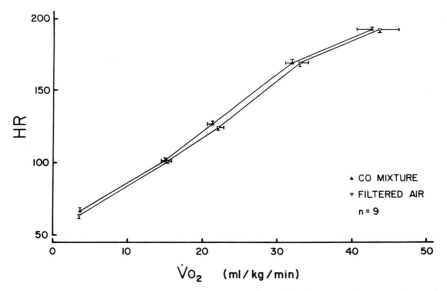

Fig. 2. Effect of CO exposure on the absolute heart rate (HR) during an incremental exercise
test.

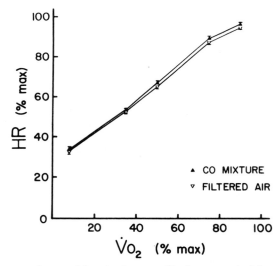

Fig. 3. CO exposure increased heart rate, normalized as percent of the maximal HR during
work expressed as percent of the highest $\dot{V}_{O_2}$ max.

difference in maximal aerobic work rate suggests that CO exposure provoked a physiological compensation such that a greater cardiorespiratory adaptation occurred at a lower work rate during CO breathing.

At a given submaximal work rate, a greater heart rate response occurred during CO exposure. Fig. 2 shows the effect of work rate, in units of $\dot{V}O_2$ cost, upon heart rate. At the submaximal work rates, the heart rate response curve is shifted to the left. To minimize individual differences in both maximal aerobic power and maximal heart rate, the data were reanalyzed, being expressed as a percent of both the highest achieved $\dot{V}O_2$ and heart rate in either the FA or CO condition, and is shown in Fig. 3. Exposure to CO increased the heart rate significantly at any work rate from rest to 75% $\dot{V}O_2$max. At 89% $\dot{V}O_2$max (highest common work rate for both the FA and CO conditions) and at $\dot{V}O_2$max, no differences in heart rate between the FA and CO conditions were found.

The heart rate adaptation to work was found to be increased when CO impaired work capacity. Our results demonstrated a small, although statistically significant, increase in heart rate during submaximal work. Other investigators who examined lower CO exposure levels have reported no differences in submaximal work heart rate (11,15,16). In the present study and at higher HbCO levels, the submaximal heart rate has been found to be significantly greater as HbCO is increased (7,11,13, 17). This suggests that the decrease in O_2 delivery produced by CO is at least in part compensated by an increase in the heart rate control of cardiac output.

Cardiac function appeared to be altered by low-level CO exposure. Systolic time intervals were measured 3 to 4 min and 19 to 22 min after the exhaustive exercise test in six subjects and are shown in Fig. 4. Heart rates at these times were not statistically different between the FA and CO conditions. During FA exposure, left ventricular ejection time (LVET) was shortened relative to LVET at rest, when the heart rate was still elevated at 4 min post-exercise. At this time, during CO exposure, LVET was significantly less shortened. A corresponding, impaired shortening of the electro-mechanical systole (QS_2) resulted, although no change in the pre-ejection period (PEP) or the PEP/LVET ratio was found. These changes were not present when STIs were measured at 20 min post-exercise.

Following maximal exercise in healthy young men at sea level, LVET was shortened to a similar extent whether subjects breathed CO or FA (1). In the present study, LVET was not shortened as much during CO breathing. A similar lack of LVET shortening at 4 min post-exercise has been reported in patients having chest pain upon exertion (8,10,12,14).

Ventilatory responses to submaximal exercise during CO exposure were also altered. Minute ventilation was significantly greater at 75% and 89% $\dot{V}O_2$max as shown in Fig. 5. Both respiratory rate and tidal volume were slightly but not significantly elevated during CO exposure. The anaerobic threshold (19) was lower

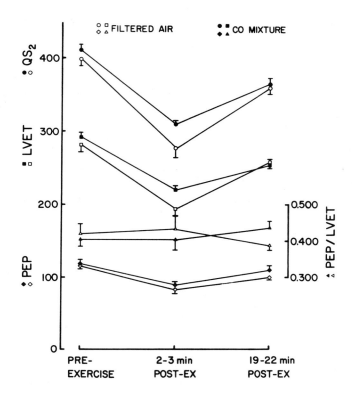

Fig. 4. CO exposure (closed symbols) lessened the post-exercise shortening of systolic time intervals observed while breathing filtered air (open symbols). Electro-mechanical systole (QS_2), left ventricular ejection time (LVET), pre-ejection period (PEP).

during CO exposure than during FA exposure, which resulted in the increased minute ventilation at heavy work rates. During FA exposure, the anaerobic threshold was 50.7 ± 2.6 $\dot{V}O_2$ max, whereas when breathing CO, the anaerobic threshold was reduced to 46.0 ± 2.2% $\dot{V}O_2$ max. At work rates above the anaerobic threshold, the venous lactate concentration was also increased significantly. Although lactate levels were higher during heavy exercise, maximal post-exercise lactate levels were not significantly different.

CO exposure appears to result in a greater metabolic lactacidosis during submaximal work, leading to hyperventilation. This hyperventilation has been observed at CO exposure levels above 7% HbCO (7,17,18) during heavy exercise, but at lower HbCO levels, such hyperventilation has not been reported (11). Metabolic lactacidosis during exercise that resulted in hyperventilation has also been found in subjects with 20% HbCO; however, no differences were observed in maximal lactate values post-exercise (17).

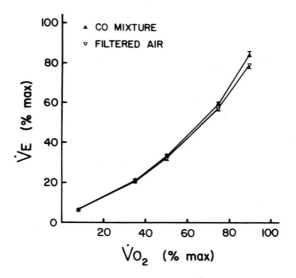

Fig. 5. CO exposure increased minute ventilation ($\dot{V}_E$), expressed as a percent of the maximal $\dot{V}_E$ observed, during work expressed as a percent of the highest $\dot{V}_{O_2}$max.

SUMMARY

Low-level CO exposure resulting in 5.1% HbCO at an altitude of 1,610 m impaired work performance to the same extent as that reported for sea level studies. Small but significant submaximal exercise changes in cardiorespiratory function were found during CO exposure. Work heart rate was elevated, and post-exercise left ventricular ejection time did not shorten to the same degree as with FA exposure. CO exposure resulted in a lower anaerobic threshold, and a greater minute ventilation at work rates heavier than the anaerobic threshold due to an increased blood lactate level.

REFERENCES

1. Anderson, E.W., J. Strauch, J. Knelson, and N. Fortuin. Effects of carbon monoxide (CO) on exercise electrocardiogram (ECG) and systolic time intervals (STI). *Circulation* 44, Suppl. II: 135, 1971.
2. Aranow, W.S., and J. Cassidy. Effect of carbon monoxide on maximal treadmill exercise. A study in normal persons. *Ann. Intern. Med.* 83: 496-499, 1975.
3. Balke, B. and R.W. Ware. An experimental study of "physical fitness" of Air Force personnel. *U.S. Armed Forces Med. J.* 10: 675-688, 1959.

4. Buskirk, E. Work and fatigue in high altitude. In: *Physiology of work capacity and fatigue.* Edited by E. Simonson. Springfield, Ill.: Charles C. Thomas, 1971, pp. 312-322.
5. Dahms, T.E. and S.M. Horvath. Rapid, accurate technique for determination of carbon monoxide in blood. *Clin. Chem.* 20: 533-537, 1974.
6. Dahms, T.E., S.M. Horvath, and D.J. Gray. Technique for accurately producing desired carboxyhemoglobin levels during rest and exercise. *J. Appl. Physiol.* 38: 366-368, 1975.
7. Ekblom, B. and R. Huot. Responses to submaximal and maximal exercise at different levels of carboxyhemoglobin. *Acta Physiol. Scand.* 86: 474-482, 1972.
8. Gillilam, R.E., W.D. Parnes, B.E. Mondell, R.J. Bouchard, and J.R. Warbasse. Systolic time intervals before and after maximal exercise treadmill testing for evaluation of chest pain. *Chest* 71: 479-485, 1977.
9. Hartley, L.H. Effects of high-altitude environment on the cardiovascular system of man. *J. Am. Med. Assoc.* 215: 241-244, 1971.
10. Hoffman, A., M. Sefidpar, and D. Burckhardt. Systolische Zeitintervalle in Ruhe und unter Belastung bei unterschiedlichem Schweregrad von Linksherzinsuffizienz. *Z. Kardiol.* 63: 768-777, 1974.
11. Horvath, S.M., P.B. Raven, T.E. Dahms, and D.J. Gray. Maximal aerobic capacity at different levels of carboxyhemoglobin. *J. Appl. Physiol.* 38: 300-303, 1975.
12. Lewis, R.P., D.G. Marsh, J.A. Sherman, W.F. Forester, and S.F. Schaal. Enhanced diagnostic power of exercise testing for myocardial ischemia by addition of postexercise left ventricular ejection time. *Am. J. Cardiol.* 39: 767-775, 1977.
13. Pirnay, F., J. Dujardin, R. Deroanne, and J.M. Petit. Muscular exercise during intoxication by carbon monoxide. *J. Appl. Physiol.* 31: 573-575, 1971.
14. Pouget, J.M., W.S. Harris, B.R. Mayron, and J.P. Naughton. Abnormal responses of the systolic time intervals to exercise in patients with angina pectoris. *Circulation* 43: 289-298, 1971.
15. Raven, P.B., B.L. Drinkwater, S.M. Horvath, R.O. Ruhling, J.A. Gliner, J.C. Sutton, and N.W. Bolduan. Age, smoking habits, heat stress, and their interactive effects with carbon monoxide and peroxyacetyl nitrate on man's aerobic power. *Int. J. Biometerol.* 18: 222-232, 1974.
16. Raven, P.B., B.L. Drinkwater, R.O. Ruhling, N. Bolduan, S. Taguchi, J. Gliner, and S.M. Horvath. Effect of carbon monoxide and peroxyacetyl nitrate on man's maximal aerobic capacity. *J. Appl. Physiol.* 36: 288-293, 1974.
17. Vogel, J.A. and M.A. Gleser. Effect of carbon monoxide on oxygen transport during exercise. *J. Appl. Physiol.* 32: 234-239, 1972.
18. Vogel, J.A., M.A. Gleser, R.C. Wheeler, and B.K. Whitten. Carbon monoxide and physical work capacity. *Arch. Environ. Health* 24: 198-203, 1972.
19. Wasserman, K. B.J. Whipp, S.N. Koyal, and W.L. Beaver. Anaerobic threshold and respiratory gas exchange during exercise. *J. Appl. Physiol.* 35: 236-243, 1973.
20. Winer, B.J. *Statistical principles in experimental design.* 2nd Ed. New York: McGraw-Hill, 1971., pp. 171-175.

ENVIRONMENTAL STRESS

RESPIRATORY AND BIOCHEMICAL ADAPTATIONS
IN MEN REPEATEDLY EXPOSED TO OZONE

Jack D. Hackney
William S. Linn
Ramon D. Buckley
Clarence R. Collier
John G. Mohler

Rancho Los Amigos Campus
School of Medicine
University of Southern California
Downey, California

INTRODUCTION

Ozone (O_3) in ambient air is now recognized as a widely occurring environmental stress potentially capable of injuring human health. Stratospheric ozone reaches the earth's surface to produce clean-air background concentrations near 0.05 parts per million (ppm),* and the gas may be formed photochemically in the lower atmosphere as a result of natural emissions of hydrocarbons or oxides of nitrogen; thus, it is not entirely foreign to man's environment. However, with the massive emissions from fuel combustion characteristic of modern urban environments, atmospheric concentrations of photochemically produced ozone may reach 0.6 ppm or higher (i.e., at least an order of magnitude greater than the background level) in areas with unfavorable atmospheric conditions such as suburban Los Angeles. Widespread acute adverse respiratory effects in the population experiencing such pollutant

*Concentrations mentioned herein are in terms of the neutral buffered potassium iodide calibration method (15).

111

episodes are not commonly reported, although controlled laboratory studies on volunteers have repeatedly shown toxic respiratory effects at ozone concentrations of 0.4 to 0.5 ppm under conditions simulating ambient exposures (1,8,9,11). This might suggest that residents of ozone-polluted areas become adapted to the exposures to the extent that they do not experience acute health effects. Many laboratory animal studies indirectly support this possibility (3,12). It has been suggested that Los Angeles residents are less affected by photochemical smog exposure than visitors (4), but until recently this possibility has not been seriously investigated. We describe here a series of experiments that support the possibility that exposure to ozone-containing Los Angeles air pollution (or to comparable concentrations of ozone in the laboratory) may produce an adaptive response in humans.

We define the term "adaptation" for present purposes as a biological response to an inhaled toxic substance that prevents or attenuates the expected adverse clinical and physiological responses to subsequent exposures to the same substance. In animal toxicology, a prior exposure to a sublethal ozone concentration allows rodents to survive subsequent exposures to higher concentrations that are lethal to animals not previously exposed (14). This phenomenon is commonly known as "tolerance." We do not use the latter term in reference to humans, since the ozone concentrations and the severity of the toxic responses in question are much less than in animal studies. Adaptations to "natural" environmental stresses (such as temperature, altitude, and exercise) are generally considered beneficial to the individual; but in the case of ozone or other toxic pollutants, one should keep in mind the possibility that the apparent adaptive response might be harmful in the long term. The observations discussed here relate only to short-term, integrative responses of the human organism in the presence of pollutant challenge. An understanding of the responses at the cellular and molecular levels, and of their long-term consequences, must await further research.

BACKGROUND

Studies of human responses to ozone, giving careful attention to realistic laboratory simulation of ambient exposure conditions, were first reported by Bates and coworkers (1). These investigators exposed volunteers in a controlled-environment chamber to concentrations up to 0.75 ppm. Exposures lasted two hours, during which time the subjects exercised on a bicycle ergometer for the first 15 min of each half hour at a light work load (sufficient to approximately double the minute volume of ventilation relative to the resting level). This protocol realistically models typical ambient exposure conditions during pollution episodes. Marked pulmonary physiological and clinical changes were found in exposures to 0.75 ppm. Less severe but still significant changes were found in exposures to 0.37 ppm. Subjects in these studies were young, healthy Canadians, living in urban environments not often subject to severe ozone pollution.

   Initial ozone studies in our laboratory used an experimental protocol comparable to that of Bates' group, except for the addition of heat stress, which usually accompanies ambient oxidant exposure in the Los Angeles area. Volunteers (all residents of metropolitan Los Angeles) were exposed to 0.50, 0.37, or 0.25 ppm at 31°C (88°F) and 35% relative humidity. In subjects with clinical respiratory hyperreactivity, as well as those with normal respiratory health, measurable physiological or clinical responses were not found at 0.25 ppm and were rarely found at 0.37 ppm. At 0.50 ppm, some subjects showed marked adverse responses, but many did not (8,9). Overall (as illustrated in Fig. 1), the dose-response behavior of Los Angeles residents seemed to indicate a mean pulmonary physiological response, as measured in terms of maximum one-sec forced-expiratory volume (FEV_1), about half as severe at a given ozone concentration as had been seen in Canadians exposed under presumably similar conditions. Adaptation in Los Angeles residents seemed a likely explanation for their apparent difference from Canadians, but many alternative explanations were also plausible, e.g., differences in exposure temperature, ozone monitoring, physiological measurement, undetected air contaminants interacting

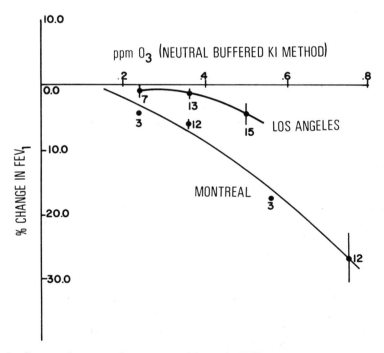

Fig. 1. Comparative ozone dose-response (change in FEV_1 vs. ozone exposure) curves for previously studied Los Angeles and Montreal subjects. Number of subjects exposed at given concentration is indicated adjacent to each point. For larger subject groups, vertical bars indicate standard error of response.

with ozone, or subject selection. To investigate some of these, a cooperative study was undertaken with the Canadian investigators. A group of four Canadian subjects was compared with a group of four Los Angeles subjects, matched as closely as possible, in a two-hour exposure to 0.37 ppm ozone at 21°C and 50% relative humidity. Experimental methodology was also compared. The Rancho Los Amigos (Los Angeles) environmental chamber was used for the exposure, since it had the most extensive air-purification equipment, capable of quantitatively removing all common air contaminants other than low-molecular-weight saturated hydrocarbons (6). The Canadian subjects reacted more severely on the average than did the Los Angeles subjects, and the difference in mean pulmonary function change following exposure was comparable to that expected from previous studies. No methodological differences capable of explaining the apparent differences in response were found; thus, the hypothesis of adaptation in Los Angeles residents was supported (7). In an additional study addressed to the same question, a group of nine preprofessional students newly arrived in Los Angeles was compared with a group of six from the same class who had resided in the area for three years or more. Exposures were as in the Canadian-Los Angeles study, except that temperature was 31°C and relative humidity, 35%. The exposures took place near the end of the summer photochemical smog season, when adaptation (if any) in Los Angeles residents should have been at maximum. Again, the non-residents appeared more reactive than the residents (5).

The distribution of FEV_1 responses for non-residents and residents from both the above studies is shown in Fig. 2. While individuals vary considerably, as expected from previous findings, the non-residents show a significantly greater loss in performance ($p < 0.05$). Since these observations provide only indirect support for the hypothesis of ozone adaptation, another exposure study was undertaken with the goal of producing and demonstrating an adaptive response directly.

METHODS

Six male volunteers from the project staff and other hospital employees were selected for study. Individual characteristics are given in Table 1. All were in good general health and had normal base line lung-function tests, but all showed some evidence of "hyperreactivity" to pollutant exposure — history of allergy or wheezing, subjective sensitivity to Los Angeles smog exposure, or marked clinical response to 0.50 ppm ozone in a previous exposure study. These "hyperreactive" subjects were chosen in preference to "normals" in order to obtain clear-cut physiological and clinical responses to ozone exposure (often lacking in "normals"). The subjects were exposed on the first day of the study to purified air (control), and on the second, third, fourth, and fifth days to 0.50 ppm ozone in purified air. It was hypothesized that adverse responses would develop after the initial ozone exposures, but these would tend to reverse even in the face of later ozone exposures because of adaptation.

Exposures took place in April near the end of the winter-spring low smog season to minimize the possibility of pre-existing adaptation due to ambient exposure. Air-monitoring data for the areas frequented by the subjects were examined to verify the absence of significant intercurrent ambient exposures. During the five-day study period, recorded ambient ozone levels never exceeded 0.09 ppm and typically remained below 0.05 ppm.

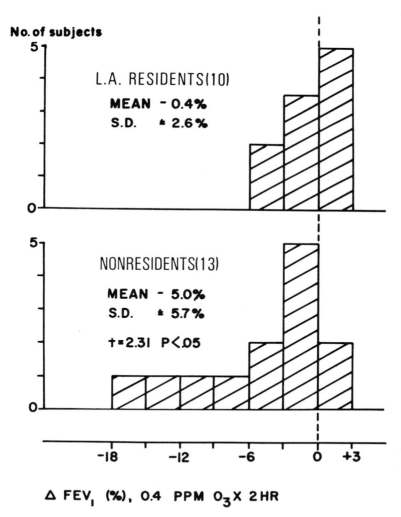

Fig. 2. Frequency distributions of response to 2-hr ozone exposure at 0.4 ppm in long-time Los Angeles residents and new arrivals previously studied.

Table 1. Subject Characteristics

I.D. No.	Age	Ht, cm	Wt, kg	Clinical Characteristics
5	57	170	68	(a,b)
9	43	183	91	(a,b,c,d)
11	31	180	70	(a,d)
16	32	178	70	(d)
24	56	178	77	(a,b,c,d)
25	23	185	77	(a,b,c,d)

(a) = allergic history. (b) = respiratory symptoms subjectively associated with ambient smog exposure. (c) = wheezing history. (d) = previously found reactive to ≤ 0.5 ppm ozone in controlled exposure.

Each exposure was for two hours, during which time exercise was performed at a work load of 150 to 200 kg • m • min^{-1} for the first 15 min in every 30 min. Exposure temperature was 31°C and relative humidity, 35%. Subjects entered and left the exposure chamber at 45-min intervals, each individual's exposure beginning about the same time each day in order to eliminate possible effects of circadian rhythm. Details of the exposure protocol and health-effect testing have been previously described (2,6). We obtained forced vital capacity (FVC), FEV_1, and maximum flow rates with 50% and 25% FVC remaining from the maximum expiratory flow volume curve performed at the beginning and end of each exposure. Other tests were done only at the end of exposure; these included the single-breath nitrogen test (from which was obtained the slope of the alveolar nitrogen plateau, or ΔN_2), measurement of total respiratory resistance by forced oscillation at 6 Hz applied at the mouth, a clinical evaluation (expressed as a semi-quantitative score based on a standardized interview concerning symptoms relatable to ozone effects), and biochemical assays of venous-blood substances expected to be sensitive to ozone effects. Other lung-function data (including closing volumes, frequency dependence of resistance, and indices of multiple-breath nitrogen washouts) are not discussed here, since they showed no obvious ozone-related changes.

Health-effect data for the group were tested for significant variation attributable to ozone exposure or adaptation by repeated-measures single factor analysis of variance. The null hypothesis tested was that no significant variation in health measures over the five exposure days would be found. In cases for which the null hypothesis was rejected, all possible pair-wise comparisons between test days were made by the Newman-Keuls test (16) to determine whether or not the time course of the response was consistent with the alternative hypothesis of adaptation.

A more rigorous experimental design would have required a separate control study incorporating five successive days' exposure to pure air alone, since repeated heat and exercise stresses alone might plausibly produce changes in some of the health-effect measures. Practical considerations prevented such a control experiment; however, evidence was available from other exposure studies which indicated that the relevant physiological and clinical measures would likely remain stable in the absence of ozone challenge. A previous similar study covering four successive days, two pure-air exposures followed by two exposures to 0.25 ppm ozone, including some of the subjects in the present study, showed no substantial lung-function or symptom changes (8). Biochemical measurements, made only on the first and last days, also remained stable. In another study, eight normal subjects were exposed similarly to purified air on three successive days and biochemical measurements were made after each exposure (R.D. Buckley, K.W. Clark, C.A. Posin, and M.P. Jones, unpublished results). In contrast to the, previous observation, some indices showed significant changes. These are discussed further in the following section.

RESULTS

Physiologic findings are discussed in detail elsewhere (10). Group mean values for the forced-expiratory measures are plotted in Fig. 3. Fig. 4 shows the other lung-function measures and the mean symptom scores. All these variables show the trend expected in an adaptive response—a decrement (relative to the day-one control value) on days two and/or three, followed by improvement on days four and five. Individual responses varied as to magnitude and time course (consistent with previous findings here and elsewhere); thus the group changes did not attain statistical significance in many cases. The FVC change from pre- to post-exposure measurements and the post-exposure $\dot{V}_{25}$ did show significant variation ($p < 0.05$). In both cases the poorest performance was after the second ozone exposure (day three)—significantly different from control (day one) as determined by the Newman-Keuls post-hoc test ($p < 0.05$). Values measured on the third and fourth ozone exposure days, however, were not significantly different from control. Delta nitrogen showed a tendency to increase with initial ozone exposures, indicating less uniform ventilation distribution within the lung, and respiratory resistance tended to increase. These changes approached statistical significance ($0.05 \leqslant p < 0.10$). Again, the maximum function decrement was seen after the second ozone exposure.

Selected function measurements for the clinically and physiologically most reactive individual (subject 11) are shown in Fig. 5. His forced-expiratory performance decreased dramatically in the second ozone exposure, and ΔN_2 increased at the same time. Respiratory resistance increased only modestly. These changes were largely reversed on the last two exposure days.

Group mean values for blood biochemical measures showing changes are plotted in Fig. 6. Hemoglobin concentration and the concentration of reduced glutathione (GSH) in red cells behaved similarly, remaining near control values after one ozone exposure, decreasing significantly after the second exposure, and showing little additional change thereafter. Red-cell 2,3-diphosphoglycerate (2,3-DPG) activity increased gradually with successive ozone exposures, but appeared to have stabilized by the third exposure. Red-cell acetylcholinesterase activity showed the most consistent change: steadily downward with successive exposures. The rate of change lessened with each successive day, and the decrease from day four to day five was not significant by the Newman-Keuls test; thus the enzyme activity may have been approaching a stable level. Red-cell fragility as determined by hydrogen peroxide challenge was increased on the first three ozone exposure days, but fell below the control level on the last exposure day. The changes in fragility did not attain statistical significance $(0.05 < p < 0.10)$.

Other biochemical measures expected to be ozone-sensitive showed no statistically significant changes or obvious trends over the five days. These included hematocrit; serum vitamin E concentration; and activity levels of glucose-6-phosphate

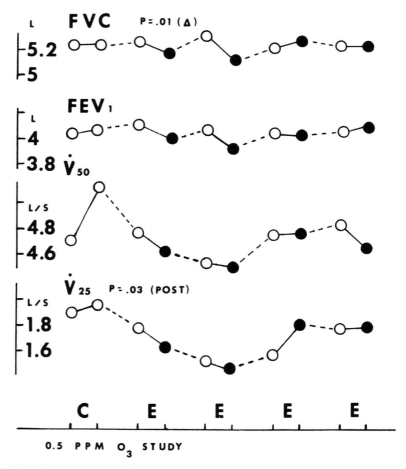

Fig. 3. Group mean forced-expiratory function values. C = control study. E= 0.5 ppm ozone exposure. Pre-exposure (on left) and post-exposure (on right) measurements on same day are connected by solid lines. Solid circles indicate measurements after ozone exposure.

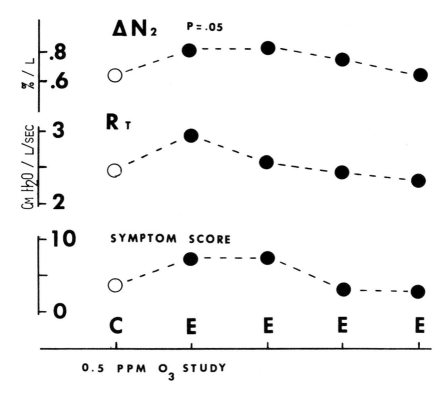

Fig. 4. Group mean values for pulmonary function and symptom measures made only post-exposure (cf. Fig. 3).

dehydrogenase, lactate dehydrogenase, glutathione reductase, and glutathione peroxidase.

The extent to which the foregoing biochemical observations are related to ozone exposure is somewhat in doubt in light of the results from eight normal subjects receiving three successive days' purified-air exposures. Mean values for this group are plotted in Fig. 7 with the same format as the present group's results in Fig. 6. For hemoglobin and 2,3-DPG, the changes in the pure-air group were similar to those in the present group over days one, two, and three. For acetylcholinesterase, the directional trends were similar, but the losses in activity tended to be larger and more consistent (giving a higher level of significance) in the ozone-exposed subjects. For GSH, there was no evidence of change in the pure-air group, in contrast to the ozone group who showed a decrease by day three. Red-cell fragility showed a large decrease on day two only in the pure-air group, as opposed to a marginal increase on all but the last ozone exposure day in the present group. However, the range of values for the pure-air group was well above that for the present group.

DISCUSSION

Results of the present and previous experiments taken together provide fairly convincing evidence that at least some normal people are capable of adapting clinically and physiologically to ozone exposure at ambient concentrations. Other important evidence in favor of this possibility comes from recent work of Parent *et al.* (13). These investigators exposed healthy Canadian subjects (presumably not frequently exposed to ambient ozone) to 0.6 ppm on two different days. Exposure-related lung-function decrement was less in the second exposure, provided it took place two, three, or four days after the initial exposure. There was no evidence of adaptation in subjects re-exposed on the day immediately following the initial exposure, however.

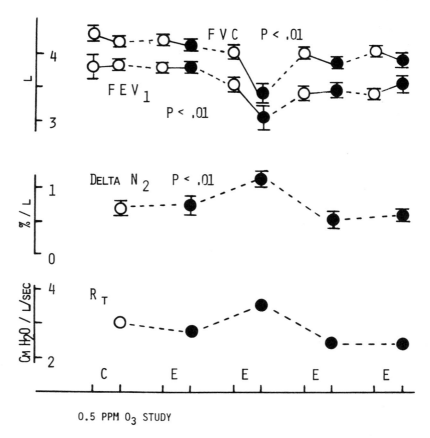

Fig. 5. Mean pulmonary function values for subject 11, the most reactive individual (cf. Figs. 3 and 4). Standard deviations are indicated by vertical bars where repeated measurements are available.

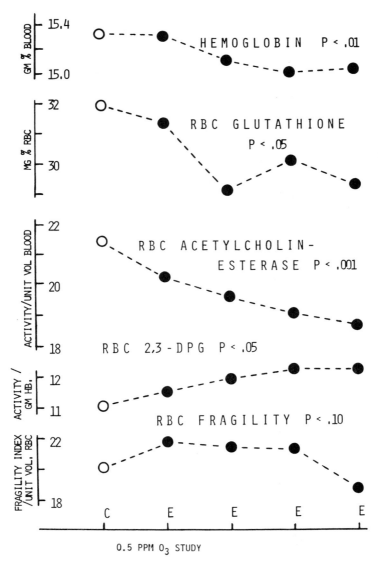

0.5 PPM O$_3$ STUDY

Fig. 6.   Group mean post-exposure values for blood biochemical variables showing apparently meaningful changes. Symbols as in Figs. 3 to 5.

122 Jack D. Hackney *et al.*

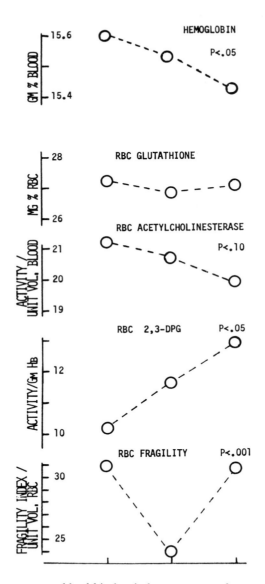

Fig. 7. Mean post-exposure blood biochemical measurements for a separate subject group exposed to purified air only on three successive days (cf. Fig. 6).

The biochemical responses seen in this study differ from the clinical and physiological responses in that the biochemical indices showing changes generally do not return quickly to control levels in the face of repeated ozone exposures. As indicated previously, the biochemical effect of ozone *per se*, as opposed to that of heat, exercise, or other accompanying stresses, cannot be fully elucidated from these data. In some of the observed responses, ozone may be relatively unimportant. Whatever the stimulus producing the biochemical responses, the changes are not inconsistent with the development of adaptation, since the measures generally appear to stabilize with only a small displacement from control values (relative to the large interindividual differences observed in normal populations). However, a contrasting view is also possible: small biochemical changes in blood cells may reflect larger changes in respiratory tract tissues more directly exposed to ozone. Such changes might represent significant harm over the long term, since they appear not to reverse quickly.

Further research efforts should be directed not only toward better understanding of the short-term biochemical response to ozone, but also toward finding out the long-term consequences—clinical, physiological, pathological, or biochemical—in repeatedly exposed "adapted" individuals. This would require comparative short-term controlled exposure studies of humans and laboratory animals, long-term studies of repeatedly exposed animals to relate short- and long-term effects, and epidemiologic studies of human populations to look for evidence of important long-term effects if any are found in animals. It is also important to find out whether or not some people fail to adapt, since this could have important implications in control of ambient air pollution and possibly in understanding etiology of respiratory diseases.

SUMMARY

Adaptation or tolerance to ozone and other irritant gases is well known in laboratory animals, but little studied in man. We investigated the ability of men to adapt to ozone exposure under conditions simulating severe prolonged air pollution episodes. Six volunteers with clinical respiratory hyperreactivity but normal base line pulmonary function were exposed two hours per day to purified air (control) on one day, then to 0.5 ppm ozone on four successive days, with secondary stresses of heat and intermittent exercise. Decreases in pulmonary function performance and increases in respiratory symptoms were maximal after two ozone exposures and largely reversed after four exposures. Several red-cell biochemical measures showed small but significant displacements from control values after one or two ozone exposures. These appeared to stabilize with further exposures, but did not reverse. The results suggest that humans can adapt to high ambient ozone concentrations. The mechanism and long-term consequences of this response remain unknown.

All human experimentation discussed here conforms to the recommendations of the Declaration of Helsinki.

REFERENCES

1. Bates, D.V., G. Bell, C. Burnham, M. Hazucha, J. Mantha, L.D. Pengelly, and F. Silverman. Short-term effects of ozone on the lung. *J. Appl. Physiol.* 32: 176-181, 1972.
2. Buckley, R.D., J.D. Hackney, K. Clark, and C. Posin. Ozone and human blood. *Arch. Environ. Health.* 30: 40-43, 1975.
3. Fairchild, E.J. Tolerance mechanisms. Determinants of lung responses to injurious agents. *Arch. Environ. Health.* 14:111-126, 1967.
4. Falk, H.L. Chemical definitions of inhalation hazards. In: *Inhalation carcinogenesis: AEC symposium series no. 18.* Division of Technical Information. U.S. Atomic Energy Commission, Oak Ridge, Tennessee, 1970.
5. Hackney, J.D., W.S. Linn, R.D. Buckley, and H.J. Hislop. Experimental studies of health effects of ozone. *Environ. Health Perspect.* (in press).
6. Hackney, J.D., W.S. Linn, R.D. Buckley, E.E. Pedersen, S.K. Karuza, D.C. Law, and D.A. Fischer. Experimental studies on human health effects of air pollutants: I. Design considerations. *Arch. Environ. Health.* 30:373-378, 1975.
7. Hackney, J.D., W.S. Linn, S.K. Karuza, R.D. Buckley, D.C. Law, D.V. Bates, M. Hazucha, L.D. Pengelly, and F. Silverman. Effects of ozone exposure in Canadians and Southern Californians. Evidence for adaptation? *Arch. Environ. Health.* 32:110-116, 1977.
8. Hackney, J.D., W.S. Linn, D.C. Law, S.K. Karuza, H. Greenberg, R.D. Buckley, and E.E Pedersen. Experimental studies on human health effects of air pollutants: III. Two-hour exposure to ozone alone and in combination with other pollutant gases. *Arch. Environ. Health.* 30:385-390, 1975.
9. Hackney, J.D., W.S. Linn, J.G. Mohler, E.E. Pedersen, P. Breisacher, and A. Russo. Experimental studies on human health effects of air pollutants: II. Four-hour exposure to ozone alone and in combination with other pollutant gases. *Arch. Environ. Health.* 30: 379-384, 1975.
10. Hackney, J.D., W.S. Linn, J.G. Mohler, and C. R. Collier. Adaptation to short-term respiratory effects of ozone in men exposed repeatedly. *J. Appl. Physiol.* 43:82-85, 1977.
11. Kerr, H.D., T.J. Kulle, M.L. McIlhany, and P. Swidersky. Effects of ozone on pulmonary function in normal subjects: an environmental chamber study. *Am. Rev. Resp. Dis.* 111: 763-773, 1975.
12. Morrow, P.E. Adaptations of the respiratory tract to air pollutants. *Arch. Environ. Health.* 14: 127-136, 1967.
13. Parent, C., M. Hazucha, and D.V. Bates. Adaptation to ozone as studied by dynamic lung function tests in man. *Clin. Res.* 25: 60A, 1977.
14. Stokinger, H.E., W.D. Wagner, and P.G. Wright. Studies on ozone toxicity: I. Potentiating effects of exercise and tolerance development. *AMA Arch. Industr. Health.* 14: 158-160, 1956.
15. U.S. Public Health Service, Division of Air Pollution. Selected methods for the measurement of air pollutants. PHS publication 999-AP-11. Cincinnati, 1965.
16. Winer, B.J. *Statistical principles in experimental design.* New York: McGraw-Hill, 1963, Chapter 4.

THE INFLUENCE OF EXERCISE
ON THE PULMONARY FUNCTION CHANGES
DUE TO EXPOSURE TO LOW CONCENTRATIONS OF OZONE

L. J. Folinsbee
B. L. Drinkwater
J. F. Bedi
S. M. Horvath

Institute of Environmental Stress
University of California, Santa Barbara
Santa Barbara, California

The acute effects of ozone on the respiratory and other body systems have received some attention in the past several years. Stokinger *et al.* (24) first reported that exercising rats were more susceptible to the toxic effects of ozone than resting animals. That such an effect might also be observed in man was suggested by an inverse correlation between cross-country running performance and oxidant pollution levels prior to competition (27). It was subsequently demonstrated (4) that if men exercised during ozone (0.75 ppm) exposure, the observed decrement in pulmonary function was considerably greater than in a resting exposure at the same concentration. In other studies in which subjects exercised during exposure, pulmonary function decrements were observed at ozone concentrations as low as 0.37 ppm (17, 23), a level that caused no effect in resting exposures. In two particularly sensitive subjects, respiratory effects have been reported after breathing 0.15 ppm O_3 through a mouthpiece (1 hr) during heavy exercise (6); such effects, however, were not significant for the group as a whole. In a recent series of experiments we (11) have demonstrated that the enhanced pulmonary effect of ozone, which results from exercise during exposure, occurs during the exercise period. In subjects who exercised for a single 30-min period during a 2-hr ozone (0.50 ppm) exposure, the maximum impairment of pulmonary function was seen immediately following exercise. Despite continued exposure to ozone at rest, pulmonary function either improved or showed no further impairment.

Ozone also influences exercise performance, inducing an alteration in the normal ventilatory pattern (12) and a reduction in maximum work performance (13). The reduction in maximum work performance was apparently due to respiratory discomfort, caused by irritation and reduction of the maximum tidal volume, and not a true limitation of $\overset{\bullet}{V}O_2$ max.

While previous studies have examined the effects of various concentrations of ozone on pulmonary function (14,15,17,19), Silverman *et al.* (23) made an attempt to define a "dose:response" relationship , as did Hackney *et al.* (15). They exposed resting and lightly exercising subjects to three different concentrations of ozone (0.37, 0.50, 0.75 ppm). Significant linear or second order correlations were found between relative change in various pulmonary function measurements and the "effective dose" of ozone (determined by the product of ozone concentration, mean ventilation volume, and the duration of exposure). The present investigation was designed to evaluate the effects of various concentrations of ozone coincident with those found in "smoggy" urban areas at several different levels of activity from rest through heavy exercise. Because the impairment of lung function was expected to occur at lower ozone concentrations as the level of exercise was increased, we planned to obtain an estimate of a "no-effect" level, a concentration of ozone at which no significant pulmonary effects are observed on exposure in "normal" individuals.

METHODS AND PROCEDURES

Subjects. Forty young adult male nonsmokers participated[a] as subjects; each was medically screened (Table 1). A medical history questionnaire, a resting 12-lead electrocardiogram, a battery of clinical spirometric tests, and a physical examination were used in evaluating each subject. Half the subjects had previously resided in an area with high ambient ozone concentrations; 14 reported symptoms associated with irritation or breathing difficulty on high-pollution days. Five of the subjects who had not resided in a polluted area reported similar symptoms upon visiting a high-pollution region. Nine subjects had a history of allergy, 11 were former smokers, and one had "asthma" as a child.

Experimental design. Forty paid volunteers, divided into four equal groups designated as A, B, C, and D, served as subjects in this experiment. Two subjects in Group C and four in Group D failed to complete the sequence of exposures and were replaced by new subjects. Each subject was exposed four times, in random

[a] The nature and purpose of the study and the risks involved were explained verbally and given on a written form to each subject prior to his voluntary consent to participate. The protocol and procedures for this study have been approved by the Committee on Activities Involving Human Subjects of the University of California, Santa Barbara.

Table 1. Physical Data of Subjects

	Age (yr)	Height (cm)	Weight (kg)	Forced Vital Capacity (ml)	% Predicted[a] FVC	FEV$_{1.0}$/FVC (%)
GROUP A	22.1 (18-28)	182.5 (173-186)	76.8 (69-92)	6049 (5223-7168)	106 (93-120)	83.5 (76-94)
GROUP B	21.6 (20-28)	1179.0 (169-193)	75.2 (62-88)	5663 (4340-6739)	102 (78-115)[b]	84.6 (70-96)
GROUP C	23.0 (21-26)	180.0 (171-186)	72.4 (59-84)	5899 (4536-7647)	106.2 (89-138)[b]	83.6 (76-90)
GROUP D	22.3 (19-26)	180.4 (170-185)	72.7 (64-85)	5835 (5002-6480)	104.6 (89-118)	86.1 (79-92)

[a] Predicted according to Anderson et al. (2). [b] The subjects with FVC at 78% of predicted (FEV$_{1.0}$/FVC = 0.96) and FVC at 138% of predicted (FEV$_{1.0}$/FVC = 0.76) have no apparent health problems.

sequence, to filtered air, 0.10 ppm O_3, 0.30 ppm O_3, and 0.50 ppm O_3 for two hours. Group A rested throughout the exposure. Groups B, C, and D exercised at four intervals throughout the exposure; each 15-min exercise period was followed by a 15-min rest period. Exercise consisted of walking on a treadmill at a load sufficient to produce a ventilation of approximately 30, 50, and 70 ℓ/min (BTPS) for Groups B, C, and D, respectively. Group A's ventilatory volumes were approximately 10 ℓ/min. The ambient environmental conditions were 25°C, 45% relative humidity.

Protocol. Experiments were performed in the morning or afternoon at the same time of day for each subject. There was a minimum interval of one week between exposures. Although the influence of adaptation to ozone (from these successive exposures) is unknown, the effect on the results would be minimal because of the randomized design. Additionally, adaptation has been demonstrated experimentally only when subjects have been exposed to high (0.50 ppm) ozone levels, on two successive days (see Hackney *et al.*, this volume).

Upon arrival at the laboratory, the subject was weighed and examined by a physician. Chest leads for the electrocardiogram were then attached. Pre-exposure pulmonary function tests were then performed, consisting of forced vital capacity, functional residual capacity, maximum voluntary ventilation, closing volume, and airway resistance. Following these preliminary tests, the subject entered the exposure chamber. Ventilation, respiratory frequency, and cardiac frequency were measured during each of the exercise periods or at similar intervals for Group A. At the end of exposure, the subject left the chamber and was examined by a physician. The pulmonary function tests were then repeated. On several occasions, two subjects were present in the chamber simultaneously; in these cases, one subject began the exposure 45 min ahead of the other.

If the subject showed changes in post-exposure pulmonary function (arbitrarily defined as a change in FVC, $FEV_{1.0}$, or FEF 25-75% of greater than 10%), he was reexamined two hours later.

Methods. Pulmonary function tests were performed in the standing position. A 13.5-ℓ Benedict Roth spirometer was used for the determination of lung volumes (VC, IC, ERV), forced expiratory parameters (FVC, $FEV_{1.0, 2.0, 3.0}$, FEF 50% FEF 75%, and FEF 25-75%), and maximum voluntary ventilation (MVV) (abbreviations according to Ref. 1). Functional residual capacity was determined by a helium dilution technique. The spirometer was interfaced with a PDP-12 laboratory computer that provided on-line analysis of the pulmonary function tests. Thoracic gas volume (TGV) and airway resistance (R_{aw}) were measured in a constant volume body plethysmograph using the method of DuBois *et al.* (9). All flows and volumes were corrected to BTPS. Closing volume was determined by a helium bolus technique (25) using an acoustic helium analyzer (Quintron). Ventilation volume was measured by having the subject breathe through a low-resistance breathing valve into a chain-compensated 120-ℓ spirometer. Respiratory frequency was taken from a continuous recording of expired carbon dioxide concentration. Cardiac frequency was calculated from the electrocardiogram.

A double-walled acrylic environmental chamber, 1.8 m wide by 2.4 m long by 2.6 m high, was used for exposures. Inlet air was dried (Drierite, silica gel) and passed through activated charcoal and particulate filters at 0.5 m³/min; chamber air turnover time was 22 min. A dehumidifier and steam inlet were used to maintain relative humidity at 45 ± 5%, and an electric heater and air conditioner were used to maintain temperature at 25 ± 1.0°C. Ozone was generated from 100% oxygen using an ultraviolet ozone generator (Ozone Research, Inc.) or an electric discharge ozone generator (3). The ozone concentration was monitored continuously using a chemiluminescent ozone meter (McMillan). The monitor was calibrated with a factory-calibrated ozone generator (Monitor Labs), which was checked periodically against the neutral buffered potassium iodide method (21). It has been demonstrated recently that neutral buffered KI methods give results that are up to 20% higher than those determined by ultraviolet spectrometry (7). We have not attempted to correct our values to these standards, since the specific method we used was not employed in the above comparisons. (As an approximation, the nominal concentrations 0.10, 0.30, and 0.50 ppm O_3 may be more nearly 0.11, 0.27, and 0.43 ppm O_3.) Measured ozone concentrations averaged 0.107 ± 0.01, 0.299 ± 0.01, and 0.487 ± 0.01, respectively, for all groups.

Data analysis. Data were treated by a three-factor analysis of variance with repeated measures across ozone concentration and time. Treatment conditions were (a) subject group (determined by ventilation volume), (b) ozone concentration, and (c) time. Levels of factor (c) were either pre- and post-test measurements or intervals within the exposure period. When interactions were significant, a test for simple main effects was applied, followed by a Newman-Keuls test of ordered means when appropriate (29). All hypotheses were tested for significance at an alpha level of 0.05. Wherever a change is indicated in the text as being significant, it is significant at this level; significance is generally not indicated in the tables.

Table 2. Effective Dose of Ozone (ml)[a]

	0.10 ppm O_3	0.30 ppm O_3	0.50 ppm O_3
GROUP A [b] (10 ℓ)	0.125	0.362	0.599
GROUP B (20.35 ℓ)	0.264	0.73	1.20
GROUP C (29.8 ℓ)	0.383	1.07	1.72
GROUP D (38.65 ℓ)	0.496	1.37	2.22

[a] Calculated from estimated mean ventilation volume · time of exposure · ozone concentration. Elevated ventilation during the recovery period was not included, but at most would account for only a 5% increase in effective dose at the highest work load used. [b] Mean 2-hr ventilation volume during exposure in parentheses.

Data were also analyzed by polynomial regression analysis (8). The analyses were performed first on mean data from each subject group (i.e. A, B, C, or D) and second on all subject groups together after computing an "effective dose" for each group at each ozone concentration. The "effective dose" is calculated as the product of ozone concentration · mean ventilation during exposure · duration of exposure (Table 2). A stepwise multiple regression (8) was performed using the mean percent change in each pulmonary function variable with the ozone concentration and the ventilation volume to estimate the contributon of each variable to the change in pulmonary function.

RESULTS

The analysis of variance revealed significant changes related to subject group (based on ventilatory volume), code (ozone concentration), and time of measurement (pre-, post-, or time periods during exposure).

Pulmonary function. There were no significant differences in any of the pulmonary function measurements in the pre-exposure period when compared across group or code (Table 3).

Following ozone exposure, forced vital capacity (FVC) was significantly decreased at the two highest concentrations, 0.30 and 0.50 ppm. Although the decrease at 0.50 ppm was significantly greater than at 0.30 ppm, there was no significant difference due to volume of ventilation. Inspiratory capacity (IC) was reduced only following ozone exposure with exercise, the 0.50 ppm exposure producing a significantly greater decrease. Total lung capacity (TLC) was significantly reduced only at 0.50 ppm. Other lung volumes (FRC, RV) were not significantly affected by ozone.

A reduction in maximum expiratory flow also occurred following the ozone exposure as measured by changes in $FEV_{1.0}$, FEF 50%, FEF 75%, FEF 25-75%, and

Table 3. Pulmonary Function Measurements Before and After Ozone Exposure[a]

		GROUP A[b]				GROUP B[c]			
		Filtered Air	0.10 ppm O_3	0.30 ppm O_3	0.50 ppm O_3	Filtered Air	0.10 ppm O_3	0.30 ppm O_3	0.50 ppm O_3
FVC	PRE	6157 ± 671	6152 ± 615	6095 ± 661	6100 ± 710	5831 ± 792	5747 ± 844	5797 ± 846	5813 ± 848
	POST	6061 ± 684	6058 ± 721	6016 ± 632	5764 ± 742	5776 ± 722	5773 ± 841	5692 ± 905	5353 ± 961
	POST 2H								5466 ± 810[d]
$FEV_{1.0}$	PRE	5016 ± 629	5008 ± 686	4972 ± 732	4972 ± 745	4837 ± 574	4741 ± 571	4885 ± 621	4884 ± 666
	POST	4966 ± 702	4929 ± 702	4924 ± 675	4602 ± 791	4807 ± 536	4831 ± 569	4659 ± 755	4263 ± 802
	POST 2H								4484 ± 431[d]
IC	PRE	3828 ± 535	3807 ± 391	3760 ± 462	3879 ± 479	3642 ± 494	3588 ± 698	3641 ± 642	3643 ± 594
	POST	3842 ± 500	3796 ± 409	3826 ± 545	3586 ± 572	3529 ± 563	3558 ± 689	3449 ± 618	3224 ± 631
	POST 2H								3296 ± 646[d]
TLC	PRE	8371 ± 1131	8322 ± 864	8468 ± 933	8467 ± 893	7694 ± 1252	7955 ± 1142	7752 ± 1470	7804 ± 1139
	POST	8235 ± 951	8523 ± 787	8331 ± 884	8195 ± 983	7618 ± 1130	7902 ± 1275	7726 ± 1430	7418 ± 1217
FEF 25-75%	PRE	5.14 ± 1.08	5.15 ± 1.41	5.18 ± 1.26	5.34 ± 1.26	5.57 ± 1.87	5.58 ± 1.93	5.73 ± 1.93	5.76 ± 1.86
	POST	5.14 ± 1.26	4.97 ± 1.34	5.26 ± 1.35	4.97 ± 1.33	5.65 ± 2.17	5.79 ± 1.87	5.28 ± 1.96	4.74 ± 1.60
	POST 2H								5.08 ± 1.93[d]
MVV	PRE	194 ± 33	207 ± 26	202 ± 24	200 ± 26	200 ± 31	194 ± 40	196 ± 34	203 ± 33
	POST	193 ± 35	198 ± 34	197 ± 29	190 ± 26	203 ± 34	207 ± 45	184 ± 43	167 ± 35
FEF 50%	PRE	5.30 ± 0.99	5.67 ± 1.35	5.30 ± 1.32	5.67 ± 1.28	5.35 ± 1.90	5.88 ± 2.43	5.73 ± 2.05	5.91 ± 2.11
	POST	5.49 ± 1.20	5.31 ± 1.26	5.62 ± 1.44	5.06 ± 1.10	5.57 ± 1.75	5.89 ± 2.29	5.48 ± 2.27	4.97 ± 1.86
	POST 2H								5.15 ± 1.79[d]
FEF 75%	PRE	2.54 ± 0.75	2.61 ± 0.89	2.42 ± 0.72	2.74 ± 0.66	2.61 ± 0.86	2.94 ± 1.19	3.01 ± 1.38	3.10 ± 1.06
	POST	2.63 ± 0.79	2.37 ± 0.68	2.47 ± 0.69	2.48 ± 0.50	2.97 ± 0.98	3.15 ± 1.46	2.94 ± 1.21	2.46 ± 0.89
	POST 2H								2.61 ± 1.21[d]
R_{aw}	PRE	2.25 ± 0.98	2.07 ± 0.91	1.80 ± 0.60	2.00 ± 0.96	1.47 ± 0.39	1.60 ± 0.75	1.53 ± 0.56	1.42 ± 0.44
	POST	2.21 ± 1.16	1.99 ± 0.91	1.72 ± 0.69	2.23 ± 0.99	1.49 ± 0.45	1.47 ± 0.45	1.67 ± 0.36	1.73 ± 0.43

		GROUP C[e]				GROUP D[g]			
		Filtered Air	0.10 ppm O_3	0.30 ppm O_3	0.50 ppm O_3	Filtered Air	0.10 ppm O_3	0.30 ppm O_3	0.50 ppm O_3
FVC	PRE	5828 ± 773	5909 ± 789	5863 ± 680	5915 ± 801	5968 ± 532	5965 ± 561	5944 ± 507	5981 ± 610
	POST	5822 ± 799	5900 ± 803	5571 ± 968	5330 ± 967	5876 ± 583	5851 ± 586	5540 ± 550	4963 ± 1104
	POST 2H				5755 ± 1062[f]			6009 ± 398[h]	5534 ± 500[i]
$FEV_{1.0}$	PRE	4899 ± 693	4946 ± 706	4921 ± 605	4933 ± 757	4968 ± 527	4898 ± 642	4972 ± 528	5021 ± 685
	POST	4939 ± 702	4971 ± 695	4613 ± 874	4234 ± 972	4953 ± 652	4773 ± 850	4589 ± 604	3826 ± 524
	POST 2H				4609 ± 908[f]			5627 ± 520[h]	4560 ± 561[i]
IC	PRE	3850 ± 658	3892 ± 585	3789 ± 575	3887 ± 630	3615 ± 319	3640 ± 338	3617 ± 425	3607 ± 254
	POST	3693 ± 621	3821 ± 643	3580 ± 752	3315 ± 688	3510 ± 280	3466 ± 371	3335 ± 367	2712 ± 785
	POST 2H				3781 ± 930[f]			3894 ± 353[h]	3403 ± 263[i]
TLC	PRE	8364 ± 1032	8348 ± 1211	8451 ± 1198	8485 ± 1441	8273 ± 872	8277 ± 1142	8461 ± 740	8180 ± 964
	POST	8406 ± 1229	8327 ± 1257	8341 ± 1214	8111 ± 1391	8261 ± 845	8356 ± 1185	8266 ± 847	7534 ± 1394
FEF 25-75%	PRE	5.37 ± 1.24	5.38 ± 1.38	5.29 ± 0.99	5.50 ± 1.50	5.63 ± 1.17	5.15 ± 1.48	5.43 ± 0.98	5.20 ± 1.00
	POST	5.51 ± 1.18	5.37 ± 1.09	4.82 ± 1.34	4.46 ± 1.64	5.28 ± 1.36	5.15 ± 1,52	4.89 ± 1.15	3.45 ± 0.88
	POST 2H				4.96 ± 1.70[f]			7.23 ± 0.26[h]	4.72 ± 0.94[i]
MVV	PRE	206 ± 32	206 ± 32	200 ± 39	205 ± 27	208 ± 34	203 ± 27	204 ± 26	196 ± 23
	POST	209 ± 27	209 ± 32	193 ± 38	168 ± 32	207 ± 33	204 ± 36	172 ± 37	142 ± 33
FEF 50%	PRE	5.88 ± 1.29	5.84 ± 1.58	5.59 ± 0.91	6.00 ± 1.50	6.09 ± 1.27	6.16 ± 1.56	6.02 ± 1.21	5.97 ± 1.55
	POST	5.72 ± 1.51	5.75 ± 1.46	5.18 ± 1.55	4.85 ± 1.78	5.91 ± 1.42	5.95 ± 1.76	5.41 ± 1.02	4.30 ± 1.19
	POST 2H				5.17 ± 2.24[f]			7.05 ± 0.30[h]	5.84 ± 1.87[i]
FEF 75%	PRE	2.68 ± 0.85	2.62 ± 1.06	2.70 ± 0.86	2.89 ± 1.09	2.61 ± 0.82	2.79 ± 1.02	2.80 ± 0.66	2.76 ± 0.94
	POST	2.87 ± 0.86	2.67 ± 0.76	2.47 ± 1.02	2.23 ± 1.05	2.74 ± 0.91	2.56 ± 0.65	2.43 ± 0.67	1.56 ± 0.43
	POST 2H				2.18 ± 1.41[f]			3.88 ± 1.30[h]	2.35 ± 0.68[i]
R_{aw}	PRE	1.73 ± 0.31	1.81 ± 0.33	1.71 ± 0.42	1.68 ± 0.54	1.77 ± 0.71	1.55 ± 0.42	1.62 ± 0.32	1.56 ± 0.41
	POST	1.66 ± 0.45	1.62 ± 0.22	2.17 ± 0.59	2.03 ± 0.42	1.82 ± 0.72	1.74 ± 0.50	1.85 ± 0.42	2.19 ± 0.44

[a] See text for discussion of significant changes. N = 10 except as noted. Values are mean ± SD. [b] $\dot{V}_E$ = 10.3 ℓ/min. [c] $\dot{V}_E$ = 30.7 ℓ/min. [d] N = 6. [e] $\dot{V}_E$ = 49.6 ℓ/min. [f] N = 5. [g] $\dot{V}_E$ = 67.3 ℓ/min. [h] N = 2. [i] N = 8.

MVV. Only $FEV_{1.0}$ and MVV showed significant interactions across group (ventilation), ozone level, and time (PRE-POST). Neither measure was significantly altered following exposure to either filtered air or 0.10 ppm O_3 at any work load. In the resting group (A), $FEV_{1.0}$ decreased at 0.50 ppm, but MVV was unchanged. In the exercising groups (B,C,D), $FEV_{1.0}$ was decreased following exposure to either 0.30 or 0.50 ppm. At the two highest work loads, significant differences in $FEV_{1.0}$ were found between ozone concentrations; 0.50 ppm exposure produced the greatest decrease, followed by 0.30 ppm, while 0.10 and filtered air exhibited no change. The greatest decrease in $FEV_{1.0}$ in any group always occurred during 0.50 ppm O_3 exposures. MVV showed a similar pattern of change, but was apparently a less sensitive indicator of pulmonary function impairment. MVV did not change after resting exposures (A), and in groups B and C only showed a significant decrease after the 0.50 ppm exposure. At the highest work load (D), MVV fell after both the 0.30 and the 0.50 ppm exposure.

The general pattern of responses showed few significant differences among the groups based on ventilation volume. Pulmonary function impairment was generally greatest at the highest concentration, 0.50 ppm O_3. When the work load (i.e. ventilation) was increased, significant pulmonary function changes also occurred at 0.30 ppm O_3. No effects were observed following filtered air or 0.10 ppm O_3 exposure.

We calculated the mean pre-exposure value across all four codes for each pulmonary function measure and then determined the percentage change post-exposure for each ozone concentration. Regression equations (Table 4) were then calculated for each group (A,B,C,D), using the percentage change (PRE-POST/PRE) of each pulmonary function parameter and the ozone concentration (some examples are shown in Fig. 1). Clear differences emerge between Groups A and D with regard to the overall response to the ozone exposures.

Regression equations have also been determined using the concept of "effective dose" (described in *Methods*). These equations (Table 5) predicting percentage change in each pulmonary function parameter were calculated from pooled data from all groups (Fig. 2). Significant correlations between predicted and actual changes were obtained for many of the pulmonary function variables indicating a quantitative relationship between the degree of pulmonary function impairment and the volume of ozone inspired.

Exercise measurements. The major effect that could be attributed to ozone exposure was a change in respiratory pattern (Table 6).

Tidal volume (V_T) decreased with ozone exposure while respiratory frequency increased. Tidal volume decreased with time of exposure in Groups B, C, and D, being most pronounced at the highest work load and at the highest ozone concentration (0.50 ppm). However, tidal volume also decreased during the filtered air exposure, although there was no change during exposure to 0.10 ppm.

The difference in V_T among Groups B, C, and D was not always significant, confirming a wide range of respiratory patterns among individuals. There was at no time a significant difference in V_T between Groups C and D, which suggests that many of the Group D subjects had reached their maximum tidal volume.

Table 4. Linear and Polynomial Regression Equations for Predicting Percentage Change[a] in Pulmonary Function from Ozone Concentration for Each Subject Group

Variable	GROUP A (10 ℓ/min)	GROUP B (30 ℓ/min)	GROUP C (50 ℓ/min)	GROUP D (70 ℓ/min)
FVC	$Y' = 1.249 - 7.643x + 33.668x^2$ [b]	$Y' = 0.559 - 9.247x + 46.82x^2$	$Y' = -0.454 + 18.7972x$	$Y' = 3.935 - 18.74x + 89.79x^2$
$FEV_{1.0}$		$Y' = 0.508 - 8.64x + 62.99x^2$	$Y' = 2.093 + 30.525x$	$Y' = -1.01 - 42.95x$
IC		$Y' = 2.52 - 7.965x + 50.50x^2$		$Y' = 3.71 - 21.06x + 126.62x^2$
MMF		$Y' = -2.07 + 27.98x$	$Y' = -2.72 + 40.66x$	$Y' = -2.346 + 65.20x$
MVV	$Y' = 3.64 - 22.79x + 53.02x^2$	$Y' = -5.06 + 40.51x$	$Y' = -2.66 + 2.75x + 76.48x^2$	$Y' = 4.53 + 67.37x$
FEF 50%			$Y' = 3.30 + 33.12x$	$Y' = 2.34 - 12.79x + 133.02x^2$
FEF 25%		$Y' = -2.83 - 51.10x + 177.47x^2$	$Y' = 4.17 + 45.10x$	$Y' = -2.86 + 80.51x$
R_{aw}		$Y' = 2.66 - 37.03x$		$Y' = -11.45 + 56.28x - 206.32x^2$
f_R (P7)[c]			$Y' = 1.94 - 68.49x$	$Y' = 0.68 - 64.69x$
V_T (P7)[c]			$Y' = -0.66 + 44.2x$	$Y' = -3.45 + 39.56x$

[a] Change is calculated as a decrease. Therefore, a negative value indicates an increase, as in the case of f_R and R_{aw}. Equations that were not statistically significant were omitted. [b] Y' = % decrease change. x = ozone ppm. [c] P7 = min 105.

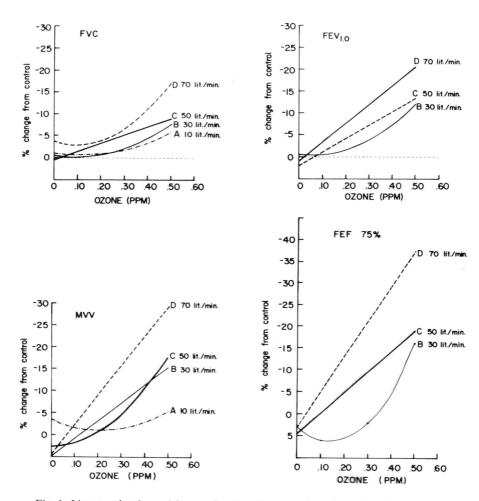

Fig. 1. Linear and polynomial regression lines for percentage change in pulmonary function [forced vital capacity (FVC), 1-s forced expired volume (FEV$_{1.0}$), maximum voluntary ventilation (MVV), and forced expiratory flow at 75% of FVC expired (FEF 75%)] as a function of ozone concentration during a 2-hr exposure with intermittent exercise. Letters indicate groups in which exercise levels were adjusted to achieve a given ventilation volume: A (rest), B (30 ℓ / min), C (49 ℓ /min), and D (67 ℓ /min). Nonsignificant lines were omitted.

Respiratory frequency was increased with time of exposure in Groups C and D. This increased frequency was significant during the second hour of exposure. The highest respiratory frequencies were observed at the highest ozone concentrations (0.50 ppm) in both Groups C and D.

Both cardiac frequency and total ventilatory volume were increased at the higher work loads, but neither was further altered by ozone exposures. Regression equations were calculated for tidal volume and respiratory frequency vs. ozone exposure in the last exercise period (Table 4). Significant linear correlations were found between ozone concentration and respiratory frequency (min 105) and between ozone concentration and tidal volume (min 105) in both Groups C and D. These were not seen in Groups A and B. The prediction equations for changes in tidal volume and respiratory frequency relative to the effective dose of ozone are shown in Table 5, and the regression line for respiratory frequency is shown in Fig. 3.

Multiple regression. To determine whether the decrement (%) observed in several pulmonary function variables could be predicted from the ozone concentration and ventilatory volume during exercise, a multiple regression analysis was performed using group means (Table 7). The highest multiple correlation ($R = 0.89$) was found for $FEV_{1.0}$, indicating that approximately 80% of the variation in $FEV_{1.0}$ could be explained by the two predictors, O_3 concentration (ppm) and $\dot{V}_E BTPS$. With the exception of fR, ozone concentration was selected as the primary predictor, with $\dot{V}_E BTPS$ adding from 7% to 23% to the predictable variation. For fR the largest proportion of the variation in response was predicted by the ventilatory volume, although inclusion of O_3 in the prediction equation improved this prediction by 28%.

The equations given in Table 7 allow prediction of the mean change in pulmonary function to be expected with a group of subjects within the given range of ozone concentration and under the same type of exposure pattern.

Table 5. Linear and Polynomial Regression Equations for Predicting[a] Relative Percentage Change[b] in Pulmonary Function from the "Effective Dose" of Ozone

Variable[c]	Prediction Equation[b]	Correlation (YY')	S.E. of Regression Coefficients	
FVC	$Y' = 1.34 + 0.037x + 0.029x^2$	0.94	± 0.19	± 0.010
$FEV_{1.0}$	$Y' = 0.19 + 0.38x + 0.028x^2$	0.96	± 0.24	± 0.012
IC	$Y' = 1.86 + 0.001x + 0.045x^2$	0.96	± 0.23	± 0.011
MVV	$Y' = -3.01 + 1.31x$	0.93	± 0.14	
FEF 25-75%	$Y' = 0.51 + 0.01x + 0.066x^2$	0.96	± 0.33	± 0.016
FEF 50%	$Y' = 0.81 + 0.17x + 0.049x^2$	0.96	± 0.31	± 0.015
FEF 75%	$Y' = -0.42 - 0.05x + 0.084x^2$	0.93	± 0.59	± 0.029
R_{aw}	$Y' = 1.61 - 1.42x$	-0.76	± 0.32	
f_R (P7)[d]	$Y' = 4.5 - 1.65x$	-0.85	± 0.27	
V_T (P7)[d]	$Y' = -2.72 + 0.95x$	0.71	± 0.25	

[a] To use these equations for prediction, enter "effective dose" of ozone in $\ell\, O_3 \cdot 10^{-4}$. See text for calculations. All equations significant at the 0.01 level. [b] The equation computes the change as a decrease. Therefore, the percentage changes calculated for f_R and R_{aw} are negative, indicating an increase in these values. [c] Abbreviations according to Ref. 1. [d] P7 = min 105.

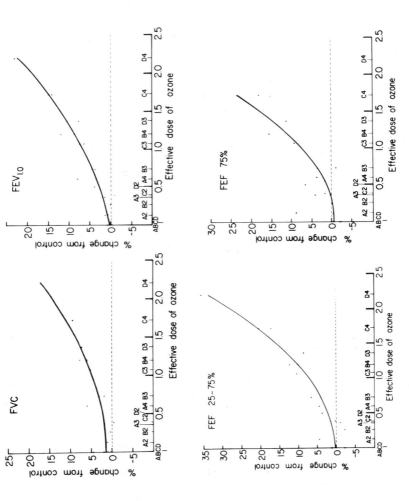

Fig. 2. Polynomial regression lines for percentage change in pulmonary function [FVC, FEV$_{1.0}$, FEF 75%, and forced expiratory flow over mid-half of the FVC (FEF 25-75%)] as a function of "effective dose" of ozone. Numbers and letters indicate group and code. See text for further explanation. All lines significant at 0.01 level of confidence.

Table 6. Tidal Volume and Respiratory Frequency at Exercise During Exposure

	Min	GROUP A Filtered Air	0.10 ppm O$_3$	0.30 ppm O$_3$	0.50 ppm O$_3$	GROUP B Filtered Air	0.10 ppm O$_3$	0.30 ppm O$_3$	0.50 ppm O$_3$
V$_T$ BTPS	15	712 ± 207	750 ± 185	605 ± 127	712 ± 261	1611 ± 324	1545 ± 309	1715 ± 373	1547 ± 281
	45	744 ± 196	769 ± 304	565 ± 142	695 ± 245	1539 ± 326	1485 ± 330	1605 ± 393	1689 ± 316
	75	691 ± 177	728 ± 204	697 ± 220	703 ± 235	1463 ± 305	1471 ± 329	1515 ± 216	1492 ± 367
	105	673 ± 258	671 ± 135	689 ± 222	779 ± 392	1518 ± 416	1423 ± 410	1542 ± 333	1361 ± 303
f$_R$	15	16.3 ± 6.5	14.6 ± 5.2	16.3 ± 4.9	16.4 ± 6.6	0.7 ± 5.3	21.2 ± 5.5	18.4 ± 5.3	20.4 ± 5.0
	45	15.8 ± 6.5	15.2 ± 6.9	15.6 ± 3.9	16.5 ± 7.0	1.2 ± 4.4	21.9 ± 5.3	19.8 ± 5.6	19.0 ± 5.7
	75	15.4 ± 5.6	16.2 ± 6.9	13.9 ± 4.6	16.6 ± 8.8	2.0 ± 4.5	21.5 ± 4.5	20.3 ± 5.1	21.5 ± 5.5
	105	16.1 ± 7.2	15.2 ± 5.9	14.5 ± 5.4	15.6 ± 7.5	1.3 ± 5.4	22.6 ± 5.5	19.7 ± 5.5	23.9 ± 5.7

	Min	GROUP C Filtered Air	0.10 ppm O$_3$	0.30 ppm O$_3$	0.50 ppm O$_3$	GROUP D Filtered Air	0.10 ppm O$_3$	0.30 ppm O$_3$	0.50 ppm O$_3$
V$_T$ BTPS	15	2355 ± 420	2182 ± 363	2074 ± 654	2396 ± 360	2239 ± 312	2268 ± 274	2328 ± 358	2255 ± 365
	45	2108 ± 289	2169 ± 334	2039 ± 562	1947 ± 288	2163 ± 246	2298 ± 342	2249 ± 318	1900 ± 562
	75	2109 ± 409	2073 ± 371	1854 ± 450	1785 ± 417	2039 ± 234	2255 ± 319	2035 ± 348	1647 ± 451
	105	2105 ± 286	2042 ± 375	1842 ± 437	1649 ± 454	2053 ± 259	2117 ± 320	1910 ± 432	1685 ± 472
f$_R$	15	22.5 ± 4.9	22.7 ± 4.4	28.3 ± 13.4	20.9 ± 4.2	29.3 ± 6.0	30.9 ± 4.3	29.8 ± 4.5	30.7 ± 7.0
	45	23.4 ± 4.7	23.7 ± 3.2	27.8 ± 10.3	26.5 ± 7.8	30.9 ± 5.3	29.9 ± 5.8	29.1 ± 5.9	39.8 ± 13.8
	75	23.0 ± 5.0	23.5 ± 3.4	29.9 ± 12.1	28.9 ± 9.9	32.8 ± 5.2	31.7 ± 5.4	32.6 ± 6.9	44.6 ± 16.2
	105	24.1 ± 5.5	24.3 ± 4.4	29.4 ± 8.5	31.6 ± 9.3	33.0 ± 5.1	35.2 ± 11.6	38.0 ± 13.3	44.1 ± 17.2

DISCUSSION

On exposure to moderate levels of ozone, the subjects in this study demonstrated a decrease of maximum voluntary lung inflation and a reduction of maximum expiratory flow. Such changes have been observed previously by others (4,11,13,15,17, 19,23).

The decrease in forced vital capacity was due primarily to a decrease in maximum inspiratory position, indicated by a decreased inspiratory capacity, although a small but nonsignificant increase in residual volume may have contributed to this decreased FVC. We (11) and others (15,23) have noted that a decrease in inspiratory capacity is responsible for the decrease in vital capacity consequent to ozone exposure. However, the decrease in FVC may result from an increase in residual volume (17) or through the combination of a reduced total lung capacity and increased residual volume (5). An increase in RV appears to occur only at higher ozone concentrations. This would be reasonable if we were to postulate that an increase of RV resulted from gas trapping and premature airway closure due to a direct effect of ozone on small airway smooth muscle or due to interstitial pulmonary edema. Dungworth *et al.* (10) observed that morphological damage in monkey lungs was restricted to the most proximal respiratory bronchioles at low ozone concentrations (0.20 ppm), but reached the alveolar duct level at high concentrations (0.80 ppm). Inhibition of inspiration following ozone exposure is generally considered to be due to a vagally mediated reflex initiated by stimulation of irritant

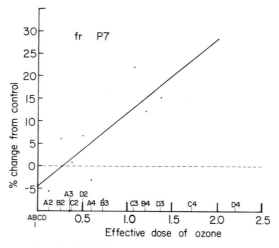

Fig. 3. Linear regression (p < 0.01) for percentage change in respiratory frequency in the last exercise period (P7) as a function of "effective dose" of ozone. Numbers and letters indicate group and code. See text for further explanation.

Table 7. Prediction of Percentage Change in Pulmonary Function
From Ozone Concentration and Exercise Ventilation

	Equation[a]	Multiple R	Increment in R^2 Due to Second Variable	Overall R^2	S.E. of Estimate
$FEV_{1.0}$	$Y' = 27.45\ (O_3) + 0.098\ (V_E) - 4.88$	0.894	0.11	0.799	± 3.16
IC	$Y' = 23.54\ (O_3) + 0.144\ (V_E) - 5.18$	0.864	0.23	0.747	± 3.52
FVC	$Y' = 17.19\ (O_3) + 0.086\ (V_E) - 3.15$	0.864	0.17	0.746	± 2.44
FEF 50%	$Y' = 30.18\ (O_3) + 0.165\ (V_E) - 7.08$	0.847	0.19	0.717	± 4.72
MVV	$Y' = 38.22\ (O_3) + 0.119\ (V_E) - 7.66$	0.842	0.07	0.709	± 5.52
MMF	$Y' = 33.95\ (O_3) + 0.172\ (V_E) - 8.12$	0.809	0.16	0.655	± 6.03
f_R	$Y' = 0.359\ (V_E) + 35.10\ (O_3) - 15.72$	0.801	0.28	0.642	± 8.45

[a] Y' = % decrease for all terms (except f_R, which is % increase). O_3 = ppm ozone during exposure. V_E = ventilation volume (ℓ /min BTPS) during exercise.

receptors (23). These receptors are also stimulated by lung deflation (22), and it is therefore conceivable that an increase in residual volume may be due to a reflexly induced inhibition of maximum expiration

A decrease in maximum expiratory flow, which was observed following exposure to certain higher concentrations of ozone, could result from a reflex bronchocon-striction (4,12), a decrease in maximum inspiratory position (18), or a direct effect of ozone on pulmonary (airway smooth muscle) tissues (26). A decrease in maxi-mum inspiratory position (as indicated by a reduction in inspiratory capacity without a concomitant change in FRC) will lead to a reduction in maximum expiratory flow, as normally calculated. However, we have shown previously (11) that this mechanism can account for only a portion of the reduction in flow. It is well established in experimental animals that irritant gases cause a reflex bronchoconstriction upon stimulation of irritant receptors and that this is a vagal reflex (28). Vagal mediation of such a reflex has not been established in man, although it seems a reasonable presumption. Bronchoconstriction consequent to irritant receptor stimulation is therefore an additional mechanism that may impede maximum expiratory flow. Watanabe et al. (26) have also demonstrated a nonreflex component of ozone-induced bronchoconstriction (i.e. bronchoconstriction due to ozone, which can be only partially reversed by atropine). This may be due to a direct effect of ozone on peripheral airways that are not innervated by the vagus nerve. Other factors may contribute to reduced air flow, such as diffuse interstitial pulmonary edema (10) and mucus plugging of the airway. However, neither mechanism seems to be an explanation because of the time frame of the response; such effects may take several hours to develop.

Some evidence suggests that the small airways are not affected by ozone at low "effective dose" levels. However, a significant decrease in flow at low lung volume (FEF 75%) after exposure to high levels of ozone may indicate an influence of ozone at the peripheral airway level. Because closing volume showed a large interindividual variability and thus no significant change related to ozone exposure, this measure did not provide additional information regarding small airway closure.

A major aim of this study was to determine the effect of a wide range of activity levels on the pulmonary toxicity of ozone exposure, providing data that could be used to predict pulmonary function changes related to ozone exposure. We also hoped to establish "no effect" levels of exposure at different activity levels. The data established that the increased ventilation consequent to activity results in increased pulmonary function impairment in the presence of ozone. The level of ventilation is apparently not as important as the concentration of ozone in predicting the degree of pulmonary function alteration. For all but one variable (f_R), the majority of the variance is described by the ozone concentration rather than the ventilation during exposure. Watanabe *et al.* (26) found that the concentration of ozone was more important than the duration of exposure in determining the pulmonary resistance changes in ozone-exposed cats. Whereas the total quantity of ozone inspired may be an important factor, it appears that the concentration of ozone must be sufficient to provoke a response.

Effective dose.   The effective dose was calculated as the product of ozone concentration, ventilation, and time. Although significant correlations were obtained with effective dose of ozone and changes in various pulmonary function variables, the apparent importance of the ozone concentration suggests that this simplified approach should perhaps be modified. Tidal volume does not appear to be a differentiating variable, since Groups C and D (50 ℓ and 70 ℓ, respectively) had similar tidal volumes; yet Group D had a greater degree of pulmonary function impairment.

We did not measure expired levels of ozone and therefore the retained dose of ozone (which is presumably the component of interest) is not known. Some of the ozone must be expired; the amount may depend on inspired ozone concentration and ventilation volume. DeLucia and Adams (6) noted that some 30% of inspired ozone was subsequently expired by one subject during exercise at 55% $\dot{V}O_2$ max.

It has been pointed out (23) that one additional complication in calculating effective dose is the transition from nose to mouth breathing. The nose is more efficient at removing ozone from inspired air than is the mouth (30), although the relative effectiveness of this procedure depends on the ventilation. At rest and during recovery from exercise, breathing is primarily nasal. At the two higher work loads (Groups C and D), breathing is likely to be primarily oral. However, ozone may induce an increase in nasal resistance (20) that could lead to mouth breathing at a lower work load. Nevertheless, the concept of "effective dose" in this simplified form is supported by the significant linear and second order relationships between the "effective dose" and the decrement in the various measures of pulmonary function.

We have compared the pulmonary function changes predicted from these data with those found in other investigations in which the "effective dose" concept can be applied. $FEV_{1.0}$ data, which are routinely obtained in pollution exposure studies (11,13,15,17,23), are presented in Fig. 4 (Table 8) with the data for the present

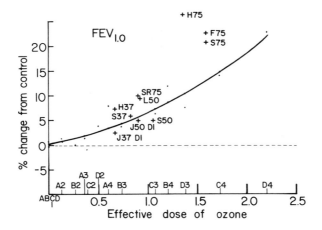

Fig. 4. Polynomial regression line of percentage change in $FEV_{1.0}$ as a function of "effective dose" of ozone in the present study. Letters correspond to those in parentheses preceding each study listed in Table 8. Accompanying numbers indicate ozone concentration in pphm. D1 after J 37, J 50 refers to the first of two exposure days. SR75 (resting exposure to 0.75 ppm — study S). See text and Table 8 for further detail.

Table 8. Comparisons of $FEV_{1.0}$ Predicted from Polynomial and Multiple Regression Equations with Actual Data from Other Similar Ozone Exposure Studies

Study	Assumed Exercise V_E ℓ/min BTPS	Ozone Concentration ppm	Ozone Effective Dose ℓ O_3/h·10^{-3}	Multiple Regression Prediction %	Polynomial Prediction %	Actual %
(H) Hazucha et al. (17)	20	0.75	1.35	17.7	10.4	26.8
		0.37	0.67	7.2	4.0	7.4
(S) Silverman et al. (23)	25	0.75	1.58	18.2	13.2	21.0
		0.50	1.05	11.3	7.3	5.0
		0.37	0.78	7.7	4.9	5.0
(F) Folinsbee et al. (13)	25	0.75	1.58	18.2	13.2	22.8
(J) Hackney et al. (15)	20	0.50	0.90	10.8	5.9	5.0
		0.37	0.67	7.2	4.0	2.3
(L) Folinsbee et al. (11)	24 [a]	0.50	1.0	11.2	6.79	9.6
MEAN				12.79	8.20	12.83

Multiple regression prediction vs. actual $r = 0.91$ [b] $t = 0.02$ [c]
Polynomial regression prediction vs. actual $r = 0.87$ $t = 2.08$

[a] Actual V_E was 37 ℓ/min, but for only 30 min total rather than 60 min. [b] Correlation coefficient of actual data given in this table with data predicted, using equations derived from present study. [c] t value for paired t test.

study. Changes in $FEV_{1.0}$ predicted from the equation (Table 5) in this study agree with the experimental findings in other investigations up to ozone levels of 0.50 ppm. $FEV_{1.0}$ changes following exposure to 0.75 ppm ozone are not predicted well from the polynomial regression equations derived from these data. The data were well correlated ($r = 0.87$), but the difference between the means of the predicted and actual values approached significance ($p < 0.10$). When the $FEV_{1.0}$ change was predicted using the multiple regression equation, the actual and predicted values were more highly correlated ($r = 0.91$) and the means were nearly identical ($t = 0.02$). This indicates that the multiple regression equation, which gives a greater weight to the ozone concentration, is applicable over a wider range of exposure levels. The simplified "effective dose" concept we used apparently needs further modification giving more consideration to ozone concentration, or perhaps should be discarded in favor of a multiple regression approach.

"No effect" levels.   The concentration of ozone at which no pulmonary function change occurred varied according to the level of activity. At rest (Group A), no significant effects were observed at concentrations up to and including 0.30 ppm O_3. In other studies, minimal pulmonary effects have been observed at rest with 0.50 ppm and 0.75 ppm (23), in accordance with our observations. At the highest work load, the "no effect" level was 0.10 ppm. The threshold concentration for subjects exercising at moderately severe energy expenditures lies somewhere between 0.10 and 0.30 ppm O_3. DeLucia and Adams (6) have observed pulmonary function changes at concentrations as low as 0.15 ppm when sensitive subjects were exposed to ozone administered via a mouthpiece during one hour of continuous exercise at 65% of maximum aerobic power.

The prediction equations developed from the present data are limited in application to the O_3 concentrations or effective doses employed in this study. Additionally, these predictions are applicable for only resting or intermittent exercise exposures. Continuous exercise during exposure may induce more complex effects than the brief (15-min) exercise bouts used here, since there would be no opportunity for recovery between exercise periods. We have previously shown (11) that pulmonary function changes, which may be greatest following exercise during exposure, tend to show improvement during the subsequent rest period despite continued exposure. Furthermore, the present data originated from healthy young male non-smokers. Responses of females, children, and elderly subjects remain unknown. It is not clear whether smokers or non-smokers are more sensitive to ozone. Kerr *et al.* (19) found smokers as a group to be less sensitive to ozone exposure, whereas Hazucha *et al.* (17) found smokers to be slightly more sensitive to ozone at higher concentrations (0.75 ppm). The subjects in this study were residents in a low-pollution area (Santa Barbara) at the time of exposure. Residents of a polluted area may have a lower sensitivity to ozone (16), and these predictions may not therefore be applicable to them. Possible application of our prediction equations to diseased subjects also would not be warranted until such subjects are studied under similar exposure conditions.

SUMMARY

Ozone, a major component of photochemical smog, can (if present at certain concentrations) cause impairment of pulmonary function and exercise performance. In the present investigation, we studied 40 subjects who were exposed for two hours to each of four different concentrations of ozone (0.0, 0.10, 0.30 and 0.50 ppm) at one of four different work loads designed to produce specific ventilatory volumes in each subject (VE = 10, 30, 49, and 67 ℓ/min). The purpose of this study was to determine if low levels of ozone would induce pulmonary function changes if the subject exercised at different intensities during exposure. "No effect" levels for different levels of activity could also be discerned with this experimental design. Pulmonary function tests were performed prior to and following exposure; standard spirometric and plethysmographic methods were used. Subjects exercised during exposure on a treadmill at a level sufficient to reach the target ventilation; 15-min exercise periods were alternated with 15-min rest periods.

At rest (10 ℓ/min), pulmonary function changes, if any, were confined to the exposure to 0.50 ppm. At the lowest exercise work load (30 ℓ/min), some changes were apparent at 0.30 ppm, but effects were more marked at 0.50 ppm. At the two highest work loads (49 and 67 ℓ/min), pulmonary function changes occurred at both 0.30 and 0.50 ppm, with the changes at 0.50 ppm usually significantly greater than at 0.30 ppm. There was a decline in vital capacity and forced expiratory flow.

Respiratory frequency (f_R) increased and tidal volune (VT) decreased with ozone exposure. These effects were not observed in the resting group. The change in ventilatory pattern was progressive and was most striking in the heavy work load groups (49 and 67 ℓ/min) and at the highest ozone concentration (0.50 ppm). Percentage change in pulmonary function was highly correlated with "effective dose" (ml O_3 inspired). A multiple regression approach established that ozone concentration was the variable that described the majority of variance in pulmonary function, suggesting that such an approach is preferable to the "effective dose" concept in predicting pulmonary function changes that will result from ozone exposure.

We conclude that for subjects engaged in vigorous exercise, 0.30 ppm of ozone is an unacceptably high level of exposure. However, 0.10 ppm of ozone did not produce measurable pulmonary function impairment in these normal subjects even at the highest work load, approximately 75% to 80% of the subject's $\dot{V}O_2$ max. There is no assurance, however, that more sensitive subjects would not be affected by this relatively low (0.10 ppm) level of ozone.

ACKNOWLEDGMENT

We are grateful for the skilled technical assistance of A. Evers, C. Craig, K. Mayeda, L. Weir, D. Brown, J. Delehunt, and R. Ebenstein.

This work was supported in part by NIH Grant ESO-1143 and California ARB Grant 4-1266. The investigators complied with the recommendations of the Declaration of Helsinki in these experiments.

REFERENCES

1. ACCP-ATS Committee on Pulmonary Nomenclature. Pulmonary terms and symbols. *Chest* 67: 585-593, 1975.
2. Anderson, T., J. Brown, J. Hall, and R. J. Shephard. The limitations of linear regressions for the prediction of vital capacity and forced expiratory volume. *Respiration* 25: 140-158, 1968.
3. Bates, D., G. Bell, C. Burnham, M. Hazucha, J. Mantha, L.D. Pengelly, and F. Silverman. Problems in studies of human exposure to air pollutants. *C.M.A. Journal* 103: 833-837, 1970.
4. Bates, D.V., G. Bell, C. Burnham, M. Hazucha, J. Mantha, L. Pengelly, and F. Silverman. Short term effects of ozone on the lung. *J. Appl. Physiol.* 32: 176-181, 1972.
5. Clamann, H. and R. Bancroft. Toxicity of ozone in high altitude flight. *Adv. in Chem.* 21: 352-359, 1959.
6. DeLucia, J.A. and W. C. Adams. Effects of O_3 inhalation during exercise on pulmonary function and blood biochemistry. *J. Appl. Physiol.* 43: 75-81, 1977.
7. DeMore, W. B., J. Romanovsky, M. Feldstein, W. Hamming, and P. Mueller. Interagency comparison of iodometric methods for ozone determination. In: *Calibration in air monitoring.* American Society for Testing and Materials, Technical Pub. 598. Philadelphia: ASTM, 1976.
8. Dixon, W. J., ed. *BMD biomedical computer programs.* University of California Press, 1975.
9. Dubois, A., S. Botelho, and J. Comroe. A new method for measuring airway resistance in men using a body plethysmograph. *J. Clin. Invest.* 35: 327-335, 1956.
10. Dungworth, D., W. Castleman, C. Chow, P. Mellick, M. Mustafa, B. Tarkington, and W. Tyler. Effect of ambient levels of ozone on monkeys. *Fed. Proc. Fed. Am. Soc. Exp. Biol.* 34: 1670-1673, 1975.
11. Folinsbee, L., S. M. Horvath, P. Raven, J. Bedi, A. Morton, B. Drinkwater, N. Bolduan, and J. Gliner. The influence of exercise and heat stress on pulmonary function during ozone exposure. *J. Appl. Physiol.* 43: 409-413, 1977.
12. Folinsbee, L. J., F. Silverman, and R. H. Shephard. Exercise responses following ozone exposure. *J. Appl. Physiol.* 38:996-1001, 1975.
13. Folinsbee, L., F. Silverman, and R. Shephard. Decrease in maximum work performance following exposure to ozone. *J. Appl. Physiol.* 42: 531-536, 1977.
14. Goldsmith, J.R. and J. A. Nadel. Experimental exposure of human subjects to ozone. *J. Air Pollut. Control Assoc.* 19: 329-330, 1969.
15. Hackney, J., W. Linn, D. Law, S. Karuza, H. Greenberg, R. Buckley, and E. Pedersen. Experimental studies on human health effects of air pollutants III. *Arch. Environ. Health* 30: 385-390, 1975.
16. Hackney, J. D., W. Linn, S. Karuza, R. Buckley, D. Law, D. Bates, M. Hazucha, L. Pengelly, and F. Silverman. Effects of ozone exposure in Canadians and Southern Californians. Evidence for adaptation? *Arch. Environ. Health* 32:110-116, 1977.
17. Hazucha, M., F. Silverman, C. Parent, S. Field, and D. V. Bates. Pulmonary function in man after short-term exposure to ozone. *Arch. Environ. Health* 27: 183-188, 1973.

18. Horvath, S. M. and L. J. Folinsbee. Effects of low levels of ozone and temperature stress: Environmental health effects research series. EPA-600/1-76-001. Research Triangle Park, N.C.: U. S. Environmental Protection Agency, 1976.

19. Kerr, H. D., T. J. Kulle, M. McIlhany, and P. Swidersky. Effects of ozone on pulmonary function in normal subjects. *Am. Rev. Respir. Dis.* 111: 763-773, 1975.

20. Pengelly, L. D., J. Leon, K. Henry, and E. Rebuck. Effect of ozone on nasal and pulmonary function in man. *Physiologist* 18: 348, 1975. (Abstr.)

21. Saltzman, B. Determination of oxidants (including ozone): neutral buffered-potassium iodide method. In: *Selected methods for the measurement of air pollutants.* Cincinnati, Ohio: U.S. Public Health Service Publ. 999-AP-11, 1965.

22. Sellick, H., and J. Widdicombe. Vagal deflation and inflation reflexes mediated by lung irritant receptors. *Q. J. Exp. Physiol.* 55: 153-163, 1970.

23. Silverman, F., L. Folinsbee, J. Barnard, and R. Shephard. Pulmonary function changes in ozone – interaction of concentration and ventilation. *J. Appl. Physiol.* 41: 859-864, 1976.

24. Stokinger, H., W. Wagner, and P. Wright. Studies of ozone toxicity. 1. Potentiating effects of exercise and tolerance development. *AMA Arch. Ind. Health* 14: 158-162, 1956.

25. Travis, D., M. Green, and H. Don. Simultaneous comparison of helium and nitrogen expiratory "closing volumes." *J. Appl. Physiol.* 34: 304-308, 1973.

26. Watanabe, S., R. Frank, and E. Yokoyama. Acute effects of ozone on lungs of cats. I. Functional. *Am. Rev. Respir. Dis.* 108: 1141-1151, 1973.

27. Wayne, W., P. Wehrle, and R. Carroll. Oxidant air pollution and athletic performance. *J. Am. Med. Assoc.* 199: 901-904, 1967.

28. Widdicombe, J.G. Reflex control of breathing. In: *Respiratory physiology.* Edited by J. G. Widdicombe. Physiology Series 1, Vol. 2. MTP Int. Rev. Sci. Baltimore: University Park Press, 1974.

29. Winer, B. J. *Statistical principles in experimental design.* 2nd ed. New York: McGraw-Hill, 1971.

30. Yokoyama, E. Comparison of the uptake rate of inspired O_3, NO_2, and SO_2 gases in the upper airway. *J. Jpn. Soc. Air Pollut. (Taiki Osen Kenk).* 7: 192, 1972. (Abstr.)

III

Work Physiology
Chairperson

Erling Asmussen
August Krogh Institute
Copenhagen, Denmark

AEROBIC WORK PERFORMANCE, A REVIEW

Irma Åstrand

Work Physiology Unit
National Board of Occupational Safety and Health
Stockholm, Sweden

Per-Olof Åstrand

Department of Physiology III
Karolinska Institute
Stockholm, Sweden

INTRODUCTION[a]

The basal metabolic rate in humans is relatively high; 5500 to 7000 kJ [b] during a 24-hr period is a realistic figure for an adult and is equivalent to the energy demand of walking about 18 miles (30 km)! The single factor which in a significant way will increase the energy output above resting level is muscular activity. The working skeletal muscles may actually increase the rate of the oxidative processes to more than 50 times their resting level.

One consequence of the unique ability of the skeletal muscles to vary their metabolic rate is that their activity can seriously interfere with the *milieu interne*, i.e., change the composition and property of the fluid within cells and of their surroundings. When the consumption of fuel and oxygen increases up to fifty-fold, the rate

[a] Portions of this chapter are reprinted with permission from *Progress in Cardiovascular Diseases* 19: 51-67, 1976. [b] 1 kJ (kilojoule) = 1000 J (joule); 1 kcal = 4.2 kJ.

of removal of heat, carbon dioxide, water, and waste products must be increased similarly. To maintain the chemical and physical equilibrium of the cells, there must be a tremendous increase in the exchange of molecules and ions between intra- and extracellular fluid; "fresh" fluid must continuously flush the exercising cells. The maintenance of a relatively stable environment within the body during exercise represents a tremendous challenge to various "service functions," primarily respiration and circulation. The human body is built for action; the circulatory system is normally dimensioned to provide optimal service not only at resting conditions, but also in connection with vigorous physical activity. It should also be emphasized that optimal function can only be achieved by regularly exposing the heart, circulation, muscles, skeleton, and nervous system to some loading, i.e. physical training.

An individual's ability to perform muscular work for minutes or longer periods of time must necessarily depend on his capacity to transport oxygen from the atmosphere into the mitochondria; oxygen is the key to large energy stores. The higher the maximal oxygen uptake ($\dot{V}O_2$ max, maximal aerobic power), the larger is the yielded amount of energy; thus, the rationale behind the quite common mapping out of an individual's $\dot{V}O_2$ max. An exact measurement of the power of the anaerobic machinery is presently not possible (except in a few well standardized conditions), which is a good excuse for excluding the anaerobic energy yield as a main theme for this discussion of physical work capacity.

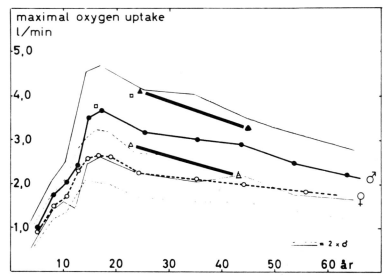

Fig. 1. Mean values for maximal oxygen uptake (maximal aerobic power) measured during exercise on treadmill or bicycle ergometer in 350 female and male subjects, 4 to 65 years of age. Included are values from a group of 86 trained students in physical education and measured maximal oxygen uptakes in a longitudinal study (21 years) of 35 female and 31 male subjects from the original group of 86 (2).

In our laboratories we have collected data over the years on maximal oxygen up-take and its variation with age and sex (Fig. 1). Similar curves have been published from various countries, but the absolute values may vary depending on the selection of subjects, their body sizes, etc. (1,17,19,28). Before beginning a more detailed discussion of the sex and age aspects of physical performance, it is of interest to discuss which factor(s) may limit the maximal aerobic power. *What is the limiting factor(s) for the oxygen transport during exercise?* (See reference 3 for additional detail.) There are many studies showing that the pulmonary function does not limit the oxygen uptake in normal individuals. During maximal exercise there is a hyper-ventilation due to an additional respiratory drive caused by a lowered blood pH. The alveolar oxygen tension increases and the PCO_2 becomes reduced. There may be a slight reduction in the arterial PO_2, but this is not necessarily caused by a limitation in the diffusion of oxygen across the alveolar-red cell membranes. Voluntarily, one can further increase the pulmonary ventilation during maximal exercise, so at least the respiratory muscles are not normally taxed to their maximal power.

When discussing the links between the capillaries in the lungs and those in the working muscles, it is wise to consider Fick's formula for O_2 transport:

$$\dot{V}O_2 = HR \times SV \times C(a-\bar{v})O_2$$

where: HR = heart rate; SV = stroke volume; HR x SV = $\dot{Q}$ (cardiac output); $C(a-\bar{v})O_2$ = arterio-venous oxygen content difference.

During exercise. the increased oxygen demand is met by an elevated cardiac output as well as a gradually wider $a-\bar{v}\ O_2$ difference.

Studies have indicated that the cardiac output at a given oxygen uptake is essentially identical in trained and untrained subjects, in arm and leg exercise respectively, in running and bicycling, and during running and swimming (3,14,18,21,30).

There are individual variations in the absolute values, but we can conclude that *the oxygen uptake gives an indirect evaluation of the cardiac output.* When looking at the situation during maximal exercise, there is in young male subjects a linear relation between data on maximal oxygen uptake and maximal cardiac output. (Subjects with a maximal oxygen uptake of 3.0 $\ell \cdot min^{-1}$ will attain a cardiac output of about 20 $\ell \cdot min^{-1}$; with a $\dot{V}O_2$ max of 6.0 $\ell \cdot min^{-1}$, the $\dot{Q}$ max will be about 40 $\ell \cdot min^{-1}$!)

During exercise there is a hemoconcentration of the blood and therefore a slight increase in the oxygen content of the arterial blood. During maximal work, the blood leaving the muscles has a very low oxygen content, and in mixed venous blood the oxygen content will be around 20 ml $\cdot \ell^{-1}$.

In upright exercise the stroke volume will increase some 50% above the "resting" value, but a maximum is reached at an oxygen uptake corresponding to about 40% of the individual's maximal $\dot{V}O_2$. Therefore, the main factor behind the rise in cardiac output during exercise is an increase in the heart rate. The increase in heart rate with oxygen uptake is essentially linear, at least for a wide range of submaximal exercises. However, in work with smaller muscle groups (e.g., in arm exercise) and isometric (static) exercises the heart rate at a given oxygen uptake is significantly higher than in dynamic leg exercise.

Despite a reduced peripheral resistance in the vascular bed during dynamic exercise, there is an increased intra-arterial blood pressure during exercise in young, healthy subjects. The aortic systolic pressure will reach about 175 mmHg, but the increase in diastolic pressure is usually less than 10 mmHg (3). In older normotensive individuals, the systolic pressure may go up to about 225 mmHg. Also in this respect, arm exercise and static work will elevate the blood pressure more than dynamic exercise.

In several studies at the GIH laboratory, we have experimentally manipulated several of the factors in Fick's formula. The results, partly confirming other reports, partly of an original nature, will be briefly summarized:

Oxygen Content in Arterial Blood

An acute hypoxia (altitude about 4,000 m) will certainly reduce the oxygen content of the arterial blood. During submaximal work there is a compensatory increase in the cardiac output (thanks to an elevated heart rate), but during maximal efforts the cardiac output is not different from control values. There are no means for better utilization of oxygen in the blood, and therefore the maximal oxygen uptake and physical performance are reduced (5).

With part of the hemoglobin blocked by carbon monoxide (up to 20%), the oxygen transport at a given submaximal rate of work can still be maintained. The heart rate is increased and the cardiac output is at control level or somewhat higher. During maximal work, the oxygen uptake is reduced more or less in proportion to the varied oxygen content of the arterial blood. With 15% HbCO, the cardiac output was on average 15% lower than in the control experiments (11).

An increased oxygen tension of the inspired air will increase the maximal oxygen uptake and improve performance. Recent studies by Ekblom *et al.* (11) on eight subjects breathing 50% oxygen in nitrogen at sea level showed an average 12% increase in maximal aerobic power. The cardiac output was only slightly elevated, but the a−v̄ oxygen difference became significantly wider.

By blood loss and reinfusion of red cells, the effect of acute variations in the hematocrit can be studied. The effect of blood loss is a deterioration of physical performance, which is related to a reduced maximal oxygen uptake. A reinfusion of red cells (equivalent to 800 ml of blood) in subjects who have recovered after blood loss could dramatically ("overnight") improve the maximal oxygen uptake and the performance to supernormal values (an average increase in $\dot{V}O_2$ max of 9%). In five subjects running at the maximal speed that could be maintained for about 5 min, the oxygen content of the arterial blood was on average 13% higher after reinfusion of red cells than after blood loss. The difference in maximal oxygen uptake was about 13%, but the individual variations were rather large. The maximal heart rate and stroke volume were more or less identical in the different experiments (12).

Stroke Volume

The heart rate during submaximal and maximal exercise can be varied markedly by various drugs. The parasympathetic activity can be diminished by infusion of atropine, and the sympathetic (adrenergic) β -receptors by propranolol. In one set of experiments, at a given oxygen uptake, heart rate varied about 40 beats • min⁻¹ (taking the extremes), but the cardiac output was similar in the four situations; the stroke volume compensated for the changes in heart rate. (In the propranolol experiments, the cardiac output was reduced an average of 1.5 to 2 ℓ• min⁻¹.) It should be noted that the subjects reached their normal maximal oxygen uptake despite a reduction in maximal heart rate from 195 to about 160 beats • min⁻¹. The performance time was, however, significantly shorter after the β -blocking, and the intra-arterial blood pressure was reduced (10).

There are large individual differences in the size of the stroke volume during exercise; the extreme values can be from 40 up to about 200 ml. There will be an almost proportional variation in maximal cardiac output (8 up to 40 ℓ• min⁻¹) and maximal oxygen uptake (from about 1 up to 6 ℓ• min⁻¹). Genetic factors are decisive for the size of the stroke volume, but the individual's habitual physical activity also comes into the picture. Longitudinal studies have shown that physical conditioning will increase the stroke volume, and prolonged inactivity (e.g. bed rest) will reduce it (9,14,25,27,29).

In the well-controlled study by Saltin et al. (30), the $\dot{V}O_2$ max decreased by 28% (mean value) in five subjects bed-ridden for three weeks (three previously very sedentary subjects and two well-trained ones). This drop was entirely due to a decrease in stroke volume. During two months of daily physical training, the maximal oxygen uptake for three of the sedentary subjects increased 33% above the pre-bed-rest control level, an increase that was apportioned about equally between increased stroke volume and a—v̄ O₂ difference. (Their improvement in $\dot{V}O_2$ max above the post-bed-rest control was about 100% !)

A 15% to 20% increase in $\dot{V}O_2$ max is a common report from training studies. The lower the maximum to start with, the greater is the increase with physical conditioning. In many studies the improved $\dot{V}O_2$ max was due entirely to an increase in stroke volume, particularly in older subjects. In others, up to 50% of the increase in $\dot{V}O_2$ max with training was caused by an increment in the a—v̄ O₂ difference, thus augmenting the increased oxygen delivery due to an increased stroke volume. (The effect of cardiac disease will in many cases be an impaired stroke volume, but that topic will not be analyzed in this review.)

The maximal heart rate is unchanged or slightly reduced (< 10 beats • min⁻¹) during a period of bed rest or physical conditioning, respectively. (Over the years there is a reduction in maximal heart rate in most individuals.)

Muscle Mass Working

Working at maximal rate with one leg on a bicycle ergometer in one study (8) brought the oxygen uptake to 2.4 $\ell \cdot min^{-1}$. With both legs working, the maximum increased to only 3.5 $\ell \cdot min^{-1}$, a relatively small increment.

In swimming with arms only and the legs tied together, a well-trained girl swimmer attained an oxygen uptake of 2.7 $\ell \cdot min^{-1}$; during swimming with leg kicks (with the arms placed on a cork plate), 3.4 $\ell \cdot min^{-1}$. In the normal stroke, she did not reach the sum of these values (2.7 + 3.4 - 0.3 = 5.8 : 0.3 = the resting metabolic rate), but "only" 3.6 $\ell \cdot min^{-1}$ (21).

Working with arms and legs simultaneously on two bicycle ergometers increased the $\dot{V}O_2$ max by less than 10% compared with maximal leg exercise, but the maximum was no higher than when the same subjects were running uphill on a treadmill (6).

This summary illustrates that the maximal oxygen uptake (maximal aerobic power) in exercise engaging large muscle groups is apparently not limited by the capacity of the muscle mitochondria to consume oxygen. Slight variations in the volume of oxygen offered to the tissue ($\dot{Q}$ x C_aO_2) will produce almost proportional changes in the oxygen consumed. In experiments 1-4, the exercise was standardized, the "manipulations" were induced acutely, and the quantity of the muscle enzymes therefore not modified. When more oxygen was transported to the working muscles (experiment 3), there were mitochondria available to consume it.

The exercise with arms (in swimming) as well as with one leg does include some muscle groups that are also engaged in the two-leg work. It is remarkable, however, that the combined exercise does not increase the $\dot{V}O_2$ max more markedly. This finding supports the hypothesis that the central circulation is the limiting factor.

PERIPHERAL ADAPTATIONS TO EXERCISE

It has been noted that over a long period of time, swimmers may attain similar maximal oxygen uptakes in a treadmill test, but when measured during swimming there may be variations in the maximum due to the intensity of swim training (21). Therefore, the ability to utilize the potential of the oxygen transporting system is probably dependent on some adaptation at the tissue level. It is also well documented that a person can work at a high percentage of his maximal oxygen uptake for longer periods of time when trained than when he is untrained. The lactate accumulation sets in at a higher relative rate of work, and the trained subject has a relatively higher energy yield from free fatty acid. Therefore, the glycogen stores will last longer. Before going into detail, it should be recalled that the human skeletal muscles are composed of slow and fast twitch fibers. Slow twitch fibers (Type I) are less fatigable due to their high potential for an aerobic energy yield and are recruited at a relatively low frequency of contraction. The fast twitch fibers (Type II) are faster in contraction time and produce more tension, but are less resistant to fatigue. Their enzymes favor anaerobic energy yield. Type II fibers have been subdivided, with Type II-a having a better potential for aerobic energy yield than Type II-b.

It should be emphasized that a period of physical conditioning will increase the volume of mitochondria in trained muscles, increasing their aerobic energy potential (15,20,22,31). Gollnick *et al.* (13) have, however, concluded from their studies of enzyme systems in skeletal muscles of untrained and trained men that the metabolic capacity of both the conditioned and unconditioned muscles normally exceeds the actual oxygen uptake of the muscles. The increase in enzymes noticed with conditioning also far exceeds the noticed improvement in maximal oxygen uptake.

Succinate dehydrogenase (SDH) analyses on isolated muscle fibers show a higher activity in slow twitch fibers. However, in highly trained cross-country runners, the SDH activity was identical in the slow and fast twitch fibers (24), which indicates an improved aerobic potential for the fast twitch fibers. As a consequence, there is an increase in the Type II-a fibers at the expense of II-b fibers, which may disappear altogether (26,31). The observed range in aerobic potential in individuals of different states of physical training is much greater than the variations in maximal oxygen uptake. As mentioned, there are studies indicating that the enzyme systems do not limit the maximal oxygen uptake. Henriksson and Reitman (16) noticed a 19% increase in maximal aerobic power in their 13 subjects who had trained for endurance for 8 to 10 weeks. The activities of SDH and cytochrome oxidase had increased 32% and 35%, respectively (muscle samples from vastus lateralis). Within two weeks' post-training, the cytochrome oxidase activity had returned to the pre-training level, and after six weeks the SDH activity was back to the control level. However, the maximal oxygen uptake was still high, some 16% above the pre-training level. The authors conclude that an enhancement of the oxidative potential in skeletal muscle is not a necessity for a high maximal oxygen uptake. Changes in the enzyme pattern may be of great importance in the utilization of various substrates in the muscles, which may have particular consequences in prolonged exercise. Henriksson (15) noted after two months of training a 27% higher SDH activity in the trained leg than in the untrained one (quadriceps femoris muscle). The maximal oxygen uptake was 11% and 4% higher in trained and untrained legs, respectively, in the post-training test compared with pre-training. In submaximal two-leg exercise at an average of about 70% of the maximal oxygen uptake (for one hour), the subjects apparently worked harder with the trained leg. However, the degree of utilization of free fatty acids was higher in the trained than in the untrained leg, indicating a difference in the oxidative capacity. These studies indicate that irrespective of what factors trigger the adaptations on the cellular level, causing effects mainly in those muscles that are being trained, there are no (or only modest) effects elicited in non-trained muscles.

Several investigators have found an increase in the number of capillaries in the muscle as a result of training. Brodal *et al.* (7), using the electron microscope, counted the number of capillaries per fiber and per mm² in biopsy samples from the vastus lateralis of the quadriceps muscle. The mean number of capillaries per fiber was 41% greater in the trained than in the untrained group. The difference was statistically highly significant (P < 0.001) and was of the same order of magnitude as the difference in maximal oxygen uptake between the two groups, i.e. 41%. Similar observations of a higher capillary density in endurance-trained athletes are reported by Saltin *et al.* (31), who also noticed an increased capillary density in a longitudinal training study on humans. An increase in the number of capillaries reduces the tissue cylinder around a capillary, increasing the capillary surface area for an exchange of materials. It would also tend to raise PO_2, lower PCO_2, and reduce the concentration of metabolites in the interstitial fluid around the muscle fiber. An increase in the myoglobin content in trained muscles will also enhance the diffusion of oxygen (5). From a teleological viewpoint, it is very efficient that the low-threshold slow twitch fibers are recruited during exercise of moderate intensity. They have the enzyme potentials for aerobic work, and the myoglobin enhances the oxygen diffusion within the cells. During very heavy exercise, the high-threshold fast twitch fibers also become activated, but the oxygen supply is deficient. The myoglobin in the slow twitch fibers directs the oxygen to those fibers preventing the fast twitch fibers from "stealing" oxygen, but the fast twitch fibers are well equipped for anaerobic work and from a mechanical point of view, they are the specialists for very intensive exercise. In other words, when the fast twitch fibers are required, a limited volume of oxygen is available for them (beginning of exercise, very vigorous exercise). It would be something of a waste of resources to have their metabolic repertoire developed as a replica of the slow twitch fibers. However, with training, the potential of the central circulation to transport oxygen out to the tissue is developed. With more oxygen available, it makes sense that fast twitch fibers can improve their aerobic potential; they will complement rather than rival the slow twitch fibers, and in fact the maximal oxygen uptake will increase.

SEX AND AGE

At a given age there is no significant difference in maximal oxygen uptake between girls and boys before puberty. From then on, females attain on average a 25% to 30% lower maximum (oxygen uptake in $\ell \cdot min^{-1}$, Fig. 1). They have a smaller maximal stroke volume than the average man, and with a similar maximal heart rate, they will inevitably attain a lower maximal cardiac output. In addition, the lower hemoglobin concentration in women will cause a further limitation in maximal aerobic power (4). When calculated per kg body weight, the females attain an average of 80% to 85% of the males' values (in Swedish subjects). The sex differences are further reduced when maximal aerobic power is related to the fat-free body weight. The fiber composition in skeletal muscles in women and men is very similar (31).

Fig. 2 summarizes data from a few countries on maximal oxygen uptake in women and men. Fig. 3 gives data on maximal isometric strength for some muscle groups. It should be noted that the degree of overlap of data for women and men is very small for maximal oxygen uptake given in liters per minute, but somewhat larger when it is calculated per kg body weight. The sex difference in isometric strength for arm muscles (flexion) is larger than for leg muscles (knee extension). It is exceptional that a female subject reaches the same value as the average male subject. In many countries there is a policy to mix girls and boys in classes in physical education. In events demanding a high aerobic energy output and prolonged vigorous exercise, this will further complicate the educational process as the range of aerobic power is already large within a sex. Women may apply for jobs within the police (as traffic officers) and fire departments, or military service. If the job demands heavy manual labor or lifting heavy items (e.g. a person from a burning car), it is reasonable that the selection of personnel is partly based on physical characteristics. It is important that one accept some biological truths. In the animal kingdom, it is more a rule than an exception that there is a difference in power between males and females. (The male is not necessarily the stronger; in the species of birds of prey, the female is usually larger and stronger than the male).

It is difficult to explain why the difference in world records set by women and men respectively in swimming and running is only about 10%, in contrast to the larger gap between mean values obtained in laboratory experiments. In the applicants for jobs in the military service studied by Vogel *et al.* (33), the women's maximal oxygen uptake was only 62% of the men's values, and when corrected for body weight the figure was 75%. The maximal isometric strength for elbow flexors was 46% lower in the female applicants; for knee flexors the figure was 34%, and for knee extensors, 23% (Figs. 2 and 3).

We know very little of the physiological background for the decline in aerobic and muscular power with age beyond about 25 years. In a longitudinal study of well-trained subjects, it was noted that at the age of 40 to 45 years, without exception, there was a decline in the maximal oxygen uptake compared with the situation 20 years earlier. The reduction in the maximal aerobic power could not be predicted from the decline in maximal heart rate (2). In the older individual there seems to be a lowering of the cardiac output at a given oxygen uptake, which means a widening of the arterio-venous oxygen difference (25,32). In a recent study we recorded a lowering of the heart rate at a given oxygen uptake with aging (unpublished data). Table 1 gives a summary of mean values of the rate of work performed on a bicycle ergometer at a given heart rate for subjects sampled at random in the Stockholm area (lower panel). Note that at a heart rate of 150, the 25-year-old women could perform 70 watts, and the 65-year-old "colleague" was able to work at 90 watts. This finding is in contrast to previous reports of a constant mean heart rate at a given submaximal oxygen uptake in subjects between the ages of 20 to about 70 years. There is definitely a need for more studies on the human biology and the effects of aging on factors of significance for physical performance.

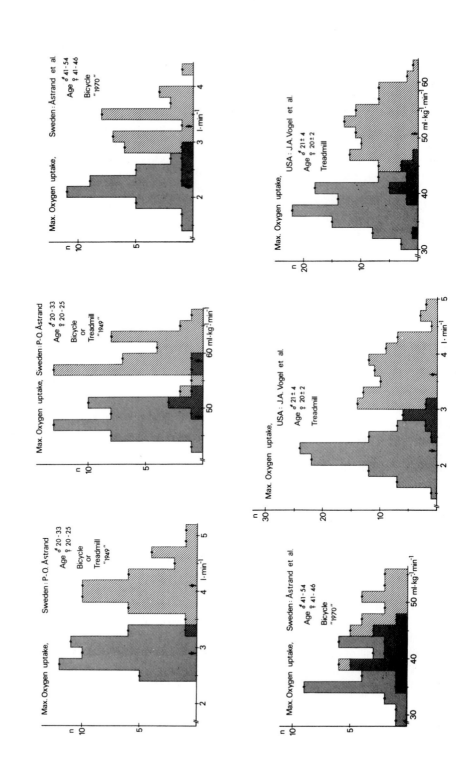

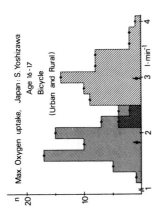

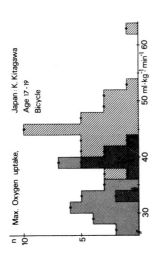

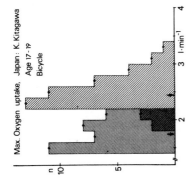

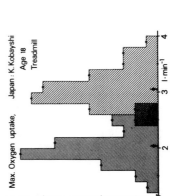

Fig. 2. Variations in maximal oxygen uptake in ℓ/min and related to body weight respectively in male and female subjects from various countries. The subjects studied by P.-O. Åstrand were students in physical education (1949) who were also studied 21 years later [1970 (2)]. The subjects of Vogel et al. (33) were studied when entering the U.S. Army. The Japanese subjects were randomly selected from school classes [Yoshizawa (34), Kitagawa (23)]. The arrow denotes the mean values for women and men, respectively (with the average for men always on a higher level). n = number of subjects in each column.

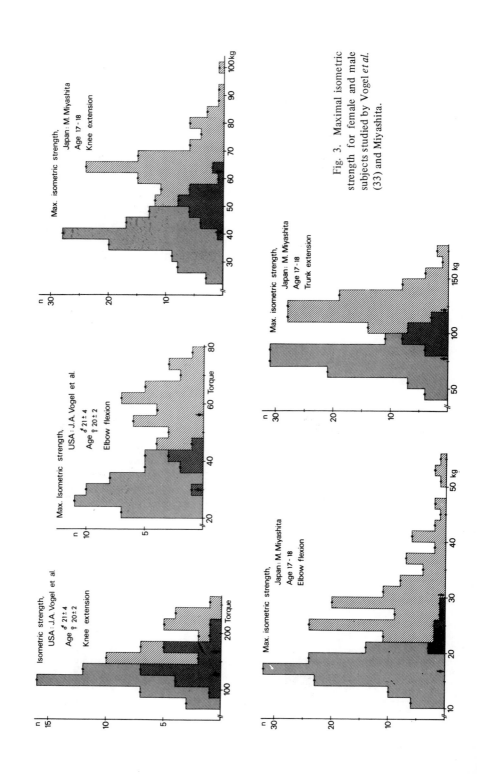

Fig. 3. Maximal isometric strength for female and male subjects studied by Vogel *et al.* (33) and Miyashita.

Table 1. Mean Values for Maximal Heart Rate, Oxygen Uptake,
and Total Energy Yield for Swedish Women (W) and Men (M)[a]

Age, yr	20 – 30 W	30 – 40 M	30 – 40 W	30 – 40 M	40 – 50 W	40 – 50 M	50 – 60 W	50 – 60 M	60 – 70 W	60 – 70 M
Max. Heart Rate beats · min^{-1}	195		185		175		165		155	
Maximal Aerobic Power										
ℓ O_2 · min^{-1}	2.4	3.4	2.2	3.1	2.0	2.8	1.8	2.5	1.6	2.1
Watts	800	1130	730	1030	670	930	600	830	530	700

Rate of Work Performed at Three Different Heart
Rates After 6 Min Exercise on a Bicycle Ergometer[b]

Heart Rate										
170	90	145	97	155	105	165	–	–	–	–
150	70	110	75	120	80	130	85	135	90	140
130	45	80	50	85	55	90	60	95	65	100

[a] Reference 1. [b] Courtesy of B. Jonsson.

SUMMARY

In vigorous exercise engaging large muscle groups, the volume of oxygen offered to the working muscles seems to be very critical for the maximal aerobic energy output. In prolonged exercise, the available carbohydrate supply may be of decisive importance; a dehydration may impair the performance. Regular physical training can modify the enzyme profiles and muscle fiber size and increase the capillary density and myoglobin concentration. These factors may be behind the specificity of training and are important for the improved ability to work close to a maximal rate of metabolism typical of endurance-trained athletes. Women and men were not born equal with regard to maximal power. There are definite differences in maximal aerobic power and muscular strength which cannot be bridged by any political, social, or educational system. Similarly, with age there are modifications of many functions. There are still many open questions about what is inevitable and what is an effect of environment, habitual activities, etc., in these modifications.

ACKNOWLEDGMENT

The authors are very grateful to Drs. Kitagawa, Kobayshi, Miyashita, and Vogel for sending the original data on their subjects.

REFERENCES

1. Åstrand, I. Aerobic work capacity in men and women with special reference to age. *Acta Physiol. Scand.* 49 (Suppl. 169), 1960.
2. Åstrand, I, P.-O. Åstrand, I. Hallbäck, and A. Kilbom. Reduction in maximal oxygen uptake with age. *J. Appl. Physiol.* 35: 649-654, 1973.
3. Åstrand, P.-O. Quantification of exercise capability and evaluation of physical capacity in man. *Prog. Cardiovasc. Dis.* 19: 51-67, 1976.
4. Åstrand, P.-O., T.E. Cuddy, B. Saltin, and J. Stenberg. Cardiac output during submaximal and maximal work. *J. Appl. Physiol.* 19: 268-274, 1964.
5. Åstrand, P.-O. and K. Rodahl. *Textbook of work physiology.* 2nd ed. New York: McGraw-Hill, 1977.
6. Bergh, U., I.-L. Kanstrup, and B. Ekblom. Maximal oxygen uptake during exercise with various combinations of arm and leg work. *J. Appl. Physiol.* 41: 191-196, 1976.
7. Brodal, P., F. Ingjer, and L. Hermansen. Number and density of capillaries in the quadriceps muscle of untrained and endurance-trained men: A quantitative electronmicroscopic study. *Am. J. Physiol.* (In press).
8. Davies, C.T.M. and A.J. Sargeant. Physiological responses to one- and two-leg exercise breathing air and 45% oxygen. *J. Appl. Physiol.* 36: 142-148, 1974.
9. Drinkwater, B.L. Physiological responses of women to exercise. In: *Exercise and sport sciences reviews.* Vol. 1. Edited by J.H. Wilmore. New York: Academic Press, 1973, pp. 125-153.
10. Ekblom, B., A.N. Goldbarg, Å. Kilbom, and P.-O. Åstrand. Effects of atropine and propranolol on the oxygen transport system during exercise in man. *Scand. J. Clin. Lab. Invest.* 30: 35-42, 1972.
11. Ekblom, B., R. Huot, E.M. Stein, and A.T. Thorstensson. Effect of changes in arterial oxygen content on circulation and physical performance. *J. Appl. Physiol.* 39: 71-75, 1975.
12. Ekblom, B., G. Wilson, and P.-O. Åstrand. Central circulation during exercise after venesection and reinfusion of red blood cells. *J. Appl. Physiol.* 40: 379-383, 1976.
13. Gollnick, P.D., R.B. Armstrong, C.W. Saubert, IV, K. Piehl, and B. Saltin. Enzyme activity and fiber composition in skeletal muscle of untrained and trained men. *J. Appl. Physiol.* 33: 312-319, 1972.
14. Hartley, L.H., G. Grimby, Å. Kilbom, N.J. Nilsson, I. Åstrand, J. Bjure, B. Ekblom, and B. Saltin. Physical training in sedentary middle-aged and older men. III. *Scand. J. Clin. Lab. Invest.* 24: 335-344, 1969.
15. Henriksson, J. Training-induced adaptation of skeletal muscle and metabolism during submaximal exercise. *J. Physiol.* (In press).
16. Henriksson, J. and S. Reitman. Time course of changes in human skeletal muscle succinate hydrogenase and cytochrome activities and maximal oxygen uptake with physical activity and inactivity. *Acta Physiol. Scand.* 99: 91-97, 1977.
17. Hermansen, L. Oxygen transport during exercise in human subjects. *Acta Physiol. Scand.* (Suppl. 399), 1973.
18. Hermansen, L., B. Ekblom, and B. Saltin. Cardiac output during submaximal and maximal treadmill and bicycle exercise. *J. Appl. Physiol.* 29: 82-86, 1970.
19. Hollmann, W. and Th. Hettinger. *Sportmedizin-Arbeits- und Trainings-grundlagen.* Stuttgart: F.K. Schattauer Verlag, 1976.
20. Holloszy, J.O. Biochemical adaptation to exercise: aerobic metabolism. In: *Exercise and sport sciences reviews.* Vol. 1. Edited by J.H. Wilmore. New York: Academic Press, 1973, pp. 45-71.
21. Holmér, I. Physiology of swimming man. *Acta Physiol. Scand.* (Suppl. 407), 1974.

22. Howald, H. and J.R. Poortmans (eds.) *Metabolic adaptation to prolonged exercise.* Basel: Birkhäuser Verlag, 1975.
23. Ikai, M. and K. Kitagawa. Maximum oxygen uptake of Japanese related to sex and age. *Med. Sci. Sports* 4: 127-131, 1972.
24. Jansson, E. and L. Kaijser. Muscle adaptation to extreme endurance training in man. *Acta Physiol. Scand.* (In press).
25. Kilbom, A. Physical training in women. *Scand. J. Clin. Lab. Invest.* 28 (Suppl. 119), 1971.
26. Nygaard, E., H. Bentzen, M. Houston, H. Larsen, E. Nielsen, and B. Saltin. Capillary supply and morphology of trained human skeletal muscle. Paper presented at the 27th International Congress of Physiological Sciences, Paris, 1977.
27. Pollock, M.L. The quantification of endurance training programs. In: *Exercise and sport sciences reviews.* Vol. 1. Edited by J.H. Wilmore. New York: Academic Press, 1973, pp. 155-188.
28. Robinson, S. Experimental studies of physical fitness in relation to age. *Arbeitsphysiol.* 10: 251-323, 1938.
29. Rowell, L.B. Human cardiovascular adjustments to exercise and thermal stress. *Physiol. Rev.* 54: 75-159, 1974.
30. Saltin, B., G. Blomqvist, J.H. Mitchell, R.L. Johnson, Jr., K. Wildenthal, and C.B. Chapman. Response to submaximal and maximal exercise after bed rest and training. *Circulation* 38 (Suppl. 7), 1968.
31. Saltin, B., J. Henriksson, E. Nygaard, P. Andersen, and E. Jansson. Fiber types and metabolic potentials of skeletal muscles in sedentary man and endurance runners. *Bull. N.Y. Acad. Med.* (In press).
32. Strandell, T. Circulation studies on healthy men. *Acta Med. Scand.* 175 (Suppl. 414), 1964.
33. Vogel, J.A., M.U. Ramos, and J.F. Patton. Comparisons of aerobic power and muscle strength between men and women entering the U.S. Army. *Med. Sci. Sports.* 9: 58, 1977.
34. Yoshizawa, S. A comparative study of aerobic work capacity in urban and rural adolescents. *J. Hum. Ergology* 1: 45-65, 1972.

CHILDHOOD UNDERNUTRITION: IMPLICATIONS FOR ADULT WORK CAPACITY AND PRODUCTIVITY

G. B. Spurr

Department of Physiology
The Medical College of Wisconsin
Milwaukee, Wisconsin
and
Research Service
Veterans Administration Center
Wood (Milwaukee), Wisconsin

M. Barac-Nieto

Department of Physiological Sciences
Universidad del Valle
Cali, Colombia

M. G. Maksud

Department of Physical Education
University of Wisconsin-Milwaukee
Milwaukee, Wisconsin

The twin spectres of hunger and malnutrition have been man's constant companions since earliest times (16,34). One of the modern concerns with this age-old problem has been directed toward the economic impact of malnutrition in adults (7,12,28,31). Because undernutrition in early life appears to have permanent effects

G. B. Spurr *et al.*

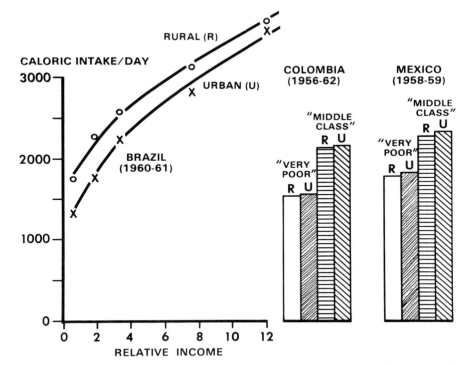

Fig. 1. Relationship of daily caloric intake to economic status of rural and urban populations in three Latin-American countries (28).

on human growth and development (26,30), there has been a tendency to concen-
trate attention on the long-term physiological (13,30), psychological (11), and
economic (18) results of malnutrition in children and to direct programs of intervention toward younger populations. Arteaga (1) pointed out that the high
incidence of undernutrition in children has led to neglect of the importance of malnutrition in adult populations whose capacity to work is the principal asset available to combat underdevelopment.

The purpose of the present paper is to review briefly the known effects of undernutrition on growth and development, and the dependency of adult work capacity
and productivity on nutrition-related parameters, and then to present the results of
some preliminary measurements of work capacity in malnourished, nutritionally
supplemented and normal 6-yr old children.

Undernutrition and Growth

Turnham (28) assembled data from a number of sources that show there is a direct
relationship between socioeconomic status and caloric intake. Some of these data are
summarized in Fig. 1 for urban and rural populations of three Latin American

countries. A similar relationship exists for daily protein intake, i.e., higher income groups enjoy more protein intake in their diets (28). Although there are difficulties in measuring individual food intake in low-income families when members are served from the same cooking pot, it is believed that within such families a hierarchy of preference exists, with the breadwinner receiving the first and best and the youngest children the least (28). Consequently, it is not unreasonable to assume that most adults who live as members of the lower socioeconomic group(s) in developing countries have at some time in their early lives been subject to dietary deprivation (6). This is borne out when one considers the heights of adults from advantaged and disadvantages groups. Fig. 2 presents the average height of men and women from six Latin-American countries (9) and the United States (25). All the Latin-American countries represented in Fig. 2 have adult heights, both for males and females, which are considerably below those for United States adults. The fact that these short statures are not entirely the result of genetic endowment is demonstrated by the data for adult Colombians from high- and low-income groups living in Bogota (Fig. 2). The genetic composition of the two socio-economic groups is presumably similar, but the lower income Colombians are shorter than the advantaged men and women. In developing countries, in general, children from the

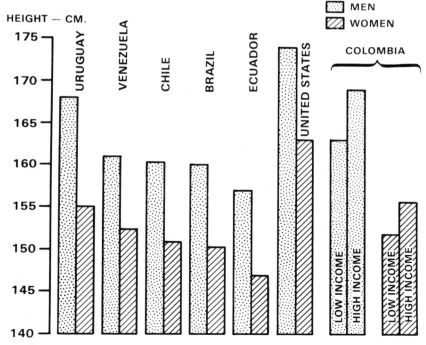

Fig. 2. Average heights of adult men and women from several Latin-American countries and the United States (9,15,25).

best-fed sections of the population approach well-fed European and North American children in their growth achievements, while those in the poor sections of the population lag very seriously behind (27). In a study of genetic and environmental determinants of growth of school-age children in rural Colombia, Mueller and Titcomb (15) established that chronic undernutrition affected the growth of rural parents and continued to affect the growth of their children. Both parents and children were smaller than their advantaged counterparts living in the capital (15). While Fig. 2 presents only heights, similar results are found for body weights (9,25).

Consequently, undernutrition early in life affects the growth of the child and results in an adult who is shorter and weighs less than those individuals who have not been exposed to malnutrition in childhood.

Physical Work Capacity, Productivity, and Nutritional Status

The physical work capacity is best measured by the maximum oxygen consumption ($\dot{V}O_2$max) (2) and is markedly depressed by undernutrition (4,5,29). Barac-Nieto *et al.* (4) have shown that the $\dot{V}O_2$max is progressively lower in subjects with increasing severity of undernutrition.

The economic impact of the decrease in $\dot{V}O_2$max of undernourished adults may be surmised from studies of productivity as related to physical work capacity (8,21). We have studied productivity in sugarcane cutters who are paid by the metric ton of cane cut each day (20,21,23). Consequently, productivity is easily measured. The men were divided into good (>4 tons/day), average (3 to 4 tons/day), and poor (<3 tons/day) cutters and referred to as Groups I, II, and III, respectively (20). Sugarcane cutting is heavy physical work that utilizes about 37% of $\dot{V}O_2$max (5 kcal/min) during the 8-hour workday (20,21,23). Some anthropometric measures of the three groups are presented in Fig. 3, together with the correlations of these values with productivity. It can be seen that while age was not related to productivity, the taller, heavier men had higher productivities. There was a tendency for men with more body fat content to produce less, although the correlation was not quite statistically significant (21).

The values for $\dot{V}O_2$max of these same men, expressed as total oxygen consumption and after normalization for body weight, are presented in Fig. 4. The data demonstrate that $\dot{V}O_2$max and productivity are directly related, although the correlation coefficients are not high. A stepwise multiple regression analysis with productivity as the dependent variable against body weight, height, surface area, percent fat, fat weight (kg), fat-free weight (kg), $\dot{V}O_2$max (liters/min and per kilogram of body weight and lean body mass), maximum heart rate, and the estimated percent $\dot{V}O_2$max sustained during 8 hours of work as independent variables gave the multiple regression equation seen in Fig. 5. The data show that productivity in Colombian sugarcane cutters is related to body size, leanness, and physical condition of the subject as measured by the $\dot{V}O_2$max (21). In the latter regard, the inverse relationship of heart rates at $\dot{V}O_2 = 1.5$ ℓ/min to productivity is further indication of the role of physical condition in productivity (20). Similar results have been reported for sugarcane loaders (22). Satyanarayana *et al.* (17) showed that the

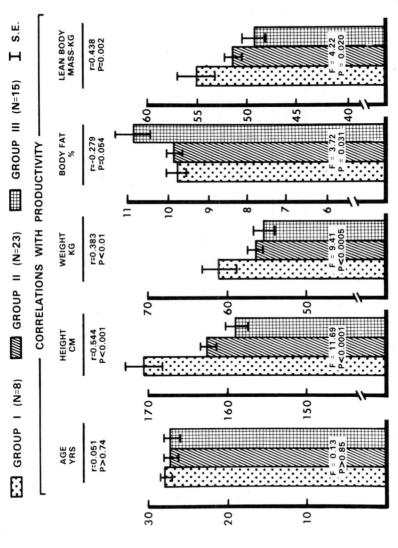

Fig. 3. Physical characteristics of good (Group I), average (Group II), and poor (Group III) sugarcane cutters (ages 18 to 34) and their relationship to productivity. F ratio values are from a one-way analysis of variance (21).

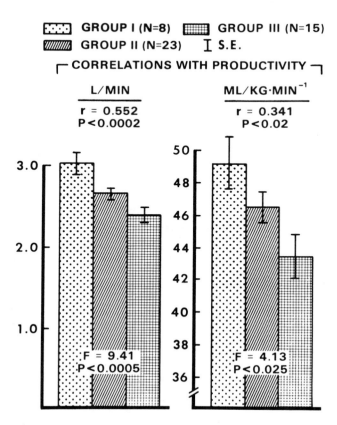

Fig. 4. $\dot{V}_{O_2}$ max (1/min) and aerobic capacity (ml/kg·min⁻¹) of good (Group I), average (Group II), and poor (Group III) sugarcane cutters (ages 18 to 34). F ratio values are from a one-way analysis of variance (21).

productivity of factory workers in India, engaged in light to moderate industrial labor, was positively correlated with their body weight and lean body weight.

The importance of these results (Figs. 3-5) lies in the relationship between productivity and indices of present ($\dot{V}_{O_2}$max, Fig. 4 and Ref. 4; percent fat, Fig. 3) or previous (height, Figs. 2 and 3) nutritional status. Consequently, it can be concluded that adults deprived of their full genetic potential of body size (height and weight) because of childhood exposure to undernutrition, or of their full working capacity because of existing undernutrition (4,5,29), are also deprived of their productive capacity (21). This exacerbates the existing vicious cycle of poverty → undernutrition → low productivity → low earning power → poverty.

Barac-Nieto *et al.* (4) have shown that approximately 80% of the low $\dot{V}_{O_2}$ max observed in severely malnourished subjects can be explained by differences in muscle cell mass, with 20% of the reduced working capacity being unexplained. The question arises about the possibility of malnutrition during the early years of life producing long-lasting cellular effects that may affect the development of adult working capacity independently of its influence on body size.

In 1975, we had the opportunity to study a small number of six-year-old children who were subjects in a larger study involving the stimulation of intellectual and social competence in Colombian preschool children affected by the multiple deprivations of depressed urban environments (14). The children had differing nutritional backgrounds and appeared to be ideal subjects to begin investigating the question of undernutrition and development of the physical condition in children. The results obtained form the basis for the remainder of this report.

METHODS

All the Colombian children studied were participating as subjects in a research program at the Human Ecology Research Station in Cali, Colombia (14). They had been solicited as volunteers when they were three years of age (± 3 mo). The 30 disadvantaged children chosen for this study belonged to the lower 10 percentile of the normative population for height and weight (25) and to the lower 25 percentile of a population of 469 three-year old children originally screened (14). At the time

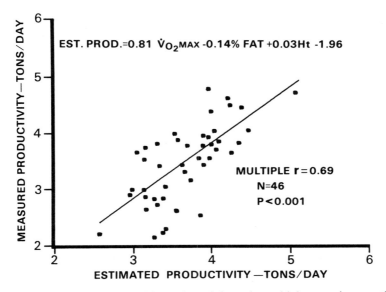

Fig. 5. Scattergram of productivity estimated from the multiple regression equation and measured productivity of sugarcane cutters (21).

of the present study, the children were six years old (± 3 mo) and were divided into the following three groups according to their nutritional history during the previous three years.

Subject Groups

Group 1 (n = 10) were six girls and four boys who received nutritional supplementation, health care, and psychoeducational stimulation in a special school operated by the program. The dietary supplement consisted of the minimum daily requirements (based on body weight) of calories, proteins (mostly vegetable), minerals, and vitamins. The children were supervised in their daily (5 days/wk) consumption of the required minimum intake, but were free to take as much additional food as desired. In addition to the nutritional intervention, they were taught personal hygiene, received excellent health care, and were stimulated educationally and socially by the programs available at the school (including supervised physical activity such as running, jumping, dancing, etc.).

Group 2 (n = 10) consisted of five boys and five girls who received the same nutritional and health care advantages as Group 1 in a home-based intervention program that necessitated feeding the child's entire family. Supervision was provided by weekly visits of a nurse who followed the food distribution and use as closely as possible and also the health of the child.

Group 3 (n = 10) was composed of seven girls and three boys who underwent the same medical, anthropometric, and psychological examinations as the subjects in Groups 1 and 2, but did not receive the nutritional supplementation. Health care was provided as needed. These children entered the program and began receiving the nutritional supplementation at school about one month prior to the measurements of $\dot{V}O_2$max, when they were six years (± 3 mo) of age.

The advantaged children in the two following groups were from the upper socioeconomic levels of society and (to our knowledge) were never exposed to nutritional or other deprivation. None was outside the limits of the normative population (25) (weight and height for age), either when they entered the program at age three (Group 4) or at the time the studies were performed at age 6.

Group 4 (n = 6) consisted of children (two girls and four boys) of Colombian parents who had all the advantages of a relatively high income, good diet, fewer siblings, etc. (14).

Group 5 (n = 6) consisted of five children of North American parents and one of Australian parents who had been living in Cali for at least one year prior to these studies. The ages of these children ranged from 72 to 87 months, with an average of 79.7 mo (6.6 yr). In what follows, the group is referred to as American (three girls and three boys). Only height and weight data obtained at the time of $\dot{V}O_2$max measurements were available for these subjects.

Exercise Testing

Prior to the $\dot{V}O_2$max test, the children were thoroughly familiarized with the laboratory, testing equipment, the treadmill (which all tried in a pretest session), and what was expected of them. During the treadmill test, the children wore a harness that was loosely attached to rigid supports as a safety device to prevent falling. Care was taken to prevent the children from supporting themselves on this device while walking on the treadmill. None of the children fell during testing.

After a resting electrocardiogram and $\dot{V}O_2$ measurement and a 3-min warm-up on the treadmill, the children walked at speeds of 2-3 mph beginning at a 2.5% grade with 2.5% grade increases every 2 min until they could continue no longer. Heart rate and $\dot{V}O_2$ measurements were made during the last 30 s of each workload until near the end, when continuous collections of expired air were made. Heart rates and $\dot{V}O_2$ were measured by standard techniques described previously (20).

Derived Data

Since heart rate (f_H), $\dot{V}O_2$, and work ([weight (kg) x treadmill speed (km/min) x % grade] /100) are linearly related to each other (2), the submaximal data obtained during treadmill walking (excluding resting values and data obtained at heart rates above 170/min) were used to construct least squares regression lines (19) for the purpose of calculating the following values:

a) Heart rate (f_H) at work = 150 kg • m • min^{-1}
b) f_H at $\dot{V}O_2$ = 300 ml • min^{-1}
c) f_H at $\dot{V}O_2$ = 30 ml • kg^{-1} • min^{-1}
d) f_H at 40% $\dot{V}O_2$max
e) $\dot{V}O_2$ (ml • min^{-1}) at work = 150 kg • m • min^{-1}
f) $\dot{V}O_2$ (ml • kg^{-1} • min^{-1}) at work = 150 kg • m • min^{-1}

Statistical Analyses

Statistical comparisons were made on the basis of unpaired student t or F ratio tests (one-way analysis of variance) as required (19). The null hypothesis was rejected at the 5% level. Except where otherwise indicated, the data are presented as means and standard deviations.

RESULTS

A χ^2 analysis of the frequency of boys and girls in the five groups gave a value of 2.31 ($P > 0.05$), indicating no significant difference in composition of the five groups by sex.

The results of the height and weight measurements and the calculations of the weight/height ratio during the approximate three-year period between ages three and six of the four groups of Colombian children are presented in Fig. 6. The results for the disadvantaged children are significantly lower (P <0.01) than those obtained in the advantaged children throughout the three-year period. Groups 1-3 were not

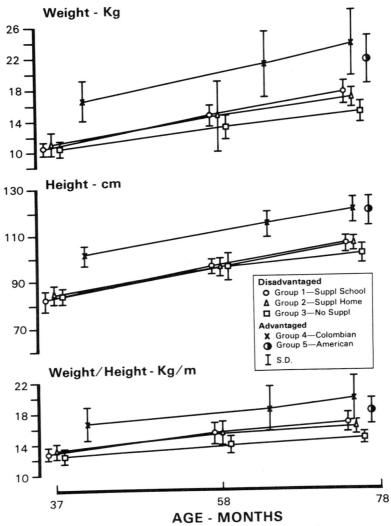

Fig. 6. Averages and standard error of the means of weight, height, and weight/height ratios of disadvantaged and advantaged children during the three years prior to the $\dot{V}O_2$max measurements made when all were approximately six years of age.

significantly different from each other in height and weight at 37 and 58 months. However, at six years of age, the differences in height, weight, and the weight/height ratio were statistically significant, with Group 3 showing lower values than Groups 1 and 2 which are not significantly different from each other (Table 1). Furthermore, the children of Group 5 were not significantly different from the advantaged Colombians, but were significantly taller and heavier than the disadvantaged children (Fig. 6, Table 1).

Table 1 summarizes the anthropometric data and the responses to the maximum exercise test. The maximum heart rates were not significantly different in the five groups of subjects. The $\dot{V}O_2max$ (ml $\cdot$ min^{-1}) values for the disadvantaged children were significantly lower than those observed in the advantaged groups. Also, the children who did not receive nutritional supplementation (Group 3) exhibited a significantly lower value for $\dot{V}O_2max$ than Groups 1 and 2 due to their low body weights, because when $\dot{V}O_2max$ was normalized for body weight (aeobic capacity), the differences between the three disadvantaged groups of children disappeared. However, the values were still significantly smaller than those measured in Groups 4 and 5. The values for maximum pulmonary ventilation ($\dot{V}_Emax$) and oxygen pulse showed a similar pattern to that seen for aerobic capacity (Table 1).

The derived data for heart rates and $\dot{V}O_2$ at a fixed submaximal workload and $\dot{V}O_2$ are presented in Fig. 7. There were no statistically significant differences between the two advantaged groups of children for any of the derived variables. The differences among the three groups of disadvantaged children were also not statistically significant for the variables presented in panels A, C, and E of Fig. 7, viz, f_H and $\dot{V}O_2$ (ml $\cdot$ min^{-1}) at 150 kg $\cdot$ m $\cdot$ min^{-1} of work, and f_H at $\dot{V}O_2$ = 30 ml $\cdot$ kg^{-1} $\cdot$ min^{-1}. The children of Groups 1-3 had significantly higher heart rates at a fixed workload (Fig. 7A) and $\dot{V}O_2$ (Fig. 7B) than Groups 4 and 5. In the latter case, the f_H of Group 3 was significantly higher than those measured for Groups 1 and 2 (Fig. 7B). In the three groups of disadvantaged children, the heart rate responses to 40% of $\dot{V}O_2max$ were significantly different from Groups 4 and 5. This was due to the high value of Group 3, since a one-way analysis of variance of Groups 1, 2, 4, and 5 showed no significant differences (P > 0.50). Finally, the $\dot{V}O_2$ (ml $\cdot$ min^{-1}) at the fixed workload (Fig. 7E) was significantly smaller in the disadvantaged children. When normalized for body weight, Groups 1-3 were not significantly different from Groups 4 and 5 (Fig. 7F). However, when Groups 1-3 were analyzed alone, there was a statistically significant F-ratio due to the high value calculated for Group 3.

DISCUSSION

The results presented in Fig. 6 show that the disadvantaged children were shorter and weighed less than children not exposed to undernutrition, and that nutritional intervention that supplied all the minimum daily requirements did not result in a "catch-up" in the three-year period studied. At 58 months, the children in Group 3 who did not receive dietary supplementation were not significantly different from

Table 1. Means and Standard Deviations of Anthropometry and Maximum Physiologic Responses of Disadvantaged Colombian and Advantaged Colombian and American Six-Year-Old Children

Group Number[e]	Disadvantaged[a]			Probability[b] (1-3)	Advantaged[c,d]		Probability[b] (1-5)
	1	2	3		4	5	
Body weight, kg	17.5 ± 1.5	16.9 ± 1.2	15.0 ± 1.3	<0.002	23.8 ± 4.2	21.6 ± 3.1	<0.0001
Height, cm	105.6 ± 3.0	105.7 ± 3.2	101.8 ± 4.3	<0.05	120.5 ± 5.3	120.0 ± 6.2	<0.0001
Weight/height, $kg \cdot m^{-1}$	16.5 ± 1.2	16.0 ± 0.9	14.7 ± 0.8	<0.002	19.7 ± 2.9	18.0 ± 1.7	<0.0001
Max heart rate, $beats \cdot min^{-1}$	197 ± 11	195 ± 9	196 ± 10	NS	197 ± 5	192 ± 6	NS
$\dot{V}_{O_2}$ max (STPD), $ml \cdot min^{-1}$	603 ± 71	584 ± 108	500 ± 97	<0.05	901 ± 112	912 ± 207	<0.0001
$\dot{V}_{O_2}$ max (STPD), $ml \cdot kg^{-1} \cdot min^{-1}$	34.6 ± 3.7	34.5 ± 5.6	33.2 ± 5.0	NS	38.1 ± 2.4	41.7 ± 4.3	<0.01
O_2 pulse, ml/beat	3.1 ± 0.4	3.0 ± 0.6	2.6 ± 0.5	NS	4.6 ± 0.7	4.8 ± 1.1	<0.0001
$\dot{V}_E$ max (BTPS), $liter \cdot min^{-1}$	29.7 ± 4.9	27.5 ± 4.7	27.1 ± 6.6	NS	44.1 ± 6.0	39.7 ± 7.1	<0.0001

[a] Undernourished at age 3. [b] One-way analysis of variance. [c] Never exposed to undernutrition. [d] No statistically significant difference between Groups 4 and 5. [e] Group 1 – dietary repletion at school (n = 10); Group 2 – dietary repletion at home (n = 10); Group 3 – no dietary repletion (n = 10); Group 4 – Colombians (n = 6); Group 5 – Americans (n = 6). NS = not statistically significant.

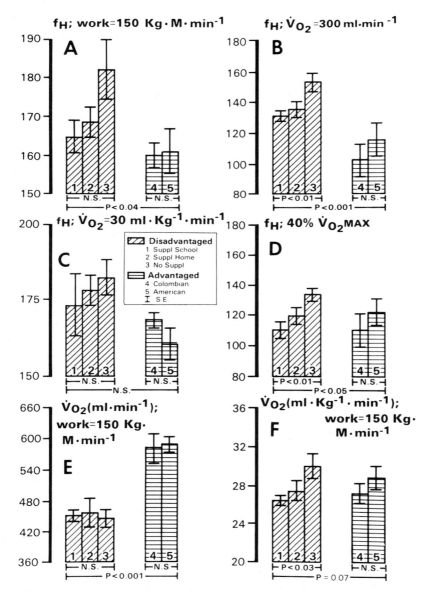

Fig. 7. Heart rates (f_H) and oxygen consumption ($\dot{V}O_2$) of advantaged and disadvantaged children at work = 150 kg·m·min⁻¹ and $\dot{V}O_2$ = 300 ml·min⁻¹. Statistical P values are derived from one-way analysis of variance or unpaired student t comparisons.

the other disadvantaged children, although the appearance of differences in weight and weight/height ratio are beginning, which (together with height) became statistically significant at age six. These same results are implied in the data presented by Young (33) for middle-class and impoverished children in Tunisia (6.5 to 12 yr of age) in comparison with United States normals. It has been suggested that the reduced size of malnourished children may be a protective device to allow for survival when there is a restriction of available food (30). The continued low weight/height ratios in the face of dietary repletion (Fig. 6) is in agreement with the statement of Viteri and Arroyave (30) that disadvantaged children "may be chronically affected by temporary periods of weight for height deficits." They offer no explanation for this observation.

The evidence for existing undernutrition in Group 3 at the time these studies were performed is limited to the height and weight data for age presented in Fig. 7 and Table 1. It is evident that more detailed studies are required.

The $\dot{V}O_2$max and maximum heart rate responses observed for the advantaged children are approximately what one would expect to obtain for normal children this age who are not in physical training. Krahenbuhl *et al.* (10) reported that even at age 8 the maximum aerobic power is lower in females (42.9 ± 5.7 ml · kg^{-1} · min^{-1}) than in males (47.6 ± 7.1 ml · kg^{-1} · min^{-1}). A combination of their (10) data for males and females gives a value of 45.4 ± 6.7 ml · kg^{-1} · min^{-1} for 38 subjects (20 males, 18 females), which is significantly higher than our values for Colombian normals and not statistically different from our American children. Also, although the maximum heart rates found in our American children were lower than the values reported by Krahenbuhl *et al.* (10), the values in the other four groups (Table 1) were not significantly different. Others (32) reported similar values for the $\dot{V}O_2$max and $f_H$max of children six years old which also are not significantly different from those measured in our advantaged children. Consequently, it would seem that the values for $\dot{V}O_2$max and $f_H$max reported here are in general agreement with those reported in the literature. Furthermore, the similarity of the maximum heart rate responses in all groups indicates a similarity in the degree of cardiovascular stress in all groups studied.

The results presented in Table 1 indicate that dietary intervention, starting at age three and continuing until age six, did not result in a significant improvement of $\dot{V}O_2$max, even though both height and weight were beginning to show significant improvement over those children who did not participate in the dietary intervention program. Our results in adults are different from those in children. Barac-Nieto *et al.* (5) demonstrated that in severely malnourished men in whom $\dot{V}O_2$max is markedly depressed, 2.5 mo of dietary intervention with adequate calories and a high protein diet (100 g/day) resulted in significant improvement in $\dot{V}O_2$max, although the values did not return to those seen in well-nourished control subjects. The early exposure of these children to malnutrition produced a serious depression in work capacity that may have effects beyond the three-year period of study described here. An explanation for the difference between adults and children is not readily apparent.

The derived data presented in Fig. 7 show interesting differences based (to some extent) on physical condition of the subjects. The heart rates at fixed submaximal workloads and $\dot{V}O_2$ (Figs. 7A and 7B) were higher in the disadvantaged children, indicating poorer physical condition (2) than the advantaged children. This was also true in Group 3 at the same relative submaximal $\dot{V}O_2$ (Fig. 7D) The effects of malnutrition (specifically, reduced physical activity and mental apathy) are well documented (11,30) and may explain in part the high heart rates at fixed workloads. The fact that the children in Group 3 have higher heart rates than those in Groups 1 and 2 may indicate existing lack of physical activity at the time measurements were made in this study. Increased heart rates at submaximal workloads and $\dot{V}O_2$ have been observed in malnourished adults (24). These responses return toward normal during dietary repletion (24).

The high $\dot{V}O_2$ (ml $\cdot$ kg^{-1} $\cdot$ min^{-1}) at 150 kg $\cdot$ m $\cdot$ min^{-1} of work for the children of Group 3 (Fig. 7F) may indicate a physical inefficiency (awkwardness) and a higher energy cost of performing tasks resulting from the previous and/or existing malnutrition of these children. If they have not been actively exploring their environment like the better nourished subjects of Groups 1 and 2, they may not have yet learned the best way to perform various tasks (in this case, treadmill walking). These derived data emphasize another result of undernutrition *viz* poor physical condition.

SUMMARY

The data presented indicate the following points:
1. Malnutrition early in life depresses growth and results in smaller adults.
2. Small adults produce less, since productivity in both light and heavy industrial work is related to body size.
3. Heavy industrial work is also related to the maximum working capacity as measured by the $\dot{V}O_2$max.
4. Undernutrition early in life depressed the body size and $\dot{V}O_2$max of six-year-old children when compared to advantaged children the same age who had not been exposed to malnutrition.
5. Three years of dietary supplementation between three and six years of age (while it improved growth) did not result in increased $\dot{V}O_2$max.

The implication of these studies is that malnutrition in young children may result in depressed work capacities as adults and further exacerbate the depressed productivity of populations in developing countries due to their smaller size. However, the studies reported here must be considered preliminary in nature. The number of participants in each of the study groups is small and not enough to describe the populations to which they belong. Furthermore, there may be a "catch-up" phenomenon later in the lives of these children with regard to physical working capacity. A more complete study of the growth and development of working capacity in advantaged and disadvantaged groups of children is needed.

ACKNOWLEDGMENTS

This work was supported in part by Contract AID/CSD 2943 with the Office of Nutrition, Agency for International Development. Also supported by the Medical Research Service of the Veterans Administration.

The authors wish to thank Drs. Harrison and Arlene McKay and Leonardo Sinisterra of the Human Ecology Research Station, Cali, Colombia, for their cooperation in giving us access to their subjects and their data on weights and heights of their subjects at ages 37 months and 58 months. The willingness of the children and their parents to participate is also gratefully acknowledged.

REFERENCES

1. Arteaga, A. The nutritional status of Latin-American adults. *Basic Life Sci.* 7:67-76, 1976.
2. Åstrand, P.O. and K. Rodahl. *Textbook of work physiology.* New York:McGraw, 1970.
3. Barac-Nieto, M., G.B. Spurr, H. Lotero, and M.G. Maksud. Body composition in chronic undernutrition. *Am. J. Clin. Nutr.* (In press).
4. Barac-Nieto, M., G.B. Spurr, M.G. Maksud, and H. Lotero. Aerobic capacity in chronically undernourished adult males. *J. Appl. Physiol.* (In press).
5. Barac-Nieto, M., G.B. Spurr, M.G. Maksud, and H. Lotero. Effects of protein-calorie repletion on maximal oxygen consumption and endurance in malnourished subjects. *Proc. Int. Union Physiol. Sci.* 11:103, 1974 (Abstr.).
6. Behar, M. Prevalence of malnutrition among preschool children of developing countries. In: *Malnutrition learning and behavior.* Edited by N.B. Scrimshaw and J.E. Gorden. Cambridge, Mass.:MIT Press, 1968, pp. 30-41.
7. Berg, A. *The nutrition factor: its role in national development.* Washington, D.C.: The Brookings Institution, 1973.
8. Hansson, J.E. The relationship between individual characteristics of the worker and output of work in logging operations. *Stud. For. Succ.* 29. Skoyshägskolen, Stockholm, 1965.
9. Interdepartmental Committee on Nutrition for National Defense, Washington D.C.: U.S. Government Printing Office. *Nutrition Survey.* (a) Ecuador: July 1959; (b) Chile: March 1960; (c) Colombia: May 1960; (d) Uruguay: March-April 1962; (e) North East Brazil: May 1963; (f) Venezuela: May 1963.
10. Krahenbuhl, G.S., R.P. Pangorazi, L.N. Burkett, M.J. Schneider, and G. Petersen. Field estimation of $\dot{V}O_2$max in children eight years of age. *Med. Sci. Sports* 9:37-40, 1977.
11. Latham, M.C. Protein-calorie malnutrition in children and its relation to psychological development and behavior. *Physiol. Rev.* 54:541-565, 1974.
12. Latham, M.C. Nutritional problems in the labor force and their relation to economic development. *Basic Life Sci.* 7:77-85, 1976.
13. Maksud, M.G., G.B. Spurr, and M. Barac-Nieto. The aerobic power of several groups of laborers in Colombia and the United States. *Eur. J. Appl. Physiol. Occup. Physiol.* 35: 173-182, 1976.
14. McKay, H., A. McKay, and L. Sinisterra. Intellectual development of malnourished preschool children in programs of stimulation and nutritional supplementation. In: *Early malnutrition and mental development.* Edited by J. Cravioto, L. Hambraeus, and B. Vahlquist. *Symp. Swed. Nutr. Found.* XII, Uppsala, Almquist and Niksell, 1974, pp. 226-233.
15. Mueller, W.H., and M. Titcomb. Genetic and environmental determinants of growth of school-aged children in a rural Colombian population. *Ann. Hum. Biol.* 4:1-15, 1977.

16. Prentice, E.P. *Hunger and history. The influence of hunger on human history.* New York: Harper, 1939.
17. Satyanarayana, K., A. Nadamuni Naidu, B. Chatterjee, and B.S. Narasinga Rao. Body size and work output. *Am. J. Clin. Nutr.* 30:322-325, 1977.
18. Selowsky, M. A note on preschool-age investment in human capital in developing countries. *Economic Development and Cultural Change* 24:707-720, 1976.
19. Snedecor, G.W. *Statistical methods.* Ames, Iowa: Iowa State Univ. Press, 1956.
20. Spurr, G.B., M. Barac-Nieto, and M.G. Maksud. Energy expenditure cutting sugarcane. *J. Appl. Physiol.* 39:990-996, 1975.
21. Spurr, G.B., M. Barac-Nieto, and M.G. Maksud. Productivity and maximal oxygen consumption in sugarcane cutters. *Am. J. Clin. Nutr.* 30:316-321, 1977.
22. Spurr, G.B., M.G. Maksud, and M. Barac-Nieto. Energy expenditure, productivity and physical work capacity of sugarcane loaders. *Am. J. Clin. Nutr.* 30: 1740-1746, 1977.
23. Spurr, G.B., M. Barac-Nieto, and M.G. Maksud. Efficiency and daily work effort in sugarcane cutters. *Br. J. Ind. Med.* 34:137-141, 1977.
24. Spurr, G.B., M. Barac-Nieto, M.G. Maksud, and H. Lotero. Heart rate response to maximal and submaximal work during dietary repletion of chronically undernourished men. *Med. Sci. Sports* 9:61, 1977 (Abstr.).
25. Stuart, H.C. and H. Meredith. *Nelson's textbook of pediatrics.* 8th Edition. Philadelphia: Saunders, 1964.
26. Thomson, A.M. The later results in man of malnutrition early in life. In: *Calorie deficiencies and protein deficiencies.* Edited by R.A. McCance and E.M. Widdowson. Boston: Little, Brown & Co., 1968, pp. 289-299.
27. Tizard, J. Nutrition, growth and development. *Psychol. Med.* 6:1-5, 1976.
28. Turnham, D. *The employment problem in less developed countries: A review and evidence.* Paris: Development Center of the Organization for Economic Cooperation and Development, 1971, pp. 73-92.
29. Viteri, F.E. Considerations on the effect of nutrition on the body composition and physical working capacity of young Guatemalan adults. In: *Amino acid fortification of protein foods.* Edited by N.S. Scrimshaw and A.M. Altschul. Cambridge, Mass.: MIT Press, 1971, pp. 350-375.
30. Viteri, F.E. and G. Arroyave. Protein-calorie malnutrition. In: *Modern nutrition in health and disease.* Edited by R.S. Goodhart and M.E. Shils. Philadelphia: Lea & Febiger, 1973, pp. 604-624.
31. Viteri, F.E. and B. Torún. Ingestión calorica y trabajo fisico de obreros agricolas en Guatemala. Efecto de la suplementación alimentaria y su lugar en los programas de salud. *Bull. Sanit. Panam.* 78: 58-74, 1975.
32. von Schmucker, B. and W. Hollman. Zur Frage der Trainierbarkeit von Herz und Kreislauf bei Kirdern bis zum 10. Lebensjahr. *Sportartz und Sport Medizin* 10:231-235, 1973.
33. Young, H.B. Measurement of possible effects of improved nutrition in growth and performance in Tunisian children. In: *Amino acid fortification of protein foods.* Edited by N.S. Scrimshaw and A.M. Altschul. Cambridge, Mass.: MIT Press, 1971, pp. 395-425.
34. Walford, C. *The famines of the world: past and present.* London: E. Stanford, 1879.

RUNNING PERFORMANCE FROM THE VIEWPOINT OF AEROBIC POWER

Mitsumasa Miyashita

Laboratory for Exercise Physiology and Biomechanics
University of Tokyo
Tokyo, Japan

Mochiyoshi Miura

Research Center of Health, Physical Fitness and Sports
University of Nagoya
Nagoya, Japan

Yutaka Murase

Department of Physical Education
Nagoyagakuin University
Seto, Japan

Keiji Yamaji

Department of Physical Education
University of Toyama
Toyama, Japan

Numerous investigators have shown that successful distance runners possess a high maximum aerobic power, and Costill and Fox (3) reported that the relationship between running speed and energy expenditure was highly predictable. Therefore, it has been accepted that aerobic power is one of the main factors contributing to distance running performance. Since it is evident that runners with nearly identical $\dot{V}O_2$ max may vary in performance and that $\dot{V}O_2$ max varies among runners of equal performance (4), other factors should be considered which might contribute to running performance.

There is limited information about the relationship between running performance and aerobic power and/or running technique among highly trained runners. The purposes of the present study were to investigate:

    1. The relationship between running performance (5,000 m) and aerobic power among trained adult male runners (Exp. 1)

    2. The effect of long-term (three years) running training on that relationship among highly trained adult male runners (Exp. 2) and trained growing boys (Exp. 3)

    3. The difference in running technique among runners with nearly identical $\dot{V}O_2$ max (Exp. 4).

PROCEDURE AND RESULTS

Experiment 1

The subjects in Exp. 1 were 45 trained middle and long distance runners of a university track team between 19 and 23 years of age. They had undertaken running training for several years.

$\dot{V}O_2$ max was measured during an exhaustive run on the treadmill. Expired air during running was collected in Douglas bags each minute of the run and analyzed by a Beckman oxygen analyzer. The heart rate was calculated from an ECG during running, and the respiratory frequency was determined with the aid of a thermistor attached to the inside of the mask. All subjects ran at least four min before they reached exhaustion.

There was not much difference in age, height, and weight among the subjects engaged in this experiment. The mean values of $\dot{V}O_2$ max are presented in the order of the best recorded time for the 5,000-m race (Table 1). There is a clear trend that the greater the $\dot{V}O_2$ max, the faster the time for the 5,000-m run.

The values obtained in this experiment are shown in Fig. 1 together with data reported on the top runners by other investigators (9; personal communication from Kagaya, 1970, and Aoki, 1970). The regression equation and its standard deviation were calculated from the mean speed (Y, m/s) and the maximum oxygen uptake (X, ml $\cdot$ kg^{-1} $\cdot$ min^{-1}) as shown below.

$$Y = 0.0431X + 2.50 \qquad (S.D. \pm 0.232)$$

These results suggest that the individual difference in performance among those runners who have the same $\dot{V}O_2$ max is within approximately 0.46 m/s, and that the increment of $\dot{V}O_2$ max by 10 ml • kg⁻¹ • min⁻¹ is equivalent to the increment of mean speed by 0.43 m/s.

Experiment 2

The subjects in Exp. 2 were nine athletes who were middle and long distance runners of a university track team, 18 to 19 years of age. Their cardiorespiratory responses to maximal work on the treadmill were observed longitudinally for three years.

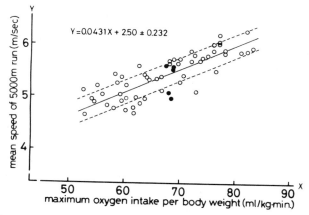

Fig. 1.   The correlation between running performance and $\dot{V}O_2$ max among the highly trained adult runners.   (●) = subjects in Exp. 4 (see text for details).

Table 1. The mean values of $\dot{V}O_2$ max in order
of a 5,000-m run for the adult runners

Record	N	Speed m/s	Height cm	Weight kg	$\dot{V}O_2$ max	
					ℓ/min	ml·kg⁻¹·min⁻¹
14:00 - 14:59	10	5.65 (0.08)	168.0 (5.2)	55.9 (3.4)	3.89 (0.23)	70.1 (4.4)
15:00 - 15-59	13	5.36 (0.07)	169.1 (5.1)	56.8 (5.0)	3.71 (0.41)	65.2 (5.1)
16:00 - 16:59	14	5.04 (0.09)	166.4 (5.7)	56.4 (4.8)	3.48 (0.32)	62.1 (5.2)
17:00 - 17:59	8	4.78 (0.08)	168.8 (6.4)	56.2 (6.6)	3.28 (0.36)	58.6 (3.4)

$\dot{V}O_2$ max was determined by the Douglas bag technique: expired gas was collected with a face mask and connecting tube (diam. = 33 mm) in the bag every minute until exhaustion. The volume of collected gas was measured by a dry gasometer and gas analysis was performed with the Scholander micro-gas analyzer.

Cardiac output was estimated by a CO_2 rebreathing method (6). The partial pressure of CO_2 in arterial blood was estimated from the mean of two determinations of partial pressure of CO_2 in the alveolar gas measured with the end-tidal method immediately after exhaustion. The partial pressure of CO_2 in the mixed venous blood was estimated by rebreathing a gas mixture of CO_2 (4% to 6%) in O_2 (7). This process was completed within 17 s after exhaustion. The values of the content of CO_2 in the arterial blood and the mixed venous blood were read on a standard CO_2 dissociation curve from the partial pressure of CO_2 in the arterial and in the mixed venous blood respectively.

Heart rate was obtained from the ECG, recorded with bipolar chest leads, and respiratory frequency was determined with the aid of a thermistor attached to the inside of the mask.

All measurements were performed in May each year. The best time for 5,000 m was chosen from all trials during that year.

Two typical trends were observed following long-term training in the present adult runners; one group with relatively high initial levels of both running performance and $\dot{V}O_2$ max showed improved performance with no change in $\dot{V}O_2$ max, while the other group improved both performance and $\dot{V}O_2$ max. Therefore, the nine subjects were divided into two groups, A and B. The mean values of $\dot{V}O_2$ max and other physiological variables are summarized in Table 2. Group A showed significantly larger initial values of $\dot{V}O_2$ max, maximal pulmonary ventilation, and arterio-venous oxygen difference in the first year than did Group B. Group A showed no improvement in $\dot{V}O_2$ max and other physiological variables, although they showed a significant improvement in running performance (+5.2%). Group B showed a significant increase in $\dot{V}O_2$ max after three years of training. The increase in $\dot{V}O_2$ max (+14.3%) was accompanied by significant increases in maximal pulmonary ventilation (+22.3%) and arterio-venous oxygen difference (+10.7%). The running perfomance was also significantly improved (+7.1%).

It has been pointed out that in sedentary men the initial level of fitness is an important factor in the increase in aerobic power with training (1). Group B, with the relatively low initial level of aerobic power, showed a significant increase in $\dot{V}O_2$ max with training. But 74.4% of the total increment of $\dot{V}O_2$ max during three years of training was gained within the first year. Therefore, it appears that highly trained and fully matured men cannot make a large improvement in aerobic power even with hard training. On the other hand, all subjects had improved their running performance with three years of training, a change that might be due to an improvement in the efficiency of running.

Table 2. Running Records and Cardiorespiratory Responses to the Maximum Work During Three Experimental Years for the Senior Runners

	Year		Height cm	Weight kg	$\dot{V}_{O_2}$max ℓ/min	$\dot{V}_{O_2}$max ml·kg⁻¹·min⁻¹	$\dot{V}_E$max ℓ/min	$\dot{Q}$max ℓ/min	$C_{a-v}O_2$ ml/ℓ	SV ml	HR beats/min	Running Record 1,500 m s	Running Record 5,000 m s
Group A (4 runners)	1971	x̄	170.5	60.9	4.31	71.9	157.0	30.3	142.0	159.4	190.5	258.0	979.8
		SD	(7.0)	(5.9)	(0.32)	(4.0)	(21.4)	(1.6)	(3.8)	(13.9)	(7.7)	(2.4)	(48.3)
	1972	x̄	171.0	59.7	4.26	71.5	171.4	29.7	137.3	152.0	197.3	251.8	942.3
		SD	(7.1)	(6.6)	(0.34)	(3.1)	(28.1)	(2.0)	(4.8)	(6.9)	(1.9)	(3.6)	(30.2)
	1973	x̄	171.0	60.5	4.10	69.7	161.6	28.1	139.6	155.4	181.5	251.8	946.5
		SD	(7.1)	(6.0)	(0.39)	(5.9)	(11.5)	(2.4)	(5.5)	(2.5)	(2.6)	(10.5)	(41.7)
	1974	x̄	171.0	60.5	4.35	71.3	159.9	30.3	143.7	159.0	190.8	246.0	931.3
		SD	(7.1)	(4.2)	(0.28)	(5.4)	(20.6)	(1.3)	(8.0)	(7.0)	(1.5)	(5.2)	(26.9)
Group B (5 runners)	1971	x̄	166.0	56.2	3.37	60.1	118.5	26.5	126.6	142.5	186.4	265.8	1001.8
		SD	(1.4)	(2.3)	(0.41)	(8.3)	(8.3)	(1.3)	(11.0)	(8.9)	(8.9)	(7.0)	(40.1)
	1972	x̄	165.8	56.2	3.73	66.4	146.8	27.2	135.8	143.7	189.2	259.2	969.0
		SD	(1.2)	(1.8)	(0.43)	(7.8)	(4.6)	(2.3)	(6.2)	(7.4)	(6.2)	(3.7)	(17.0)
	1973	x̄	166.1	56.3	3.98	71.0	136.2	28.0	142.3	149.6	187.4	252.2	953.6
		SD	(1.5)	(1.9)	(0.12)	(2.3)	(10.6)	(1.2)	(2.9)	(9.4)	(5.2)	(6.8)	(26.4)
	1974	x̄	166.0	56.3	3.87	68.7	144.9	27.7	140.1	147.8	188.0	250.8	935.4
		SD	(1.5)	(2.1)	(0.20)	(3.9)	(15.1)	(1.3)	(4.2)	(4.8)	(3.7)	(4.9)	(24.5)

Experiment 3

The subjects were six boys of 14 to 15 years of age, who were the elite runners in the National Championship of Junior High Schools. Their cardiorespiratory responses to maximal work on the treadmill were observed longitudinally for three or four years. The method for determination of $\dot{V}O_2$ max was the same as that used in Exp. 1.

Table 3 shows individual values for anthropometrical, physiological, and running data with growth and training. The increment in height and weight was less than 3 cm and 5 kg respectively during three or four years of training sessions except subject D, who was the youngest. On the other hand, a definite increase in aerobic power was observed (14.0-37.7% in $\dot{V}O_2$ max and 7.0-38.8% in $\dot{V}O_2$ max/kg). All subjects showed a rapid improvement in running performance.

Table 3. Age-Related Increase in $\dot{V}_{O_2}$ max and Improvement
in Running Performance for the Junior Runners

Subj	Age yr, mo.	Height cm	Weight kg	$\dot{V}_{O_2}$ max		Running Records
				ℓ/min	ml·kg^{-1}·min^{-1}	
A	14.8	173.4	59.0	3.63	61.5	6'7"6 (2,000 m)
	15.7	173.4	59.5	4.18	70.3	4'15" (1,500 m), 15'59" (5,000 m)
	17.8	174.5	62.5	4.67	74.6	3'59"2 (1,500 m), 14'49" (5,000 m)
	18.9	174.8	64.5	5.00	77.6	3'58"2 (1,500 m), 14'52" (5,000 m)
B	14.7	167.8	53.0	3.03	57.2	6'7"4 (2,000 m)
	15.6	168.9	57.0	3.94	69.1	16'20" (5,000 m)
	16.7	169.4	58.5	3.94	67.4	14'53"2 (5,000 m)
	17.7	169.4	59.5	4.21	70.8	4'1"6 (1,500 m), 14'52" (5,000 m)
	18.8	169.5	56.0	4.45	79.4	————————
C	14.10	161.8	53.5	3.60	67.3	6'11"6 (2,000 m)
	16.10	162.4	57.0	4.10	72.0	4'10" (1,500 m), 15'25" (5,000 m)
	17.10	162.0	55.0	4.41	80.1	4'4"2 (1,500 m), 14'51" (5,000 m)
	18.11	162.4	56.0	4.47	79.9	15'11" (5,000 m), 32'3" (10,000 m)
D	14.0	172.6	52.5	3.62	69.0	2'11"0 (800 m), 4'26"6 (1,500 m)
	15.0	175.8	55.5	4.00	72.1	2'3"2 (800 m), 4'21"0 (1,500 m)
	15.11	179.1	61.0	4.12	69.2	2'1"6 (800 m), 4'9"6 (1,500 m)
	16.11	178.6	63.5	4.46	70.2	1'59"5 (800 m), 4'12"1 (1,500 m)
	18.0	179.0	61.0	4.51	73.9	1'57"5 (800 m), 4'3"3 (1,500 m)
E	15.0	163.2	51.5	3.72	72.2	5'52"4 (2,000 m) 8'59"4 (3,000 m)
	16.0	164.5	53.5	4.03	75.3	2'1"3 (800 m), 4'8"5 (1,500 m)
	17.2	165.0	54.5	4.12	75.7	1'58"6 (800 m), 4'3"3 (1,500 m)
	18.1	165.1	54.0	4.25	78.6	1'56"8 (800 m), 3'57"4 (1,500 m)
F	14.11	168.5	52.5	3.26	62.1	6'6"0 (2,000 m)
	15.11	170.4	55.0	3.51	63.8	2'1"9 (800 m), 4'13"2 (1,500 m)
	17.1	170.8	57.0	3.93	68.9	1'59"5 (800 m), 4'4"0 (1,500 m)
	18.0	171.2	58.5	4.14	70.8	1'58"7 (800 m)

The $\dot{V}O_2$max of the present elite runners was definitely larger than the mean value of ordinary boys of the same age. Also, the increase in $\dot{V}O_2$max as related to age was greater for the elite runners than for the ordinary boys. Ekblom(5) reported that $\dot{V}O_2$max had significantly increased from 53.9 to 59.4 ml $\cdot$ kg^{-1} $\cdot$ min^{-1} in 11-yr-old boys following six months of training. There are two possible reasons that may be taken into consideration in order to explain why the present elite runners showed a definitely larger $\dot{V}O_2$max at 14 or 15 years of age; one is the genetic factor described by Åstrand and Rodahl (1) and the other is special training before 14 years of age.

Experiment 4

In Exp. 4, five subjects were selected from the university runners who had a simi-lar $\dot{V}O_2$max of about 68 ml $\cdot$ kg^{-1} $\cdot$ min^{-1} (age 18 to 20 years). Three of the subjects have best records for the 5,000-m race ranging from 14'48'' to 14'56''. The other two have poorer records of 16'10'' and 16'52''. Therefore, the subjects were divided into two groups according to their running performance. There was no difference in height, weight, and $\dot{V}O_2$max between the two groups.

All subjects were asked to run 5,000 m at their maximum effort. When the sub-jects were near the 3,000-m point, running at a nearly constant speed, their running form was filmed on a 16-mm high-speed motion picture camera, which was placed 30 m from and perpendicular to the center of the 20-m filming zone at the back stretch of the track. To obtain quantitative data from the film, we used an NAC Film Motion Analyzer which enlarged the image 15 times and projected it on a x-y coordinate screen. Analyses were made of the running speed, stride length, dura-tion of stride, leg action, and displacement of center of gravity.

The mean speed for the good runners was 5.48 m/s and that for the poor runners was 5.06 m/s. The difference between the good runners and the poor runners was 0.42 m/s. The mean value of stride length (two steps) for the good runners was 3.54 m and that for the poor runners was 3.19 m. Therefore, the good runners advanced about 0.35 m further with each stride than did the poor runners.

The force exerted to produce the running movement has two components: hori-zontal and vertical. The angle to the vertical plane of the kick leg at the instant the toe leaves the ground was compared among the runners. That is, we measured the angle between the vertical and a line drawn from the toe to the center of gravity.

The mean value of the angle for the good runner was 34.7° and that for the poor runner was 31.8°. Therefore, the body of the good runner is leaning more forward at the instant the toe leaves the ground than is that of the poor runner.

This means that the body of the good runner moves farther horizontally, while the body of the poor runners moves farther upward. Fig. 2 shows the vertical dis-placement of the center of gravity of a good runner and a poor runner. The mean difference in vertical displacement is 4 cm. Therefore, the poor runner is charac-terized by a so-called bouncing run.

$\dot{V}O_2$max per body weight for subjects employed in this experiment were almost identical, but there was a difference in physical performance in 5,000-m running. This difference was mainly based on the running pattern or style.

One of the distinguishing features between the good runner and the poor runner was the length of stride. An average step length for the good runner was 1.77 m and that of the poor runner was 1.60 m. The good runner can therefore run 5,000 m with 2,825 steps, while the poor runner requires 3,125 steps.

There was also a difference in the vertical displacement of the center of gravity between the two groups of runners. The value for the good runner was 6 cm during one step cycle, and for the poor runner, 10 cm. The body weight of the good runner was 55.5 kg, and that of the poor runner was 57.5 kg.

Assuming that the vertical work is the vertical displacement multiplied by the body weight, then the work performed in running a 5,000-m race is almost 9,407 kg • m for a good runner and 17,968.7 kg • m for the poor runner. The difference amounts to approximately 8,561 kg • m, a value which was greater than expected. The poor runner could run faster if he improved his running pattern to make more efficient use of the oxygen.

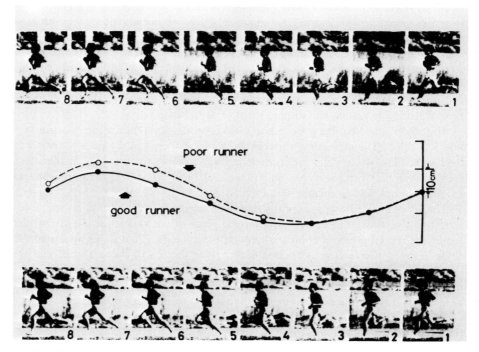

Fig. 2. Comparison of running forms and vertical displacement of center of gravity for a good runner and a poor runner.

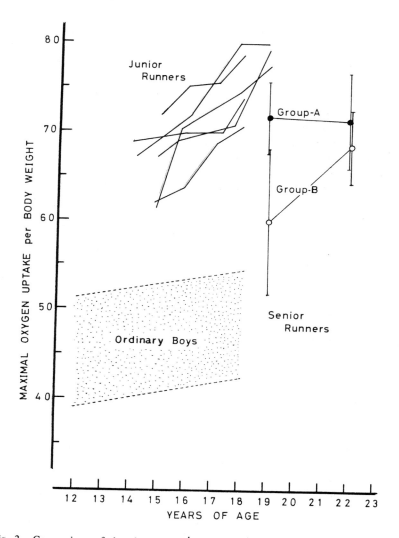

Fig. 3. Comparison of the changes in $\dot{V}_{O_2}$ max related to age for ordinary boys, junior runners (Exp. 3), and senior runners (Exp. 2).

DISCUSSION

Åstrand (2) stated that very few individuals have the inborn capacity to reach a maximum exceeding or approaching 80 ml • kg⁻¹ • min⁻¹, which is a demand for successful participation in Olympic Games and world championships in middle-

and long-distance races. In fact, one of the highest values (82.0 ml $\cdot$ kg^{-1} $\cdot$ min^{-1}) was recorded on the world record holder from Kenya (9). In the present study we found a linear relation between running performance and aerobic power among the trained runners.

Generally speaking, aerobic power can be improved by growth and training. However, it was pointed out by Saltin and Åstrand (9) that there are limits to the individual's ability to improve his $\dot{V}O_2$ max, and therefore natural endowment is important. The point that there is a definite ceiling in the increase of maximum aerobic power by training may be supported by the results obtained in the present observations on the senior runners; namely, there was no change in $\dot{V}O_2$ max during the last two years of training. The reason that five of nine runners could improve their $\dot{V}O_2$ max during the first year might be explained by the fact that they were in the last stage of growth. The present junior runners showed larger increases in maximum aerobic power related to age than those of ordinary boys, a difference that may be caused by both training and growth. They already had a larger $\dot{V}O_2$ max at 14 or 15 years of age than the ordinary boys of the same age. They must have been physically active in the preceding years. Therefore, the greater natural endowment in aerobic power of the elite junior runners is probably reflected by a higher level of physical activity early in life (Fig. 3).

Running performance in the 5,000-m race was clearly improved in the present elite junior runners with an increase in $\dot{V}O_2$ max and senior runners without an increase in $\dot{V}O_2$ max. The following factors might contribute to running performance:

1. An increase in $\dot{V}O_2$ max per body weight,

2. A decrease in the oxygen demand of running at any given velocity (increased efficiency),

3. Working closer to $\dot{V}O_2$ max for a greater portion of the race,

4. Greater anaerobic involvement (4).

In the present study the difference in running technique among the senior runners was brought into focus. Although there was a high linear relation between the mean speed of 5,000 m run and $\dot{V}O_2$ max per body weight, the mean speed was different even among runners who had a similar $\dot{V}O_2$ max . In comparing runners whose $\dot{V}O_2$ max was about 70 ml $\cdot$ kg^{-1} $\cdot$ min^{-1}, some runners ran at a speed of about 5.2 m/s, while other runners ran at a speed of about 5.6 m/s. The present analysis showed that the slower runners used their energy to perform more vertical work relative to the faster runners.

REFERENCES

1. Åstrand, P.-O. and K. Rodahl. *Textbook of work physiology.* New York: McGraw-Hill, 1970.
2. Åstrand, P.-O. Physical education in the age of post-industrialization from the viewpoint of work physiology. Proceedings of First International Seminar on Physical Education in Japan, 1973, University of Nagoya, pp. 51-56.

3. Costill, D.L. and E. L. Fox. Energetics of marathon running. *Med. Sci. Sports* 1:81-86, 1969.
4. Daniel, J. and N. Oldridge. Changes in oxygen consumption of young boys during growth and running training. *Med. Sci. Sports.* 3:161-165, 1975.
5. Ekblom, B. Effect of physical training in adolescent boys. *J. Appl. Physiol.* 27:350-355, 1969.
6. Jernerus, R., C. Lundin, and D. Thomson. Cardiac output in healthy subjects determined with a CO_2 rebreathing method. *Acta Physiol. Scand.* 59: 390-399, 1963.
7. Klausen, K. Comparison of CO_2 rebreathing and acetylene methods for cardiac output. *J. Appl. Physiol.* 20: 763-766, 1965.
8. Matsui, H., M. Miyashita, M. Miura, K. Kobayashi, T. Hoshikawa, and S. Kamei. Maximum oxygen intake and its relationship to body weight of Japanese adolescents. *Med. Sci. Sports* 4:29-32, 1972.
9. Saltin, B. and P.-O. Åstrand. Maximum oxygen uptake in athletes. *J. Appl. Physiol.* 23: 353-358, 1967.

THE INFLUENCES OF AGE, SEX, AND BODY FAT CONTENT
ON ISOMETRIC STRENGTH AND ENDURANCE

A. R. Lind
J. S. Petrofsky

Department of Physiology
St. Louis University Medical School
St. Louis, Missouri

INTRODUCTION

This report presents the results of a series of studies published in detail elsewhere (14,15,16,17,18). In consequence, we shall provide only a summary of those results, trying to emphasize those points which we believe to have importance in terms of the physiological mechanisms involved.

We are concerned exclusively with isometric handgrip contractions. Dr. Horvath was an early investigator in this field in a paper published in 1957 (23). In that study, the procedure called for the subjects to sustain a maximal effort, during which the tension fell rapidly from the maximal strength of the fresh muscles. The relationship they described between the loss of strength and the duration of the contraction supported the findings of other studies on different muscles (10,11,22). Because of the different circumstances of motor unit recruitment and rate coding and of the differences in local blood supply, we believe that the causes of fatigue during the maintenance of a maximal effort may be different than in sub-maximum tensions held constant until the target can no longer be achieved. Our own work has so far been almost exclusively concerned with sustained sub-maximal contractions.

Following the description of the influence of muscle temperature on the endurance of sub-maximal static effort, the effect of successive contractions and the recovery of both strength and endurance, we and others have described the local and central cardiovascular responses to static effort and some of the underlying physiological mechanisms (1,3,7,8,9).

Some years ago we considered that it would be worthwhile to assess the influence of some inherent factors on isometric strength and endurance and the effect of those factors on blood pressure and heart rate during fatiguing static effort.

METHODS AND PROCEDURES

The Influence of Age, Sex, and Body Fat Content on Isometric Performance

Subjects.   One hundred male and 83 female volunteers acted as subjects in these experiments. Their general characteristics have been reported previously (15,17). Each individual was first interviewed, at which time the procedures used in the experiments, as well as any possible hazards of the experiment, were explained in detail. All potential subjects were then medically examined.

Since the intention was to examine only healthy individuals, volunteers were not accepted as subjects if (1) they had a history of any form of cardiovascular disease, (2) their resting blood pressure exceeded 155/95 mmHg, or (3) there was any evidence of abnormality in their 12-lead resting ECG.

Strength and endurance.   Isometric strength and endurance were measured on a portable strain-gauge, hand dynamometer similar to the one described by Clarke *et al.* (1).   Strength was taken to be the larger of two brief (2 s to 3 s) maximal voluntary contractions (MVC); 2 min were allowed between these contractions. After a rest of 5 min, isometric endurance was measured as the duration of a sustained handgrip contraction at a tension of 40% MVC; the duration was measured to the nearest second. All measurements were recorded with the subject in the seated posture with male subjects nude above the waist and female subjects wearing halter tops; environmental temperature was kept constant at $24^{\circ}C \pm 1^{\circ}C$.

Heart rate.   Heart rate measurements were obtained before, during, and after each endurance contraction from a continuous recording of the ECG. Since the duration of the isometric contraction varied from person to person, the heart rates during exercise were measured on a relative time scale; measurements were made at 20, 40, 60, 80, and 100% of the duration of the 40% MVC; they were measured 1 min before and at 30, 60, and 90 s post-exercise.

Blood pressure.   Blood pressure was measured by auscultation on the inactive arm. Measurements were made each minute at rest, as often as possible during exercise, and at 30, 60, and 90 s following the contraction. As with the measurement of the heart rates, blood pressure measurements during exercise were converted by a computer to a relative time base by interpolation of the raw data at 20, 40, 60, and 80% of the endurance time, and by extrapolation for the values at the end (100%) of the endurance contraction.

Assessment of body fat content.   In experiments involving large numbers of subjects, it was expedient to estimate body fat content empirically. From mortality figures compiled by large insurance companies, an average weight based only on height and sex has been derived. This "ideal" weight represents the statistical assessment of the weight at which the individual will live longest (21). Coincidentally, "ideal" weight corresponds with the weight of most individuals at about the age of 20 years when the body fat content averages 10% to 20% for men and 15% to 30% for women (6,13,20). In studies involving large populations where the individual age varies, a simple ratio of "ideal" weight to actual weight has been demonstrated to be a satisfactory index of body fat content (19). Therefore, in this study we calculated the weight factor as follows:

weight factor = [(measured weight - ideal weight)/ideal weight] x 100.

A weight factor of 0 is therefore calculated for an individual whose weight is at the "ideal" value for his height, while a weight factor of 100, for example, would represent an individual whose weight was twice that of the table value. The table of "ideal" weight used here is the one published by the Metropolitan Life Insurance Company (21).

Statistical analysis of data.   In the analysis of the data, the calculations of means, variances, standard deviations, correlations, unrelated *t* tests, and regression equations by the method of least squares were performed as required on a Linc computer. When applicable, the means are reported ± the standard deviation. Regression coefficients are reported ± the standard error of the slope.

Where the data were considered to be affected by more than one independent variable, the statistical analysis was based on multiple regression. The multiple regression and corresponding standard error of the regression (reported as y ± the standard error), F tests and partial correlations will be referred to as multi-factor analysis. These calculations are part of the SPSS statistical package (12) calculated on an IBM 360 computer. Statistical significance in all cases was chosen at P values less than 0.05.

Where regression lines are drawn in figures, the upper and lower limits of the lines are set at the corresponding upper and lower levels of the actual data. Statistical analysis showed that all data distributions were sinusoidal in form.

Isometric Performance and the Menstrual Cycle

In a subsequent series of experiments, the strength and endurance of isometric contractions were measured in women with a normal menstrual cycle and in women who were taking oral contraceptives. As well as measuring the strength and the endurance of a contraction at 40% MVC throughout two menstrual cycles, the heart rate and blood pressure were measured during the endurance contractions. Also, the deep muscle temperature of some of these women was measured throughout the menstrual cycle. The details of the methods and procedures are described elsewhere (15).

RESULTS

The Influence of Age, Sex, and Body
Fat Content on Isometric Performance

The results of this investigation are described in detail elsewhere (14,17,18). They are summarized in Table 1, which shows the results of statistical analysis based on multiple regressions to assess the influence of the various independent variables which were measured.

A simple comparison of strength and endurance showed that the men were stronger than the women, but that endurance was longer, at 40% MVC, for women than for men. In the same way, men appeared to show no change of strength or endurance as aging proceeded, whereas the strength of the women decreased with age while endurance increased. When the analysis took account of the amount of body weight that was fat, aging was shown to be associated with a decrease in strength and a reciprocal increase in endurance. The men had an increase in body fat content associated with age; in turn, there was a direct relationship between body fat content and strength, while endurance decreased as body fat content increased. When the opposing influences of age and body fat content were taken into account in the analysis, we found no significant correlation between strength and endurance.

The increase in heart rate throughout a fatiguing contraction at 40% MVC was strikingly similar in men and women. Heart rate, which always increased during exercise, attained its highest magnitude during exercise in the subjects who had the highest resting heart rates. Older subjects displayed a smaller increase in exercising heart rates than the younger subjects. The blood pressure at the end of the 40% MVC was directly related to the resting blood pressure. However, aging increased the resting systolic blood pressure in men and women; this aging effect was further exaggerated during the exercise.

Isometric Exercise and the Menstrual Cycle

The results of this investigation are described in detail elsewhere (15). Briefly, maximal strength remained constant throughout the menstrual cycle in both groups of women. In the women taking the oral contraceptives, there was no change in the endurance of a contraction at a tension of 40% MVC. But in women who were not taking oral contraceptives, there was a sinusoidal relationship between the endurance and the days throughout the menstrual cycle (Table 2), with the longest endurance midway through the pre-ovulatory phase and the shortest endurance halfway through the luteal phase of the cycle.

There were no differences in the heart rate and blood pressure responses to the endurance contractions throughout the menstrual cycle, in either group of women.

In experiments to examine the possible influence of muscle temperature on the isometric endurance, the muscle temperature was measured at 40% of the distance

Table 1. Table of Correlations and Partial Correlations in Men and Women

Category	Men				Women			
	Age	Weight factor	Age [a]	Weight factor [a]	Age	Weight factor	Age [a]	Weight factor [a]
Endurance	0.110	-0.441 [c]	0.240 [b]	-0.480 [c]	0.279 [b]	-0.176	0.359 [c]	-0.290 [c]
Strength	-0.118	0.419 [c]	-0.240	0.449 [c]	-0.503 [c]	0.043	-0.545 [c]	0.247 [b]
Resting systolic BP	0.225 [b]	0.295 [c]	0.169	0.264 [b]	0.602 [c]	0.469 [c]	0.541 [c]	0.367 [c]
Resting diastolic BP	0.051	0.187	-0.094	0.205 [b]	0.283 [c]	0.313 [c]	0.227 [b]	0.246 [b]
Resting heart rate	-0.043	0.110	-0.069	0.132	-0.256 [b]	0.012	-0.274 [b]	0.102
Systolic BP at end of 40% MVC	0.325 [b]	0.116	0.278 [c]	0.003	0.639 [c]	0.281 [c]	0.604 [c]	0.107
Diastolic BP at end of 40% MVC	0.024	0.166	-0.040	0.200	0.168	0.153	0.127	0.106
Heart rate at end of 40% MVC	-0.273 [c]	-0.126	-0.254	-0.067	-0.476 [c]	-0.020	-0.495 [c]	0.157

[a] Denotes partial correlations with other variable held constant. [b] $P < 0.05$. [c] $P < 0.01$. If neither b nor c is present, then $P > 0.05$.

between the skin and the center of the forearm in two women with "normal" menstrual cycles. The results are shown in Fig. 1. Clearly, there was a variation in the muscle temperature of the brachioradialis muscle during the course of their respective menstrual cycles. Both subjects displayed the lowest temperatures at or shortly following the onset of bleeding and the highest temperatures just before the end of their menstrual cycles. In subject DL, this temperature difference was some 3.5°C during the course of the menstrual cycle, while subject ML showed a peak-to-peak variation of 2°C.

When the arm was immersed in water at 37°C in order to clamp the muscle temperature, the birth control pill had no discernible influence on isometric endurance (mean endurance = 113 s). In the "normal" subjects SS and DL, the endurance varied with the menstrual cycle (Fig. 2). However, unlike the sinusoidal variation reported from Series 1, maximum endurance was recorded at the beginning and the end of the menstrual cycle (155 and 160 s, respectively) with the lowest endurance in mid-cycle (90 and 100 s, respectively); the previous sinusoidal variation was converted into a hyperbolic relationship.

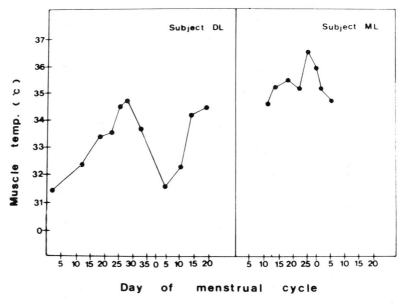

Fig. 1. The temperature of the brachioradialis muscle measured at 40% of the distance between the skin and center of the forearm for two control subjects during one menstrual cycle.

Table 2. Duration of 40% MVC
during the Menstrual Cycle

Subject	Pre-ovulatory phase (s)	Luteal phase (s)	P
Normals			
SS	226	193	< 0.01
DL	167	149	< 0.05
MS	206	161	< 0.01
Mean	200	168	
Birth Control Pill			
SF	99	105	> 0.05
SL	134	124	> 0.05
Mean	117	114	

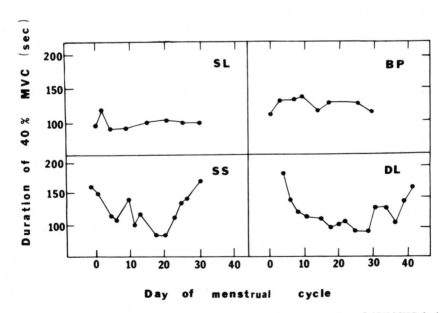

Fig. 2. The duration of a fatiguing isometric contraction at a tension of 40% MVC during one menstrual cycle in two women taking oral contraceptives (SL and BP – upper panels) and two "normal" subjects (SS and DL – lower panels) when the muscle temperature was stabilized by immersion of the forearm in water at 37°C.

DISCUSSION

The Influence of Age, Sex, and Body
Fat Content on Isometric Performance

The mechanism of the decrease in strength and increase in endurance associated with aging is open to speculation. One possibility is that this increase in endurance is merely a reflection of the lower absolute tension during the sustained isometric contraction in older people. This thesis is supported by the findings from the female subjects, who had lower isometric strength than their male counterparts and therefore exerted less absolute tension at 40% MVC, but had a longer isometric endurance. However, our multiple-regression analysis revealed that once the influence of age and body fat were removed from the data, there was no significant correlation between the MVC and endurance. Further, even after correcting for the differences in age and body fat, our men still displayed greater strength and less endurance than the women. But other factors might influence the strength and endurance. Muscle in humans is a mixture of "fast" and "slow" fiber types. The "fast" fibers, low in myoglobin, have been associated with great strength and speed but low endurance, while the "slow" fibers have been associated with low strength but great endurance. The relative proportion of these fibers in any one muscle varies with age (5). At birth, all skeletal muscle fibers appear to be "slow." During the period of growth and maturation, some of these differentiate into "fast" fibers (2). However, as aging proceeds, some of these "fast" fibers are reconverted to "slow" fibers (4). This evidence, coupled with a reduction in the number of active muscle fibers due to aging (5), might help explain the decrease in strength and increase in endurance in men and women.

The influence of body fat on isometric endurance has been recently explored in our laboratory (16). Our findings were that body fat content was directly related to deep muscle temperature in the forearm. Increases in muscle temperature result in a marked decrease in isometric endurance (1), and in our studies the changes in the temperature of muscles in the forearm were sufficient, in themselves, to explain the corresponding changes in isometric endurance (16).

While the resting systolic blood pressure of both the women and the men increased with age, the resting values for the young women were lower than those of the young men, but increased faster with age than the men's until at age 60 there was no difference due to sex. These facts are in agreement with the extensive data compiled by the Society of Actuaries (21). Multi-factor analysis revealed that at least part of this age-associated hypertension was associated with an increase in body fat content in the older individuals. In fact, the correlation of body fat content to resting systolic blood pressure was so high that in males the contribution of aging became statistically insignificant.

The systolic blood pressure at the end of the 40% MVC was affected by both the resting systolic blood pressure and age. It was particularly striking that the duration of the 40% MVC had no correlation to either systolic or diastolic blood pressure when the influences of age and resting blood pressure were delineated. It is usually considered that, at rest, the increase in systolic blood pressure is due to a decrease in the flexibility of the arteries. The decreased arterial compliance would then be responsible for the greater blood pressures in our older men during isometric contractions.

Isometric Exercise and the Menstrual Cycle

Isometric strength was unaltered during the menstrual cycle in both the "normal" subjects and the subjects who were taking oral contraceptives. Although endurance remained constant in the subjects taking oral contraceptives, the "normal" subjects displayed a sinusoidal variation in the endurance of isometric contractions during the menstrual cycle with the arm bare and held in the air. While it is reasonable to suspect the involvement of a hormonal mechanism, no sex hormone is known to cycle in phase with these endurance changes, although it remains possible that a latency exists between hormonal changes and the isometric muscular response. The matter is further confounded in that no female sex hormone is known to have a direct effect on skeletal muscle performance. While it is well established that deep muscle temperature is inversely related to isometric endurance (2,9), this explanation cannot by itself account for the changes seen here. The muscle temperature of our "normal" subjects rose continuously throughout the first three quarters of the menstrual cycle, at a time when isometric endurance first rose and then fell. When the muscle temperature was clamped at $36^\circ C$ to $37^\circ C$, the endurance first fell and then rose during this period. Obviously, these findings do not follow the expected course of relationship between the temperature of skeletal muscle and its endurance.

It follows from our results that there are least two factors in "normal" subjects (but not in the subjects taking the birth control pill) that influence isometric endurance during the course of the menstrual cycle. First, there is a variation in deep muscle temperature in the forearm during the menstrual cycle, which can explain some part of the associated variation of isometric endurance. Second, with the stabilization of muscle temperature, the resultant hyperbolic endurance response throughout the menstrual cycle must be the result of the influence of some other variable, possibly a hormone, which was beyond the scope of our investigations. It seems that this second factor is dominant over the influence of muscle temperature.

The question remains as to whether the blood flow through the forearms of women is greater than that for men at the same proportional tension. Also, it may be that forearm blood flow is altered throughout the menstrual cycle. Experiments are underway in our laboratory to assess this possibility.

There was no significant difference in either the blood pressure or the heart rate responses from either group of subjects during the menstrual cycle. Thus, neither endurance time nor the menstrual cycle appeared to modify the magnitude of the blood pressure and heart rate responses to isometric exercise.

ACKNOWLEDGMENT

This work was aided by AF Grant F33615-71C-1320 and HEW Grant HSM-099-7121.

REFERENCES

1. Clarke, R.S.J., R.F. Hellon, and A.R. Lind. Duration of sustained contractions of the human forearm at different muscle temperatures. *J. Physiol.* 143: 454-473, 1958.
2. Close, R. Dynamic properties of fast and slow skeletal muscles of the rat during development. *J. Physiol.* 173: 74-95, 1964.
3. Coote, J.H., S.M. Hilton, and J.F. Perez-Gonzalez. The reflex nature of the pressor response to muscular exercise. *J. Physiol.* 215: 789-804, 1971.
4. Drahota, Z., and E. Gutmann. Long term regulatory influence of the nervous system on some metabolic differences in muscles of different function. *Physiol. Bohemoslov.* 12: 339-348, 1963.
5. Gutman, E., and V. Hanzlekova. *Age changes in the neuromuscular system.* Bristol: Scientechnica Ltd., 1972, p. 82.
6. Hermansen, L., and W. von Dobeln. Body fat and skinfold measurements. *Scand. J. Clin. Lab. Invest.* 27: 315-319, 1971.
7. Humphreys, P.W., and A.R. Lind. The blood flow through active and inactive muscles of the forearms during sustained handgrip contractions. *J. Physiol.* 166: 120-135, 1963.
8. Lind, A.R., G.W. McNicol, and K.W. Donald. Circulatory adjustments to sustained (static) muscular activity. *Proc. Sympos. Physical Activity in Health and Disease.* Oslo: Universitets-forlaget, pp. 39-63, 1966.
9. Lind, A.R., S.H. Taylor, P.W. Humphreys, B.M. Kennelly, and K.W. Donald. Circulatory effects of sustained voluntary muscle contraction. *Clin. Sci.* 27: 229-244, 1964.
10. Merton, P.A. Voluntary strength and fatigue. *J. Physiol.* 123: 553-564, 1954.
11. Naess, K., and A. Storm-Mathisen. Fatigue of sustained tetanic contractions. *Acta Physiol. Scand.* 34: 351-366, 1955.
12. Nie, N.D., D. Bent, and C.H. Hull. *Statistical package for the social sciences.* St. Louis: McGraw-Hill, 1970.
13. Novak, L. Aging, total body potassium, fat free mass, and cell mass in males and females between ages 18 and 85 years. *J. Gerontol.* 27: 438-443, 1972.
14. Petrofsky, J.S., R.L. Burse, and A.R. Lind. Comparison of physiological responses of women and men to isometric exercise. *J. Appl. Physiol.* 38: 863-868, 1975.
15. Petrofsky, J.S., D. LeDonne, J. Rinehart, and A.R. Lind. Isometric strength and endurance during the menstrual cycle. *Europ. J. Appl. Physiol.* 35: 1-10, 1976.
16. Petrofsky, J.S. and A.R. Lind. The relationship of body fat content to deep muscle temperature and isometric endurance. *Clin. Sci. Mol. Med.* 48: 405-412, 1975.
17. Petrofsky, J.S. and A.R. Lind. Aging, isometric strength and endurance, and cardiovascular responses to static effort. *J. Appl. Physiol.* 38: 91-95, 1975.
18. Petrofsky, J.S., and A.R. Lind. Isometric strength, endurance, and the blood pressure and heart rate responses during isometric exercise in healthy men and women, with special reference to age and body fat content. *Pflügers Arch.* 360: 49-61, 1975.
19. Shephard, R., M. Kaneko, and K. Ishii. Simple indices of obesity. *J. Sports Med. Phys. Fitness.* 11: 154-161, 1971.
20. Sloan, A.W. Estimation of body fat in young men. *J. Appl. Physiol.* 23: 311-315, 1967.
21. Society of Actuaries. Addendum: weights for heights. *Society of Actuaries Stat. Bull.*, 1959.
22. Stephens, J.A. and A. Taylor. Fatigue of maintained voluntary muscle contraction in man. *J. Physiol.* 220: 1-18, 1972.
23. Tuttle, W.W. and S.M. Horvath. Comparison of effects of static and dynamic work on blood pressure and heart rate. *J. Appl. Physiol.* 10: 294-296, 1957.

LONGITUDINAL COMPARISONS
OF RESPONSES TO MAXIMAL EXERCISE

Robert A. Bruce
Timothy A. DeRouen

Department of Medicine, Division of Cardiology
Department of Biostatistics
University of Washington
Seattle, Washington

Physiological adaptations to work or muscular exercise may be defined by an increase in functional aerobic capacity and secondarily reduced cardiovascular responses to moderate levels of exertion (17). Functional aerobic capacity or maximal oxygen uptake ($\dot{V}O_2$ max) (8) equals the product of cardiac output and the difference in arterial and mixed venous oxygen contents at symptom-limited maximal exertion (15). In normals, the mechanism for the increase in $\dot{V}O_2$ max is an increase in cardiac output resulting from an increase in stroke volume (11); there is also a widening of the a-v O_2 difference. Even cardiac patients can increase $\dot{V}O_2$ max, but the major mechanism is widening of the a-v O_2 difference (10) primarily due to increased arterial O_2 content, secondary to compensatory erythrocytosis. At the same level of submaximal exertion, active cardiac patients manifest a lower cardiac output, systolic pressure, and heart rate after training, which is in proportion to the diminished *relative* aerobic requirements; i.e., ratio (%) of the same submaximal oxygen requirement to the augmented $\dot{V}O_2$ max (4).

Evaluation of adaptations to environmental stress of work or exercise requires data on direction and magnitude of variations in longitudinal changes in response to maximal exercise. It is particularly important to know the long-term changes in healthy subjects and ambulatory cardiovascular patients of middle age when the interactions of cardiovascular disease occur as age advances.

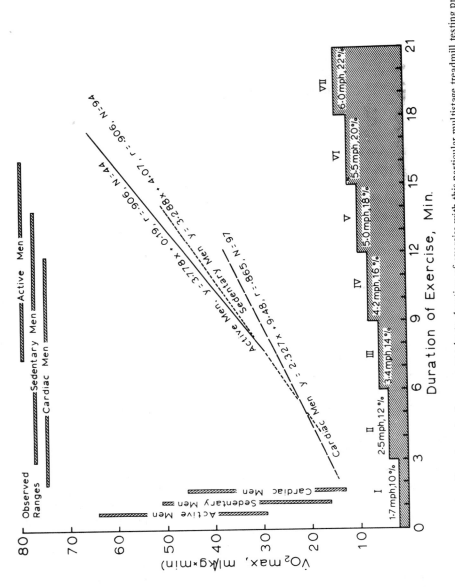

Fig. 1. Relationship of weight-adjusted maximal oxygen uptake to duration of exercise with this particular multistage treadmill testing protocol. Note the correlation coefficients ranged from +0.865 to +0.906. [Reprinted with permission from Bruce *et al.* (8).]

Two developments encouraged us to believe that the longitudinal comparisons in the same persons may be assessed non-invasively, quantitatively, and safely. The first is the demonstration that with a standardized multistage treadmill protocol, the weight-adjusted $\dot{V}O_2$ max is directly related to the duration of exercise testing [r = +0.9 in normal middle-aged men and women and +0.86 in men with heart disease (Fig. 1)] (8). The second is the extensive experience of the prospective, community Seattle Heart Watch study (7) over the past six years. Cross-sectional differences in duration and in cardiovascular response to exercise in healthy men were compared with patients with hypertension and/or coronary heart disease (7). Separation of the effects of disease vs. aging on heart rate and blood pressure responses (6), the predictive value of clinical and exercise variables (5), the variations in blood pressure responses (13), and the syndrome of exertional hypotension and post-exertional ventricular fibrillation (12) have also been reported.

To evaluate adaptations to work that may occur spontaneously in a population sample of middle-aged men, this study reports the longitudinal changes in duration and circulatory responses to maximal exercise testing.

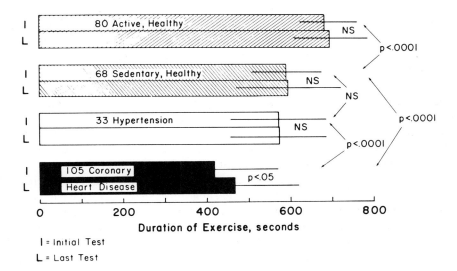

I = Initial Test

L = Last Test

Fig. 2. Variations in exercise duration (means ± standard deviation) over two to four years in 286 men with unchanged clinical status. Note the significant differences initially between physically active and sedentary men, and between hypertensive men and patients with coronary heart disease. The only significant longitudinal change occurred in the men with coronary heart disease.

Table 1. Distribution of Men with Identical Classifications
on Longitudinal Follow-Up Examinations

N	Classification	Interim Status, %		
		Unchanged	Improved	Unimproved
110	Active, Healthy	72.7	21.8	5.5
87	Sedentary, Healthy	78.2	12.6	9.2
48	Hypertension	68.8	16.7	14.6
197	Coronary Heart Disease	53.3	28.9	17.8
442	Altogether	64.7	22.9	12.4

MATERIAL AND METHODS

Only the records of 461 men, who were selected by their physicians for follow-up exercise tests more than one year after the initial examination, were drawn from the computer registry file for this study; reclassification data were available on 442.

When Each patient returned, the examining physician categorized the interim clinical status as "improved," "unchanged," or "unimproved." Men who at least once each week exercised vigorously and long enough to perspire were classified as "active;" those who did not were classified as "sedentary." Of 197 clinically healthy men, 148 (75.1%) who remained unchanged also maintained the same activity classification, and 49 (24.9%) spontaneously changed their activity habits (Table 1). Of 245 men with hypertension or coronary heart disease, 138 (56.3%) were clinically "unchanged," and 92 (46.7%) with coronary heart disease had either improved (57 patients) or were worse (35 patients) For the purpose of this study, hypertension was defined as an elevation of resting blood pressure without evidence of heart disease. Coronary heart disease was defined by the clinical syndromes of typical angina pectoris with exertion or emotion, which could be relieved by rest or nitroglycerin, by prior myocardial infarction, or by prior cardiac arrest with resuscitation by ventricular defibrillation, whether or not associated with hypertension.

The symptom-limited maximal exercise tests were administered according to the same protocol previously described (7). Resting and exercise data were paired with the initial observations for each individual; the relationships were examined by correlation, and the significance of longitudinal differences in exercise testing was ascertained by paired t tests. Longitudinal rates of change for varying amounts of elapsed time, ranging from one to five years, were normalized for one-year intervals. Ninety-five percent confidence intervals for the expected rates of change per year for an individual were determined by using the mean change plus or minus two standard deviations.

The third minute of exercise at Stage I (1.7 mph, 10% gradient) with an average normal $\dot{V}O_2$ requirement of 17.4 $\pm$ 1.4 ml · kg^{-1} · min^{-1} (8) represented a level of submaximal exertion that all could perform. This required about 4 to 4.5 "Mets" or multiples of the resting oxygen requirement. The relative aerobic requirement for each individual was calculated as the percentage of the maximal oxygen uptake, using the regression equation $y = 3.88 + 0.056x$ (x = seconds of total duration of exercise to symptom-limited capacity (8). The percentage relationships of heart rate, systolic pressure and the product of rate and pressure (x 10^{-2}) at Stage I to the individual's observed maximal rate, systolic pressure and product of rate and pressure were calculated whenever such data were reported.

RESULTS

1. Maximal Exercise Duration in Clinically Unchanged Men

Average durations of exercise to symptom-limited effort are shown in Fig. 2 for two groups of healthy men and for two other groups of men with cardiovascular disease who were classified as clinically "unchanged." The active and sedentary healthy men and the hypertensive men show no significant changes in duration of exercise over an elapse of about 3.25 years. However, the men with coronary heart disease reveal a significantly *longer* duration of exercise with retesting (+44 sec, +10.5%, P < 0.05).

Although there were differences (P < 0.001) in initial exercise duration between physically active and sedentary healthy men, and between healthy men and coronary patients, there were no differences between sedentary men with normal blood pressure and hypertensive men without heart disease.

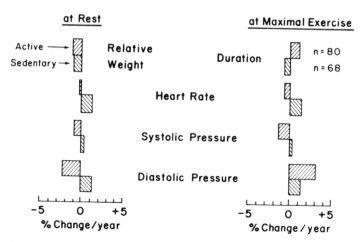

Fig. 3. Relative rates of change over two to four years in % per year of primary variables at rest and at maximal exercise in 148 healthy men with no change in status. Except for diastolic pressure, they are less than 2% per year.

2. Comparisons of Resting and Exercise Variables

Detailed analyses of the correlations between initial and final values, annual rates of change, 95% confidence intervals, and means and standard deviations of primary resting and maximal exercise variables are presented in Tables 2-5. At rest, the most reproducible variable with respect to correlation analysis was the relative body weight, which was based upon the percent difference between observed and predicted ideal body weight obtained by regression on observed height. [The correlation between relative body weight and ratio of weight to height^2 (wt/ht^2) is +0.98.] (8) With maximal exercise, the exercise duration was the most reproducible variable in each of the four groups of subjects or patients.

The relative annual rates of change or percentages of the primary resting and maximal exercise variables on the initial tests are shown in Figs. 3 and 4. With the exception of diastolic pressure, the changes average less than 2%. An exception is the 25.2 s per year or 6% annual increase in exercise duration in men with coronary heart disease. This is much less than 0.9 ±2 min prolongation in exercise duration reported by Smokler et al. in 63 patients with angina pectoris who were retested within six months (19).

For evaluation of changes in any given healthy subject, the most useful parameters of longitudinal changes are the 95% confidence intervals for the *annual* rates of change (Tables 2-5). These intervals are fairly broad for exercise duration, heart rate, and systolic pressure, which are the primary continuous variables for routine evaluation of exercise performance and cardiovascular capacity. (Should one wish to use these intervals where the elapsed time is only six months, the observed changes

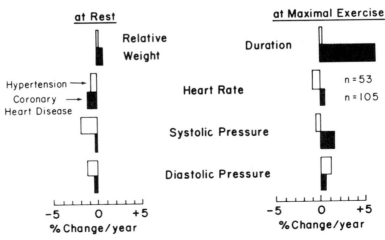

Fig. 4. Percentage changes per year in primary variables at rest and at maximal exercise in 138 hypertensive men and patients with coronary heart disease and with no change in status. Except for longer duration of exercise in the latter group, all changes are less than 2% per year.

Table 2. Comparisons in 80 Unchanged, Active Healthy Men
(Initial Age 47.6 ± 5.9 Years)

Variables	Correlations r	Annual Rates of Change	95% Confidence Intervals for Annual Rates of Change [a]	Last Test Means ± SD
Relative Weight, %	+0.847	-1.2 ± 1.6	-4.4 to 2	94.6 ± 7
Resting Heart Rate, b/min	+0.461	-0.1 ± 4.3	-9 to 9	67 ± 9
Systolic Pressure, mmHg	+0.440	-0.9 ± 5.4	-10 to 12	119 ± 11
Diastolic Pressure, mmHg	+0.133	-1.7 ± 4.4	-11 to 7	74 ± 7
Maximal Exercise				
Duration, s	+0.822	6.9 ± 22.0	-37 to 51	696 ± 88
Heart Rate, b/min	+0.782	-1.5 ± 2.8	-7 to 4	174 ± 9
Systolic Pressure, mmHg	+0.642	-2.7 ± 6.7	-16 to 7	185 ± 21
Diastolic Pressure, mmHg	+0.101	2.1 ± 7.7	-13 to 18	73 ± 11
HR x SBP x 10^{-2}	+0.701	-7.5 ± 12.8	-33 to 18	321 ± 38
FAI, %	+0.801	-1.8 ± 2.8	-7 to 4	-10 ± 12
LVI, %	+0.694	1.8 ± 3.7	-6 to 9	3 ± 11
HRI, %	+0.638	0.5 ± 1.4	-2 to 3	1 ± 5
PCI, %	-0.621	6.2 ± 1.0	-8 to -4	-14 ± 14

[a] For individuals

Table 3. Comparisons in 68 Unchanged, Sedentary Healthy Men
(Initial Age 47.5 ± 6.3 Years)

Variables	Correlations r	Annual Rates of Change	95% Confidence Intervals for Annual Rates of Change [a]	Last Test Means ± SD
Relative Weight, %	+0.793	-0.9 ± 2.3	-5.5 to 3.7	100 ± 10
Resting Heart Rate, b/min	+0.518	0.9 ± 3.8	-7 to 9	74 ± 11
Systolic Pressure, mmHg	+0.396	0.4 ± 5.8	-11 to 12	122 ± 13
Diastolic Pressure, mmHg	+0.349	1.0 ± 4.1	-7 to 9	78 ± 8
Maximal Exercise				
Duration, s	+0.786	-2.1 ± 23.1 [b]	-48 to 44	583 ± 91
Heart Rate, b/min	+0.821	-1.3 ± 3.3	-8 to 5	175 ± 14
Systolic Pressure, mmHg	+0.627	0.4 ± 5.8	-11 to 12	186 ± 20
Diastolic Pressure, mmHg	+0.498	-1.0 ± 4.0	-7 to 9	79 ± 12
HR x SBP x 10^{-2}	+0.638	-1.7 ± 11.8	-25 to 22	324 ± 40
FAI, %	+0.784	-1.3 ± 3.6	-9 to 6	-4 ± 14
LVI, %	+0.641	0.2 ± 3.4	-7 to 9	2 ± 11
HRI, %	+0.749	0.4 ± 1.8	-3 to 4	1 ± 7
PCI, %	-0.637	-1.8 ± 12.8	-27 to 24	-6 ± 17

[a] For individuals   [b] $P < 0.01$

Table 4. Comparisons in 33 Unchanged Hypertensive Men
(Initial Age 50.6 ± 6.4 Years)

Variables	Correlations r	Annual Rates of Change	95% Confidence Intervals for Annual Rates of Change[a]	Last Test Means ± SD
Relative Weight, %	+0.875	0.2 ± 3.5	-6.8 to 7.2	102.8 ± 13.4
Resting Heart Rate, b/min	+0.072	-0.7 ± 9.2	-19 to 18	72 ± 12
Systolic Pressure, mmHg	+0.418	-2.8 ± 10.4	-24 to 18	133 ± 14
Diastolic Pressure, mmHg	+0.267	-1.2 ± 5.0	-11 to 9	88 ± 8
Maximal Exercise				
Duration, s	+0.785	1.0 ± 47.7	-94 to 96	569 ± 116
Heart Rate, b/min	+0.367	-1.8 ± 8.5	-19 to 15	171 ± 16
Systolic Pressure, mmHg	+0.501	-1.4 ± 12.8	-27 to 24	199 ± 32
Diastolic Pressure, mmHg	-0.293	1.0 ± 8.9	-17 to 19	90 ± 13
HR x SBP x 10^{-2}	+0.355	-5.7 ± 25.8	-61 to 46	339 ± 61
FAI, %	+0.688	-1.5 ± 7.1	-16 to 13	1.4 ± 16.5
LVI, %	+0.411	1.2 ± 7.5	-14 to 16	-2.2 ± 18.0
HRI, %	-0.329	0.7 ± 4.7	-9 to 10	2.7 ± 8.2
PCI, %	-0.373	4.1 ± 19.4	-35 to 43	3.9 ± 23.2

[a] For individuals

Table 5. Comparisons in 105 Unchanged Coronary Heart Disease Men
(Initial Age 51.9 ± 6.8 Years)

Variables	Correlations r	Annual Rates of Change	95% Confidence Intervals for Annual Rates of Change[a]	Last Test Means ± SD
Relative Weight, %	+0.553	0.6 ± 11.0	-22.6 to 21.4	103 ± 18.7
Resting Heart Rate, b/min	+0.334	-0.9 ± 6.1	-13 to 11	71 ± 12
Systolic Pressure, mmHg	+0.526	-0.1 ± 7.8	-16 to 16	127 ± 18
Diastolic Pressure, mmHg	+0.464	-0.3 ± 5.4	-11 to 10	80 ± 14
Maximal Exercise				
Duration, s	+0.590	25.2 ± 65.6	-106 to 156	462 ± 152
Heart Rate, b/min	+0.483	0.6 ± 10.8	-21 to 22	148 ± 24
Systolic Pressure, mmHg	+0.426	2.6 ± 13.7	-25 to 27	169 ± 28
Diastolic Pressure, mmHg	+0.060	0.4 ± 6.5	-13 to 13	83 ± 14
HR x SBP x 10^{-2}	+0.150	5.4 ± 33.2	-61 to 72	251 ± 62
FAI, %	+0.497	-4.5 ± 10.0	-25 to 16	15.0 ± 21.8
LVI, %	+0.254	0 ± 11.9	-24 to 24	23.2 ± 18.5
HRI, %	+0.469	0.7 ± 6.1	-13 to 14	14.6 ± 13.2
PCI, %	-0.404	-3.5 ± 15.1	-34 to 29	8.9 ± 20.0

[a] For individuals

are multiplied by two; when the elapsed time is two years, then the observed changes are divided by two, thereby reducing all changes to equivalent values for a one-year period. The above computation implies that the rates of change are linear functions over the entire period of time, when in reality they are more likely to be curvilinear or exponential functions.)

3. Relative Responses to Submaximal Exercise

Representative differences in relative aerobic requirements for the first stage of submaximal exercise for the initial tests in healthy men and in patients with coronary heart disease are shown in Fig. 5. The relative aerobic requirements, for performance of the *same physical work* of walking 1.7 mph at 10% gradient ranged from 40 $\pm$ 5% in the active healthy men to 69 $\pm$ 22% in the sedentary men with coronary heat disease. Likewise, the *relative myocardial demand for oxygen*, represented by the percentage relationship of the product (14) of the individual's maximal heart rate and maximal systolic pressure (x 10^{-2}) to the average normal product at maximal exercise also ranged from 40 $\pm$ 9% in the active healthy men to 69 $\pm$ 16% in the sedentary men with coronary heart disease. The absence of a difference in this variable between active and sedentary patients with coronary disease was attributed to the presence of hypertension in some of the patients.

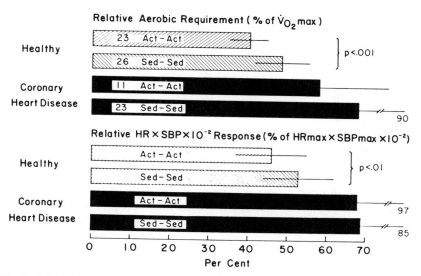

Fig. 5. Initial differences in relation to responses to submaximal exercise of Stage I, 3rd min, in healthy men and patients with coronary heart disease (means $\pm$ SD). Note that relative aerobic requirement and relative hemodynamic stress, and inferentially, myocardial aerobic requirements are least in active men and greatest in sedentary patients with coronary heart disease. See text for comments.

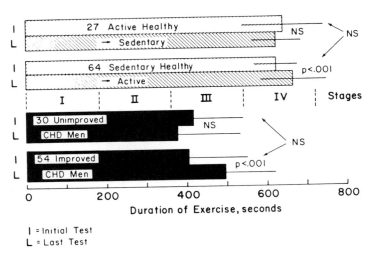

Fig. 6. Variations in exercise duration over one to four years in 175 men with a change in activity status. Note the significant increase in exercise duration in sedentary persons who become physically active, whether healthy subjects or patients with coronary heart disease.

Thus, in either healthy men or patients with coronary heart disease, the relative stress of the same amount of submaximal exercise in physical units varied inversely with the individual's total aerobic capacity and myocardial aerobic capacity indicated by the peak response in the double product at symptom-limited exercise.

4. Changes in Exercise Duration with Changes in Physical Activity or Clinical Status

Changes in duration of maximal exercise between the initial and final tests in 27 active healthy men who became sedentary, and in 64 sedentary men who became active are shown in Fig. 6. Only the sedentary men who became active show a significant prolongation of exertion of +43 s or +5.9% ($P < 0.001$), which represents a 2.4 ml • kg^{-1} • min^{-1} gain in maximal oxygen uptake.

Whereas 30 men with coronary heart disease who were clinically unimproved when retested show insignificant changes in exercise duration (Fig. 6), 54 other patients who were clinically "improved" reveal a significant increase in duration of +91 s or +22.4%, $P < 0.001$. This is equivalent to 5.1 ml • kg^{-1} • min^{-1} increase in maximal oxygen uptake, which is identical to that reported for coronary patients after three months of training (10).

5. Electrocardiographic Manifestations
of Myocardial Ischemia

Segmental ST depression of one or more millimeters, horizontal or downsloping from the J-point for at least 0.06 s in consecutive QRS-T complexes for one min or longer, was considered electrocardiographic (ECG) evidence of myocardial ischemia as a result of maximal exertion. Since this ECG sign is non-specific for disease and represents only a functional imbalance between coronary supply of oxygenated blood and hemodynamic demand of the left ventricle, it is not considered diagnostic of either coronary vascular disease or myocardial disease (3). The consistency of this reponse, whether "positive" or "negative," and changes in interpretation in these four groups of men are presented in Table 6. The prevalence of ischemic ST depression was higher in cardiovascular patients than in healthy men. Over about three years, the prevalence remained the same in active, healthy men, diminished in sedentary, healthy men and hypertensive men, but increased in coronary patients. In 70% to 89% of the men in all four groups, there is no change in interpretation. Conversely, only a few individuals in all groups have shown different classifications.

DISCUSSION

A comparative study by Shephard *et al.* (18) of three methods of eliciting maximal exercise in 24 healthy men indicated higher values of oxygen uptake, heart rate, and lactate concentration were obtained with the treadmill than with either a bicycle ergometer or step test (Fig. 7). Prior studies in Seattle demonstrated that maximal oxygen uptake could be assessed accurately and repeatedly with a single test using a standardized multistage treadmill protocol that increased the speed and gradient of walking continuously every three minutes. Although the absolute volume of oxygen consumed per kilogram of body weight at maximal exercise was significantly

Table 6. ECG Responses in Men with
Unchanged Status and Same Classifications

Classification of Men	Clinical Interpretation of Ischemic ST Depression on Last Test vs. Initial Test			Prevalences	
	No Change, i.e. Agreement %	Change + to − %	Change − to + %	Initial Test %	Last Test %
Active, Healthy	77.5	11.3	11.3	15.0	15.0
Sedentary, Healthy	88.5	8.0	3.4	12.6	8.0
Hypertension	70.8	18.8	10.4	25.0	16.7
Coronary Heart Disease	69.1	10.8	20.1	35.6	44.8
Altogether	75.1	12.0	9.3	24.5	28.9

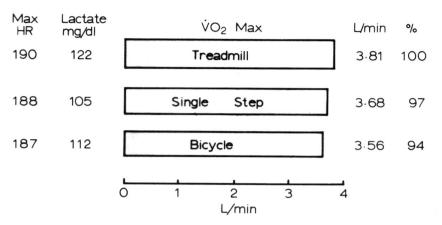

Fig. 7. Comparison of maximal exercise responses by three different methods of exercise stress testing in the same 24 healthy men of 20 to 40 years of age [adapted from Shephard *et al.* (18)]. Note the highest values for heart rate, lactate, and oxygen uptake were obtained by use of a treadmill.

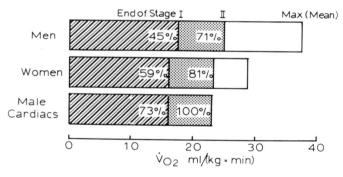

Fig. 8. Cross-sectional comparisons of maximal oxygen uptake, weight-adjusted. Note the marked differences in aerobic capacity, and as a consequence, the relative aerobic requirements of Stages I and II levels of submaximal exercise.

greater in healthy, middle-aged men than in women and lower in ambulatory men with heart disease (Fig. 8), the relative aerobic requirements for identical physical stresses of submaximal exercise varied from 45% to 73% for Stage I and from 71% to 100% for Stage II. The absolute oxygen requirements per kilogram of body weight for submaximal work loads were 15% greater in men than in women, or almost exactly equal to those reported for healthy young adults. Such differences relate to bodily dimensions and composition (16) and mechanical efficiency during rapid walking and jogging upgrade (2).

When the peak value for minute-by-minute measurements of oxygen uptake were scaled to 100%, the percentages of maximal uptake for the preceding three min were

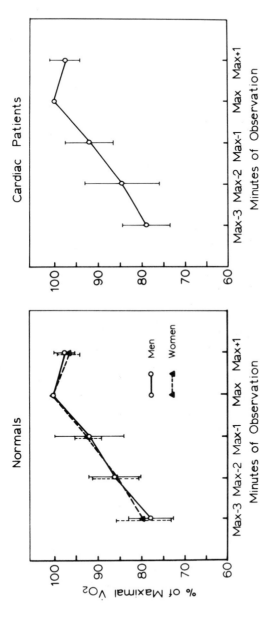

Fig. 9. Minute-by-minute approach to maximal oxygen uptake when absolute values are scaled to percentage of highest value in each subject. Note the similarity of mean values in men and women, and even in men with heart disease. [Reprinted with permission from Bruce et al. (8).]

the same for all three groups (Fig. 9). For subjects who could continue strenuous exercise for up to a minute longer, a minor fall in oxygen uptake demonstrated that the physiologic maximum had been attained. Furthermore, the weight-adjusted oxygen uptakes per min varied directly with the duration of exercise with this continuous testing protocol of progressive increments in speed and gradient of effort for strolling, walking, and jogging (Fig. 1).

Duration of symptom-limited maximal exercise testing provides, therefore, the simplest quantitative estimate of the functional cardiovascular capacity of an ambulatory individual in a single, non-invasive test. When appropriate precautions and monitoring are employed, the safety of this testing procedure is attested by no fatality in a community experience of over 20,000 tests.[a] Nevertheless, the need for professional monitoring and supervision is indicated by the occurrence of six instances of exertional hypotension and post-exertional cardiac arrest from ventricular fibrillation in patients with severe coronary vascular disease (12). Yet with prompt defibrillation, all survived and none exhibited evolving myocardial infarction. Repeated maximal exercise testing reveals longitudinal changes in exercise performance and capacity in the same individuals and provides an opportunity to detect adaptive changes. Because of the numerous interacting variables present in healthy persons, and even more so in patients with symptomatic cardiovascular disease who are undergoing one or more types of treatment, identification of adaptive changes is difficult. In this study we relied upon the clinical judgment of the examining physician as to whether or not each individual manifested changes in the interim clinical status that he could categorize. The majority of subjects, both healthy men and cardiovascular patients, exhibited an "unchanged" clinical status; only a few individuals showed clinically apparent "improvement" or "unimprovement" in functional status. When data in the other categories are examined, there is more variation in longitudinal comparisons of variables. Hence, for this longitudinal study of variations with retesting of maximal exercise, those individuals with an "unchanged" interim clinical status, which was ascertained prior to the last exercise test, were separated from those whose clinical status had changed.

The 95% confidence intervals for the annual rates of change in the primary variables of duration of maximal exercise, heart rate, and systolic pressure responses indicate that subtle differences that may be observed are likely to be within the normal limits of variation. Accordingly, only substantial changes with long-term retesting are probably significant. Therefore, these confidence intervals provide a frame of reference by which individual changes may be more readily and reliably interpreted.

Physically healthy men, whether active or sedentary, and hypertensive men show a minor and statistically insignificant increase in average duration of exercise. This obscures the expected 17 s shortening of duration per year based upon the previously

---

[a] Of several hundred men with coronary heart disease who had coronary arteriography, all survived maximal exercise testing, but six died suddenly during the invasive study.

reported small reduction in maximal oxygen uptake with aging (-0.94 ml • kg⁻¹ • min⁻¹ per year) (9). Possible mechanisms for this phenomenon include a minor reduction in body weight, a greater familiarity with testing procedure, as well as motivation to prolong post-maximal exertion by greater use of anaerobic glycolysis. In addition, several patients with hypertension but with no apparent heart disease were being medically treated for hypertension. Evaluation of the annual rate of change in maximum heart rate suggests that prior familiarity with the testing protocol or increased motivation did not significantly influence exercise performance. Maximum heart rate fell by 1.3 to 1.8 beats per min per year in both the healthy men and the hypertensive men. From cross-sectional data, the decline should be 0.6 beats per min per year in healthy men (6).

Men with coronary heart disease unexpectedly show a significant increase in exercise duration even though they were clinically classified by their physicians as "unchanged" in their interim status. Many were receiving medical treatment, a few had surgery, and several were participating in a community physical training program designed to facilitate cardiac rehabilitation of coronary patients. This suggests that until symptom-limited maximal exercise is performed, neither patients nor physicians were aware of substantial improvement in functional capacity.

Cross-sectional measurements of maximal oxygen uptakes by Åstrand (1) (Fig. 10) show progressive increments in the early years of life with growth and development, and after adolescence a progressive decline with subsequent aging. Usually greater values were found in men than in women, and in physically active persons of either sex than in sedentary persons. However, when weight-adjusted values of maximal oxygen uptake are examined, there is essentially the same rate of decline over all decades of age in males (Fig. 11).

Longitudinal studies of maximal oxygen uptake in the *same* men tested years apart show greater decrements with aging than are seen in the cross-sectional data for the population as a whole. The rate of decline in $\dot{V}O_2$ max was similar in three separate studies using different population samples, different periods of time, and different methods of measurement (Fig. 12). A more rapid decline was observed in sedentary men, whereas physically active men in all three studies had a slower rate of decrease in $\dot{V}O_2$ max (Fig. 13). Statistically, the only valid method of assessment of such functional changes with relatively prolonged periods of time is by means of longitudinal observations in the *same* person. These differences in rate of decline between active and sedentary individuals document an important adaptation to the environmental stress of physical activities. Although not based upon measurements of oxygen uptake, estimates of maximal oxygen uptake derived by regression on observed duration of exercise testing by this protocol have shown similar changes in healthy, middle-aged women (20). It is important to emphasize, however, that the subjects in these longitudinal studies were familiar with the testing protocol because of performance of one or more earlier tests to symptom-limited capacity. Since additional testing to provide familiarity with the procedure is not always feasible in cross-sectional population studies, the question remains whether or not non-specific adaptations to exercise stress occur that may influence the present test in the same persons even though the previous tests were conducted more than a year earlier.

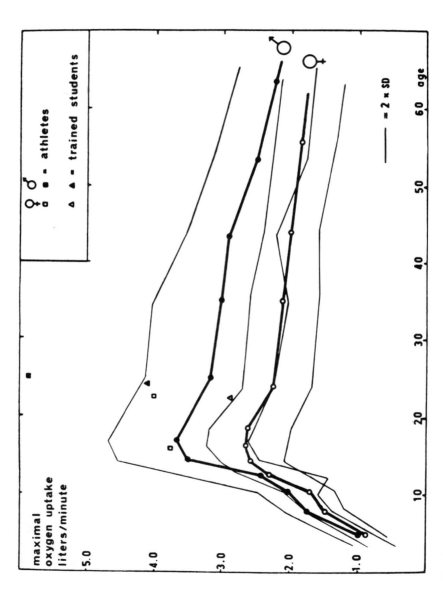

Fig. 10. Relationship of absolute volumes of oxygen uptake, not weight-adjusted, to age in healthy men and women. See text for comments. (Reprinted with permission of the *Journal of the American Medical Association*.)

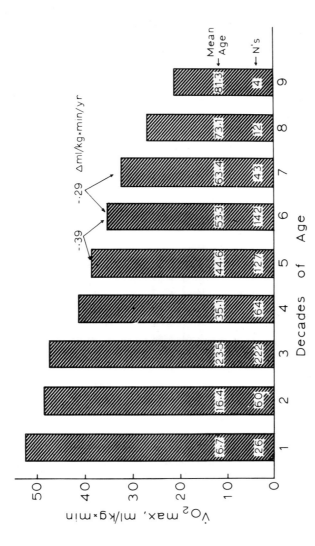

Fig. 11. Relationship of weight-adjusted maximal oxygen uptake to age, cross-sectional differences in healthy non-athletic men (n = 700). (Composite of 18 surveys.) See text for comments.

Effective treatment of hypertension and/or coronary heart disease may also either simulate or add to the effect of physical activity. It is important to note, furthermore, that invasive exercise studies of coronary patients who were participating in a physical training program demonstrate an increase in arterial oxygen content with widening of arterial-mixed venous oxygen difference as the primary physiological mechanisms for the observed 22% in $\dot{V}O_2$ max (10). As a consequence of the increase in functional aerobic capacity, the relative aerobic requirement for any given level of submaximal exertion is reduced, and heart rate and systolic pressures are lower, thereby diminishing the hemodynamic stress on the ischemic and/or infarcted left ventricle.

In conclusion, two adaptive changes are observed when multistage treadmill tests of symptom-limited maximal exercise are repeated in the same middle-aged men one to five (averaging 3 ± 1) years later. One is a minor (< 1%) per year, non-specific and statistically insignificant increase in total duration of exercise that obscures the expected small decrement associated with aging in healthy or hypertensive men who maintain the same activity classification and who manifest no changes

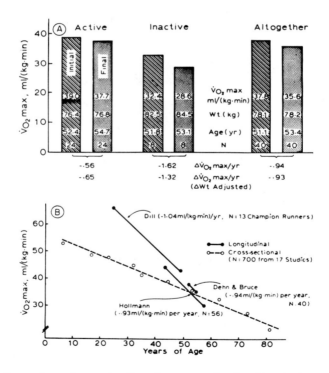

Fig. 12. Comparison of regression lines for cross-sectional and longitudinal observations of the relationship of maximal oxygen uptake to age. See text for comments. [Reprinted with permission from Dehn, M.M. and R.A. Bruce (9).]

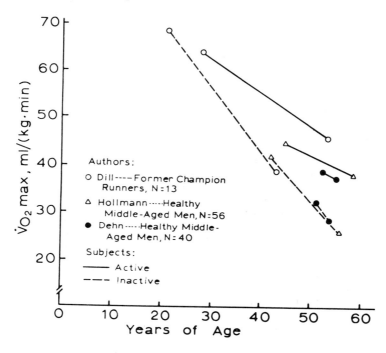

Fig.13.

over the intervening three years in their clinical status as appraised by their examining physicians. The major and significant adaptation is a substantial increase in exercise duration observed in sedentary healthy men who are motivated to participate in physical activities on a regular basis. The 6.9% gain in duration of exercise and its associated increase in functional aerobic capacity of 2.4 ml $\cdot$ kg^{-1} $\cdot$ min^{-1} is equivalent to a gain of 2.5 to 3.5 years over the expected decline from aging. In addition, effective treatment of cardiovascular disease permits partial recovery of the diminished functional capacity and possibly the life span that would otherwise be lost.

These studies have been supported in part by research grants HL-13517 from the National Heart, Lung and Blood Institute and MB-00184 from the Health Resources Administration of the Public Health Service.

REFERENCES

1. Åstrand, P.-O. Physical performance as a function of age. *J. Am. Med. Assoc.* 204: 105, 1968.
2. Booyens, J. and W. Keatinge. The expenditure of energy by men and women walking. *J. Physiol.* 138: 165, 1957.
3. Bruce, R.A. Values and limitations of exercise electrocardiography. *Circulation* 50: 1-3, 1974.
4. Bruce, R.A. The benefits of physical training for patients with coronary heart disease. In: *Controversy in internal medicine, II.* Edited by Ingelfinger, Ebert, Finland, and Relman. Philadelphia: Saunders, 1974, pp. 145-161.
5. Bruce, R.A., T. DeRouen, D. Peterson, J. Irving, N. Chinn, B. Blake, and V. Hofer. Non-invasive predictors of sudden cardiac death in men with coronary heart disease. Predictive value of maximal stress testing. *Am. J. Cardiol.* 39: 833-840, 1977.
6. Bruce, R.A., L. Fisher, and M. Cooper. Separation of effects of cardiovascular disease and age on ventricular function with maximal exercise. *Am. J. Cardiol.* 34: 757-763, 1974.
7. Bruce, R.A., G. Gey, M. Cooper, L. Fisher, and D. Peterson. Seattle heart watch: Initial clinical, circulatory and electrocardiographic responses to maximal exercise. *Am. J. Cardiol.* 33: 459-469, 1974.
8. Bruce, R.A., F. Kusumi, and D. Hosmer. Maximal oxygen intake and nomographic assessment of functional aerobic impairment in cardiovascular disease. *Am. Heart J.* 85: 546-562, 1973.
9. Dehn, M.M. and R. Bruce. Longitudinal variations in maximal oxygen intake with age and activity. *J. Appl. Physiol.* 33: 805-807, 1972.
10. Detry, J., M. Rousseau, G. Vandenbroucke, F. Kusumi, L. Brasseur, and R. Bruce. Increased arteriovenous oxygen difference after physical training in coronary heart disease. *Circulation* 44: 109-118, 1971.
11. Hartley, L.H., G. Grimby, A. Kilbom, N. Nilsson, I. Åstrand, J. Bjure, B. Ekblom, and B. Saltin. Physical training in sedentary middle-aged and older men: III. Cardiac output and gas exchange at submaximal and maximal exercise. *Scand. J.Clin. Lab. Invest.* 24: 335-344, 1969.
12. Irving, J.B. and R Bruce. Exertional hypotension and postexertional ventricular fibrillation in stress testing. *Am. J. Cardiol.* 39: 849-851, 1977.
13. Irving, J.B., R. Bruce, and T. DeRouen. Variations in and significance of systolic pressure during maximal exercise (treadmill) testing. Relation to severity of coronary artery disease and cardiac mortality. *Am. J. Cardiol.* 39: 841-848, 1977.
14. Kitamura, K., C. Jorgensen, F. Gobel, H. Taylor, and Y. Wang. Hemodynamic correlates of myocardial oxygen consumption during upright exercise. *J. Appl. Physiol.* 32: 516-522, 1972.
15. Mitchell, J.H., B. Sproule, and C. Chapman. The physiologic meaning of the maximal oxygen intake tests. *J. Clin. Invest.* 37: 538-547, 1958.
16. Quenoville, M.D., A. Boyne, W. Fisher, and I. Leitch. Statistical studies of recorded energy expenditure in men. I. Basal metabolism related to sex, stature, age, climate and race. Bucksbun, Scotland (Rowett Institute), Commonwealth Bureau of Animal Nutrition, Tech. Commun. No. 17, 1951.
17. Saltin, B. Physiological effects of physical conditioning. *Med. Sci. Sports.* 1: 50-56, 1969.
18. Shephard, R.J., O. Allen, A. Benade, C. Davies, P. diPrampero, R. Hedman, J. Merriman, K. Myhre, and R. Simmons. The maximum oxygen intake. An international reference standard of cardiorespiratory fitness. *Bull. W.H.O.* 38: 757-764, 1968.
19. Smokler, P.E., R. MacAlpin, A. Alvaro, and A. Kattus. Reproducibility of a multistage near maximal treadmill test for exercise tolerance in angina pectoris. *Circulation* 48: 346-351, 1973.
20. Voigt, A.E., R. Bruce, F. Kusumi, G. Pettet, K. Nilson, S. Whitkanack, and J. Tapia. Longitudinal variations in maximal exercise performance of healthy sedentary middle-aged women. *J. Sports Med. Phys. Fitness* 15: 323-327, 1975.

GRAVITATIONAL EFFECTS ON BLOOD DISTRIBUTION, VENTILATION, AND GAS EXCHANGE AT THE ONSET AND TERMINATION OF EXERCISE

Jack A. Loeppky
Michael D. Venters
Ulrich C. Luft

Department of Physiology
Lovelace Foundation for Medical
Education and Research
Albuquerque, New Mexico

INTRODUCTION

When man assumes the upright posture with minimal assistance from the leg muscles as when sitting or supported in the upright position on a tilt table, or when he is exposed to lower body negative pressure (LBNP), a substantial volume of blood is shifted to the lower extremities. As this posture or LBNP continues over a period of minutes, a relatively constant distribution of blood is attained, and the volume of blood in the dependent parts of the body can simplistically be considered a "pooled" or "stagnated" compartment. Cardiovascular adjustments take place to maintain the circulatory system under these conditions of a diminished central blood volume (11,34). The increment in blood volume in the lower extremities under these conditions of a superimposed hydrostatic effect of 1.0 G is in the order of 0.5 ℓ to 1.0 ℓ or 20% of the total circulating blood volume (10,23,24). Under these conditions, there is no evidence that the tissue O_2 requirements in the lower extermities are altered, and the O_2 content in the pooled blood will be significantly reduced and the CO_2 content increased as a result of the reduction in blood flow.

For example, an arterio-venous O_2 difference ($AVDO_2$) of 14 vol% has been measured in the femoral blood in the upright posture (27). A previous study from this laboratory (19) demonstrated that when human subjects were returned to the supine position after 10 min of passively standing upright on a tilt table at 60° from the horizontal, a marked increase of O_2 uptake occurred immediately at the pulmonary capillary level ($\dot{V}O_2PC$). This transient rise was 75% complete after the first min and amounted to 200 cc in total after 3 min, equaling the loss in blood O_2 stores (O_2B) incurred while upright. This repayment of O_2B was associated with a significant rise in ventilation ($\dot{V}I$). Quantitatively similar results were obtained after the release of 10 min LBNP at -40 mmHg. The rise in $\dot{V}O_2PC$ was quite similar to the tilt table study (230 cc) with $\dot{V}I$ showing a corresponding rise (20). In the latter study, leg volume was reduced approximately 2% within 10 s after LBNP release, coinciding with the rise in $\dot{V}O_2PC$ and $\dot{V}I$.

Initiating leg exercise in the upright position results in the mobilization of the pooled blood from the legs to the thorax via the muscle pump. Since this is similar in terms of blood volume shifts to the release of LBNP or the return to supine after being upright, one would anticipate a greater rise in early $\dot{V}O_2PC$ and $\dot{V}I$ in comparing responses to those of the same subjects during supine exercise requiring the same metabolic rate where no blood pooling had taken place prior to exercise. As early as 1913 Krogh and Lindhard (16) deduced certain circulatory changes at the beginning of exercise from corresponding alterations in O_2 uptake, but they did not consider the effects of prior venous pooling.

During recovery from upright exercise, a significant volume of blood presumably again pools in the lower extremities, depleting O_2B, thereby resulting in a lower $\dot{V}O_2PC$ than during recovery from supine exercise. This depletion of O_2B due to blood pooling is a much slower process than the repayment of these stores, because the depletion of venous O_2B is a function of metabolic rate of poorly perfused dependent parts of the body (19).

The breath-by-breath measurement of $\dot{V}O_2PC$ is a more valid assessment of O_2 uptake during non-steady states than is $\dot{V}O_2$ measured at the mouth by open-circuit gas exchange techniques because lung stores may change during the collection period. A number of assumptions are inherent in this report:

1. In a true steady state, O_2 transfer at the mouth ($\dot{V}O_2E$) is equal to $\dot{V}O_2PC$, which in turn is equal to the cellular O_2 consumption or metabolic rate. Since the rate of change in cellular $\dot{V}O_2$ cannot be accurately described during the onset or termination of exercise, we have assumed it to be identical in both postures so that differences in $\dot{V}O_2PC$ between the two postures will be the result of variations in O_2B resulting from shifts in blood volume or flow.

2. Changes in lung O_2 stores (O_2L) are reflected by differences between $\dot{V}O_2E$ and $\dot{V}O_2PC$, i.e. if $\dot{V}O_2E > \dot{V}O_2PC$, then O_2 is stored in the lung and vice versa.

It was the purpose of this study to compare the time courses of gas exchange transients during early onset and recovery of exercise in the upright and supine posture and to determine if any differences could be ascribed to variations in

Table 1. Physical Characteristics
and Aerobic Power of Subjects

Subject	Age yr	Height cm	Weight kg	$\dot{V}_{O_2}$ max ℓ/min
1	29	183	65.6	2.69
2	45	176	79.9	3.23
3	34	183	82.7	3.18
4	31	170	76.0	2.64
5	47	190	77.2	2.59
6	32	174	74.7	2.33
Mean	36.3	179	76.0	2.78
S.D.	7.7	7	5.9	0.35

concomitant blood volume and flow shifts. We also tried to determine whether the two postures would have different effects on the time course of pulmonary ventilation during the on- and off-transients of mild exercise.

METHODS

Subjects and exercise. The supine and upright exercises were performed by six subjects consecutively for 8 min with at least 30 min of rest between exercises. Three subjects performed supine exercise first, and the other three began in the upright position. In Table 1 are listed the physical characteristics of the subjects and their $\dot{V}_{O_2}$ max, which had been determined prior to the study by a progressive bicycle ergometer exercise in the sitting position (21). The subjects engaged in occasional endurance exercises for recreation; however, none was in regular training.

The upright exercises were done on a Von Döbeln bicycle ergometer and the supine exercise on an ENSCO (Model BE-5) ergometer attached to the foot of a bed. Work loads on the two ergometers resulting in the same steady state $\dot{V}_{O_2}$ (1.0 ℓ/min) were determined with two other subjects prior to the study. This relatively low work load (35% of $\dot{V}_{O_2}$ max) was chosen to avoid exceeding the anaerobic threshold and to assure the attainment of steady state $\dot{V}_{O_2}$ during exercise and recovery. For both exercises, subjects pedaled at a rate of 40 cycles/min to a metronome. In the supine posture the feet were elevated approximately 20° above the hips, while the upper body was approximately 10° above the hips. For both exercises the feet were taped to the pedals, and the feet were returned to the pre-exercise position on the pedals when exercise was stopped.

Equipment and calibration procedures. The instrumentation and analytical procedures were quite similar to those described in an earlier study (19). Subjects breathed continuously though a Rudolph model 2600 valve (45 ml dead space) and rubber mouthpiece. The valve was attached with corrugated plastic tubing to a 110-ℓ bag-in-box apparatus, forming a closed system wherein the subjects inspired

humidified air from the bag and exhaled into the box. The inspiratory line was connected to a 6-ℓ Krogh spirometer with electrical output. Two-way stopcocks were
inserted in the tubing to allow for periodic open-circuit gas exchange determinations. Concentrations of O_2, CO_2, and Ar were recorded continuously with a respiratory mass spectrometer (SRI-MEDSPECT MS8) from a sampling capillary in the
Rudolph valve about 3 cm from the subject's mouth. The mass spectrometer was
calibrated before and after each run with gas mixtures analyzed by the Scholander
technique. The electrical signals of the gases and spirometer were recorded with a
Visicorder (Honeywell-1508A), providing a breath-by-breath recording on a time
base. Functional residual capacity (FRC) was determined after each exercise with
a gas dilution procedure, rebreathing 9% Ar in 3 ℓ of O_2 in a 5-ℓ capacity bag with
attached mouthpiece and stopcock.

The areas encompassed by the O_2 and CO_2 curves during each exhalation were
determined by planimetry and divided by the time from beginning to end of expiration to obtain the mixed expired gas concentrations for each breath for O_2 and CO_2.
Preliminary trials showed that these values were equal to mean mixed expired gases
collected in a bag over the same period.

The volume record was corrected for time lag between spirometer and mass
spectrometer and sampling rate of the latter. The spirometer was calibrated after
each run with a 1.0-ℓ syringe. The barometric pressure and ambient temperature
during the experiments were approximately 630 mmHg and 24°C, respectively.

Leg volume (percent change) was obtained from continuous recordings of right
calf circumference by means of a mercury-in-silastic strain gauge (Model 270 plethysmograph, Parks Electronic Laboratories) which was displayed on an oscillograph
recorder (Honeywell, Model 906B). Gauge calibration and attachment, as well as
the subsequent calculation of limb volume change, closely followed the procedure
described by Holling *et al.* (15). During exercise, the large oscillations observed in
calf volume with the alternate contraction and relaxation of the leg muscles were
integrated to obtain a mean value for each cycle. This was accomplished by feeding
the strain gauge signal through a Krohn-Hite multi-function variable filter (Model
335, Cambridge, Mass.).

Heart rates were obtained from ECG recordings, utilizing leads on the right arm,
apex, and forehead, and were displayed on the Honeywell 906B recorder.

Protocol. After attaching ECG leads and the mercury strain gauge, the subject
remained at rest in the supine or upright position for 10 min before exercise began.
From 4 to 1.5 min before exercise, a 2.5-min Douglas bag of mixed expired gas was
collected to calculate steady state gas exchange. The subject was connected to the
bag-in-box 30 s before the exercise began for baseline breath-by-breath gas exchange
measurements. He began pedaling at a signal following a 5-s countdown, reaching
the prescribed rate in 2 s or less. Breath-by-breath gas concentrations and $\dot{V}I$ were
recorded through the first 3 min of exercise. The subject was then switched from
the bag-in-box and another 2.5-min Douglas bag was collected from 4.0 to 6.5 min
of exercise. After 7.5 min of exercise, the subject was again connected to the bag-
in-box (which had been refilled with humidified air) for breath-by-breath determinations. The exercise was stopped at 8 min following a countdown, and recovery

gas exchange was measured for 3 min. Another Douglas bag was collected from 4.0 to 6.5 min after exercise. Three FRC determinations were then made with the rebreathing technique, each beginning after a normal exhalation. In each case the mixed Ar concentration reached a plateau after three or four breaths with a large tidal volume. Following a rest interval, the subject assumed the other posture and the entire protocol was repeated.

Calculations. The FRC for each breath was computed from the mean value obtained by the three subsequent rebreathing maneuvers, adjusted for changes in end-tidal volume on the spirometer.

Oxygen uptake was calculated at the mouth ($\dot{V}O_2E$) and the pulmonary capillary membrane for each breath according to Auchincloss *et al.* (1):

$$(1) \quad \dot{V}O_2PC = F_IO_2 \cdot V_I - F_EO_2 \cdot V_E - (F_AO_2 \cdot V_A - \tilde{F}_AO_2 \cdot \tilde{V}_A)$$

where F_I and F_E refer to inspired (0.2094) and mixed expired gas fractions, and V_I and V_E refer to inspired and expired tidal volumes (STPD). The result of the first two terms in Equation 1 is equal to $\dot{V}O_2E$. F_A and V_A represent end-tidal gas fractions and lung volumes (FRC) at the end of the breath, and $\tilde{F}_A$ and $\tilde{V}_A$ represent the same quantities at the beginning of the breath. $\dot{V}CO_2PC$ was calculated in a similar manner:

$$(2) \quad \dot{V}_{CO_2}PC = F_ECO_2 \cdot V_E - F_ICO_2 \cdot V_I + (F_ACO_2 \cdot V_A - \tilde{F}_ACO_2 \cdot \tilde{V}_A)$$

The respiratory exchange ratio at the pulmonary capillary membrane was computed as follows:

$$(3) \quad Rpc = \dot{V}CO_2PC / \dot{V}O_2PC$$

These values were computed on a per second basis for each breath.

The respiratory frequency (f) was calculated from the time interval between successive end-expirations and was multiplied by the corresponding inspired tidal volume (V_I) to obtain $\dot{V}_I$ on a per minute basis.

Calculations were performed with a programmed digital calculator. From individual breath-by-breath values, means for the six subjects were computed at selected time intervals before, during, and after the exercise. Portions of the areas under mean curves were determined by planimetry to quantitate changes for specific time intervals. All testing for statistical significance ($p < .05$) was done with the *t*-test for paired samples.

RESULTS

Steady State Levels of Cardiorespiratory Responses

The steady state values for gas exchange were computed from the Douglas bags collected before, from 4.0 to 6.5 min during, and 4.0 to 6.5 min after exercise.

The mean FRC after recovery was 1.0 ℓ (BTPS) greater in the upright posture (3.92) than in the supine posture (2.92). This difference was significant ($p < .005$).

The other values are summarized in Table 2. The mean $\dot{V}_{O_2}$ during exercise was almost identical for the two exercises, being 952 ± 68 cc/min during upright exercise (U) and 967 ± 71 cc/min during supine exercise (S). In percent of upright $\dot{V}_{O_2}$ max the values were 34.8 ± 5.2 and 35.2 ± 3.8 for U and S, respectively. Thus, the O_2 requirements could be considered equal for the two postures. Although some differences were apparent in the various measurements between U and S in Table 2, the only ones of significance were heart rate (HR) and $P_{ET}CO_2$ before and after exercise. The HR was higher when upright in all cases, but the difference during exercise was not significant. $P_{ET}CO_2$ was 3.2 and 5.1 mmHg higher in the supine posture before and after exercise. The ventilation equivalent for O_2 was consistently but not significantly higher in the upright posture. The differences in $P_{ET}CO_2$ and $\dot{V}_I/\dot{V}_{O_2}$ with posture signify an increase in ineffective ventilation in the upright posture compared to supine, reflecting a redistribution of ventilation and perfusion within the lung. During exercise, these differences were less apparent.

Non-Steady State Responses to Exercise and Recovery

The average values for the breath-by-breath measurements are shown in Figs 1-4 for the first 3 min of exercise and recovery. Baseline mean values were obtained by averaging the mean curves for 30 s preceding exercise and recovery.

Leg volume. The mean responses are shown in Figs. 1 and 2. Although leg volume (LV) is shown to have the same baseline for U and S in order to compare relative changes with each exercise, in actual fact LV is in the order of 3% greater in the upright posture (32), so that during exercise the absolute LV values were probably nearly equal.

The LV rise to 2.0% above baseline after starting S was completed in less than 5 s and remained at this level for the first 3 min, thereafter decreasing by 0.5%. During supine recovery, LV rapidly fell below baseline within 5 s, continued to decline for 40 s, and then remained about 1.5% below the pre-exercise baseline.

With the onset of U, the LV decreased by 2.0% before 5 s and subsequently increased to -1.5% during the 8 min of exercise. When U was terminated, LV rose exponentially during the first min to -0.5% and continued gradually toward baseline during the following 2 min. The differences in LV between U and S were significant during the first 3 min of exercise and the second 2 min of recovery.

The inferences from the curves in Figs. 1 and 2 are:

1. Following the onset of exercise, blood is immediately shifted from the trunk to the legs in the supine posture while the opposite is true in the upright posture.

2. When exercise is terminated, blood shifts from the legs to the trunk after S and back to the legs after U, with both these changes being 90% complete after the first min.

Blood O_2 stores. Marked differences in O_2B fluctuations with posture were evident from changes in $\dot{V}_{O_2}PC$ (Figs. 1 and 2).

Table 2. Mean Steady State Values for HR and Respiratory Variables
for Six Subjects in the Upright (U) and Supine (S) Posture 3 min
Before and 5 min After Beginning Exercise and Recovery

		$\dot{V}_{O_2}$ cc/min	R	$\dot{V}_I$ ℓ/min	$\dot{V}_I/\dot{V}_{O_2}$	$P_{ET}CO_2$ mmHg	HR/ min
Before	U	294	.77	9.54	32.9	33.1[b]	76[a]
	S	295	.80	8.80	29.9	36.3	63
Exercise	U	952	.82	25.15	26.5	38.3	96
	S	967	.82	24.15	25.0	39.6	93
Recovery	U	312	.86	11.22	35.9	32.3[a]	83[b]
	S	282	.88	9.09	32.3	37.4	66

[a] $p < .05$. [b] $p < .01$. $\dot{V}_{O_2} = O_2$ consumption by Douglas bag.
R = respiratory exchange ratio. $\dot{V}_I$ = ventilation. $\dot{V}_I/\dot{V}_{O_2}$ = ventilation equivalent for O_2. $P_{ET}CO_2$ = end-tidal CO_2 pressure. HR = heart rate.

With the onset of U the O_2 transfer began increasing immediately, whereas with S there was a 5-s delay. After 40 s, the curves again coincided and remained relatively similar throughout exercise. During the first 40 s of U, the increase in O_2 B was 136 cc greater than for S. Despite large quantitative differences in these values between subjects (Table 3), the mean difference was statistically significant ($p < .005$). The half-times for $\dot{V}_{O_2}$PC taken from the two curves in Fig. 1 were 8 and 33 s for U and S, respectively. Both curves reached the steady-state levels between 1.5 and 2.0 min. The apparent "notch" between 45 and 60 s seen in both curves was not statistically significant, reflecting changes seen only in two or three individuals.

During recovery the difference between the $\dot{V}_{O_2}$PC curves was reversed compared to the onset of exercise (Fig. 2). In this case, $\dot{V}_{O_2}$PC after S remained higher until 100 s, where the curves converged. The significant difference in O_2 B between the curves of 152 cc (Table 3) was quantitatively quite similar to that after the onset of exercise, but was evident 2.5 times as long. During the first 15 s after S, there was a transient rise in $\dot{V}_{O_2}$PC above the exercise level equal to 20 cc of O_2. Although not statistically significant, this transient rise was noted in all but one subject. The relatively larger increase in O_2 B at the start of U thus appeared to be compensated by the relatively greater loss in O_2 B during recovery. The mean $\dot{V}_{O_2}$PC recovery half-times for U and S were 25 s and 35 s, respectively.

End-tidal gases and gas exchange. The mean $P_{ET}O_2$ values showed opposite courses during the first 15 s of exercise (Fig. 1), falling 1.7 mmHg with U and rising 2.8 mmHg with S after 10 s, the latter difference being significant ($p < .005$). The $P_{ET}O_2$ time courses for both exercises were parallel after 15 s, being offset by approximately 3 mmHg after 45 s, the same amount as during the pre-exercise baseline. $P_{ET}O_2$ for both exercises showed a significant but transient drop of 45 s, being

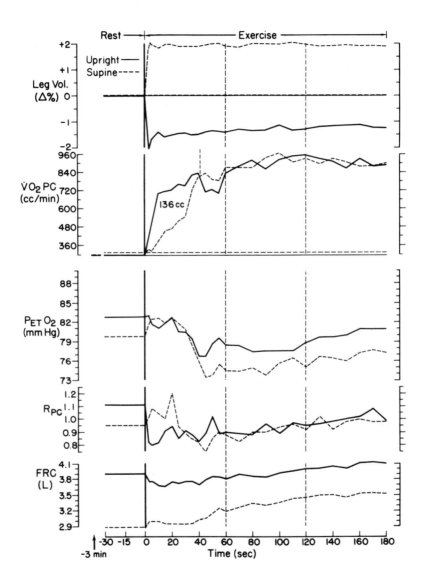

Fig. 1. Right calf volume (LEG VOL.) see text, O_2 transfer across the pulmonary capillaries ($\dot{V}_{O_2}$ PC), end-tidal O_2 pressure ($P_{ET}O_2$), respiratory exchange ratio at the pulmonary capillaries (Rpc), and functional residual capacity (FRC) at rest and during the first 3 min of exercise. Oxygen consumption was measured by Douglas bag from 4 to 1.5 min before exercise. Average barometric pressure was 630 mmHg.

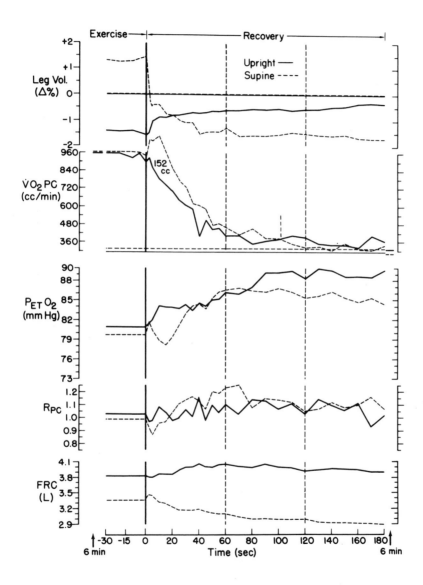

Fig. 2. Same as Fig. 1 for last 30 s of the 8-min exercise and during the first 3 min of recovery. Oxygen consumption was measured by Douglas bag from 4 to 6.5 min during exercise and 4 to 6.5 min after the end of exercise.

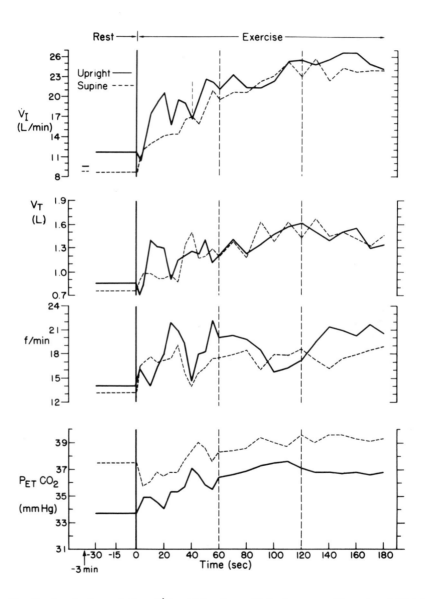

Fig. 3. Inspired ventilation ($\dot{V}_I$), tidal volume (V_T), frequency (f), and end-tidal CO_2 pressure ($P_{ET}CO_2$) corresponding to Fig. 1 in time.

6 mmHg below baseline in each case (p < .05) followed by a 2-mmHg rise during the next 10 s. Divergent trends were again apparent during the first 15 s of recovery (Fig. 2), with $P_{ET}O_2$ falling by 1.5 mmHg after S and rising 3.0 mmHg after U (p < .05). During the second min of recovery, both values plateaued about 7 mmHg above the exercise level.

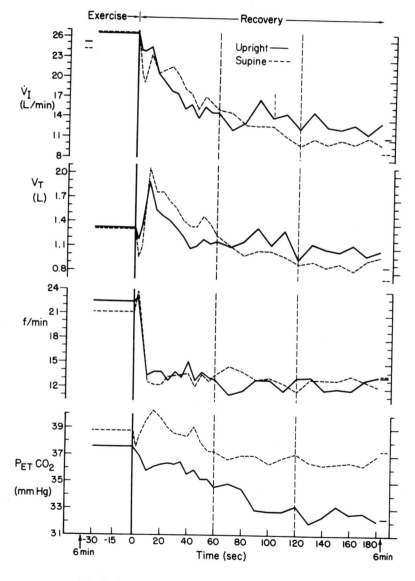

Fig. 4. Same as Fig. 3 corresponding to Fig. 2 in time.

Table 3. Differences in Pulmonary Capillary O_2 Transfer ($\dot{V}_{O_2}$PC),
Ventilation, and Heart Rate Between Upright (U) and Supine (S) Posture
During First 40 s of Exercise and First 100 s of Recovery

	EXERCISE Upright − Supine			RECOVERY Upright − Supine		
Subject	$\Delta\dot{V}_{O_2}$PC cc	$\Delta\dot{V}_I$ ℓ	(0-40 s) ΔHR / min	$\Delta\dot{V}_{O_2}$PC cc	$\Delta\dot{V}_I$ ℓ	(0-100 s) ΔHR / min
1	245	3.48	9	- 31	8.93	-13
2	109	0.75	16	-158	-1.12	-24
3	98	0.92	30	-244	-0.87	-33
4	98	3.38	12	-198	-7.86	-22
5	104	1.20	5	-147	-2.36	-16
6	160	2.88	3	-135	-0.11	11
Mean	136	2.10	13	-152	-0.57	-16
S.D.	58	1.28	10	72	5.43	15
P	< .005	< .02	< .05	< .005	NS	< .05

Values for $\Delta\dot{V}_{O_2}$PC represent the area between the curves for U and S from Figs. 1 and 2, and $\Delta\dot{V}_I$ values were obtained similarly from Figs. 3 and 4. Symbols as in Table 2.

The trends for PET_{CO_2} (Figs. 3 and 4) were reversed from those of PET_{O_2}. After 10 s of exercise, PET_{CO_2} fell 1.5 mmHg with S and rose 1.2 mmHg with U ($p < .05$). Both curves peaked at about 45 s, leveled off at 90 s, and thereafter remained parallel but offset by about 2 mmHg. After 15 s of recovery, PET_{CO_2} had risen 1.5 after S and fallen 1.8 mmHg after U, with both curves leveling off at 90 s about 3 mmHg below exercise level.

The respiratory exchange ratio at the pulmonary capillary level (Rpc) during the first 20 s of exercise (Fig. 1) was markedly affected by posture. Rpc decreased by 0.30 after 5 s of U and increased by 0.13 with S, this difference in trends being statitically significant ($p < .02$). The relatively high baseline Rpc was the result of a transient hyperventilation prior to U. After 20 s of exercise, Rpc remained fairly similar for both exercises. No marked differences were apparent during the 3 min of recovery; however, Rpc after S fell during the first 5 s and then increased for the next minute, whereas Rpc after U remained fairly level.

During the first 40 s of exercise, the average Rpc was 0.87 for U and 0.98 for S (Fig. 1). These values signify that $\dot{V}_{CO_2}$PC during U was retarded relative to $\dot{V}_{O_2}$PC, whereas during S the gas transfer of O_2 and CO_2 were nearly equal. The postural difference in $\dot{V}_{CO_2}$PC was thus about two-thirds that of $\dot{V}_{O_2}$PC (136 cc) at the start of exercise. During the first 100 s of recovery, the mean Rpc after U was again lower than after S (1.06 vs. 1.12), indicating that CO_2 transfer to the lungs recovered slower during S than $\dot{V}_{O_2}$PC. This is indicated by the upward trend in Rpc from 5 s to 70 s following S (Fig. 2). The lag in $\dot{V}_{CO_2}$PC recovery following

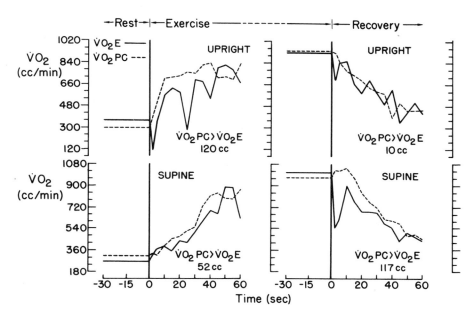

Fig. 5. O_2 intake by the lungs and pulmonary capillaries during the first min of exercise and recovery. When $\dot{V}_{O_2}PC$ exceeds $\dot{V}_{O_2}E$, then lung O_2 stores are depleted. The amount of O_2 lost is estimated from the area between the curves.

S resulted in a $\dot{V}_{CO_2}PC$ difference which was about 50% greater than the $\dot{V}_{O_2}PC$ difference of 152 cc.

Ventilation. The differences seen in $\dot{V}I$ with posture during early exercise corresponded to those of $\dot{V}_{O_2}PC$. The rise in $\dot{V}I$ was faster during U than S (half-times of 15 s and 48 s, respectively) and remained higher for 80 s (Fig. 3). The greater $\dot{V}I$ during U over the first 40 s amounted to 2.10 ℓ and in spite of large individual variations (Table 3) was statistically significant ($p < .02$). The average $\dot{V}I$ was 22% greater during the first 40 s of U. The enhanced $\dot{V}I$ was accomplished primarily by a relatively larger VT (0.11 ℓ), whereas f was only 0.8/min higher for U. However, from Fig. 3 it is apparent that VT was mainly responsible for the greater $\dot{V}I$ during the first 20 s and f during the second 20 s. After 40 s, VT and f showed some fluctuations, but were generally independent of posture.

During recovery from U, $\dot{V}I$ was generally less for the first 80 s and higher than S during the second and third min. An exception to this occurred during the first 10 s where $\dot{V}I$ after S was significantly lower ($p < .05$). Again quantitating the individual differences over 100 s, it was found that the average $\dot{V}I$ was greater after S in five of six subjects, but the mean difference was not statistically significant (Table 3). The half-times for $\dot{V}I$ from Fig. 4 for U and S were 21 s and 38 s, respectively, being quite similar to those for $\dot{V}_{O_2}PC$. The greater $\dot{V}I$ for S was the result of a larger VT from 10 s to 70 s, with f falling from the exercise to resting level within the first 10 s of recovery in both postures.

Lung O_2 stores and FRC. The changes in O_2 L were estimated for the first min of exercise and recovery for both postures by comparing mean breath-by-breath values for $\dot{V}O_2E$ and $\dot{V}O_2PC$ (Fig. 5). When $\dot{V}O_2PC$ exceeds $\dot{V}O_2E$, then O_2 L is reduced and vice versa. The net area between the two curves corresponds to the change in O_2 L.

During the first min of U, O_2 L decreased by 120 cc, while at the same time FRC fell by 0.14 ℓ (Fig. 1). With S, O_2 L was depleted by 52 cc over the first min, and FRC rose 0.13 ℓ.

After U, essentially no change in O_2 L was apparent after the first min of recovery, while the FRC (Fig. 2) was slightly greater (0.12 ℓ) than during exercise. However, after S, a 117-cc reduction occurred in O_2 L along with a reduction in FRC of 0.14 ℓ. This decrease in O_2 L was similar to that seen after the onset of U with about one-half the loss occurring during the first 10 s.

Fig. 5 clearly demonstrates that the largest losses on O_2 L were incurred after the onset of U and recovery from S. Both of these transients had in common a marked reduction in LV and a small reduction in FRC. From Figs. 1 and 2 it is evident that for the first min of exercise or recovery, the changes in FRC were directly related to changes in LV. Although variations in FRC during the transients were small, they were consistent, e.g. the mean value after 10 s of exercise was 0.23 ℓ less (U) and 0.10 ℓ more (S) than baseline, but both differences were significant ($p < .05$).

Heart rates. The HR responses were significantly different with posture during the first 40 s of exercise. The mean value in Table 3 indicates that HR increased by 12 beats/min more after 40 s of S than U. A relatively greater rise in HR with S (10 beats/min) was also seen for the steady-state values shown in Table 2. Since stroke volume usually does not change much between rest and S (5), any increase in cardiac output is more dependent on HR than in U.

The postural difference in recovery HR was also significant (Table 3). The mean HR after 100 s of recovery had dropped by 16 beats/min more after S, and this difference was maintained about the same throughout the steady-state measurement after 4 min (Table 2).

DISCUSSION

Rapid changes in LV were seen consistently in all these experiments, whereby there was an increase early in S and a decrease in recovery, while exactly the opposite phenomenon occurred with U. It is well known that blood flow increases rapidly in active muscles. Measurements of blood flow in the gastrocnemius muscle in man with an isotope clearance method (29) have shown that flow increases instantaneously with the first contraction and reaches its maximum after 30 s to 40 s of rhythmic work. Furthermore, Guyton *et al.* (13) have demonstrated an immediate increase of 40% in cardiac output when muscular activity was elicited by motor nerve stimulation in dogs with an increase of 86% within 40 s. Rising muscle perfusion is associated with regional vasodilation as well as higher perfusion pressure, so that one would expect an increase in volume of the active limbs. This was regularly the case in S in our study, but in U a shrinkage in LV was observed of about the same magnitude as it increased in S. Early experiments by Waterfield (32) with a water plethysmograph below the knee have shown that in the erect posture the relaxed LV is 2% to 3% greater than recumbent, and that brisk contraction of the muscles reduces the volume by approximately the same amount as seen in our experiments. He attributed this to the squeezing action of the muscles on the veins and venules (muscle pump). Thus the increase in muscle volume due to hyperemia in exercise, which is undoubtedly present in U as in S, is completely masked by the greater shift of pooled blood out of the legs in U. The small but consistent differences in the changes in FRC associated with exercise in the different postures appear to be closely related to the changes in LV and probably reflect the shifts of blood volume to or from the thorax with corresponding decrease or increase in lung gas volume. Sjöstrand (28) first demonstrated such effects of redistribution of blood volume on lung gas content in developing his concept of the lungs as a blood depot.

The observed reverse shifts of blood volume with exercise in the two different postures provided valuable clues as to the origin of transient discrepancies in alveolar gas exchange and ventilation seen between the two conditions. Redistribution of blood volume to or from the central blood vessels and the lungs is always associated with alterations in $O_2 B$ that must have a transient effect on $\dot{V}O_2 PC$. The following simplified calculation for the transition from rest to exercise might be appropriate to explain the observed differences in $\dot{V}O_2 PC$ in our experiments, whereby it is necessary to assume values for the $AVDO_2$ in the femoral blood. Reeves et al. (26,27) did a comprehensive study of $AVDO_2$ in the systemic and femoral blood in the supine and upright positions at rest and during exercise. We have used the following values from their data for a work load similar to ours in order to estimate changes in $O_2 B$ at the onset of exercise:

$$AVDO_2 \ (cc/\ell)$$

		Pulmonary	Femoral
S	Rest	39	42
	Exercise	95	131
U	Rest	53	139
	Exercise	101	139

It is noteworthy that the femoral $AVDO_2$ increases markedly with S while there is no change with U, where $AVDO_2$ in the legs is already large at rest due to sluggish circulation. Using these figures and assuming that an extra 0.5 ℓ of venous blood is moved from the legs to the trunk at the onset of U, the oxygenation of this blood in the lungs would increase $\dot{V}O_2 PC$ by 0.5 x 139 = 70 cc. If the same volume of arterial blood were shifted into the legs during S, as indicated by the increase in LV, the amount of O_2 extracted by the tissues, namely 0.5 x 131 = 66 cc, would be borrowed from O_2 B and would not appear as $\dot{V}O_2$PC. The assumption that the femoral AVDO$_2$ in S increases so rapidly appears justified from sequential measurements of mixed venous PO$_2$ during early exercise by a rebreathing method by Edwards *et al.* (4). A significant drop in P$\bar{v}O_2$ (> 10 mmHg) was already present after 15 s. The value of 66 cc is in the same order of magnitude as that reported by Raynaud *et al.* (25). From arterial and mixed venous blood samples and cardiac output determined with radioactive krypton, they estimated that 40 cc of excess O_2 was transported to the lungs during the first 90 s of mild supine exercise. The net result of the excess $\dot{V}O_2$PC in U and the decrement in S represents a difference of 136 cc, as was found in our measurements. The preceding calculation does not take directly into account any differences in the redistribution of blood flow with posture during exercise. In a model developed in an earlier publication from this laboratory (19), it was shown how the redistribution of flow during changes in posture at rest could account for alterations in O_2 B and $\dot{V}O_2$PC.

During recovery the relationship between changes in LV and the difference between $\dot{V}O_2$PC in the two postures was reversed. In the supine position, $\dot{V}O_2$PC actually exceeded the level recorded during exercise in the first 20 s, and this coincided with a steep drop in LV that amounted to 3% over the first min, documenting an upsurge in venous return with a corresponding rise in gas exchange in the lungs. In the upright position, LV increased by 0.6% in the first 10 s and continued to increase at a slower rate over the next 3 min as blood accumulated in the legs again. $\dot{V}O_2$PC was lower after U than S for more than a minute and the difference amounted to 152 cc (Fig. 2). Thus O_2 B is replenished again after S, while it is depleted after U.

This course of events is also reflected in the behavior of PETO_2, PETCO$_2$, and Rpc during these experiments. The fall in PETO_2 and rise in PETCO$_2$ during the first 20 s of U is compatible with a rapid influx of venous blood into the lungs, while PETO_2 rises in S and PETCO$_2$ drops coincident with the reduction in central blood volume while metabolic demands are being met in part frm O_2 B. The largest departure in the end-tidal gases from the baseline occurred at about 10 s, the time when the difference in $\dot{V}O_2$PC was also at its maximum. During recovery, the course of PETO_2 and PETCO$_2$ also differed markedly, but in the opposite direction as after the onset of exercise. PETO_2 started to rise after U and PETCO$_2$ to fall immediately, while after S, PETO_2 showed a momentary rise followed by a significant drop. PETCO$_2$ also showed a biphasic pattern, dropping for a few seconds and then rising above the preceding level during exercise. The incisure in both PETO_2 and PETCO$_2$ probably signals the arrival of additional venous blood from the extremities coincident with the depletion of LV after S. This was also indicated by the substantial loss in O_2 L following S (Fig. 5), which was similar to the loss incurred during the first min of U. The passage of additional venous blood through the lung will reduce O_2 L as O_2 B is replenished.

The sharp drop in Rpc immediately after starting U suggests that $\dot{V}CO_2$PC was not keeping up with $\dot{V}O_2PC$, which was rising rapidly at this point. This view is supported by the observation that the difference in $\dot{V}CO_2PC$ in the first 40 s between U and S was only 65% of that for $\dot{V}O_2PC$. A similar phenomenon in the opposite sense was noted after S, where an elevated Rpc indicated that $\dot{V}CO_2PC$ did not subside as rapidly as $\dot{V}O_2PC$ during recovery. In other words, the on- and off-transients for $\dot{V}CO_2PC$ were slower than for $\dot{V}O_2PC$. Similar observations have been made by others, notably Linnarsson (18), who ascribed this to the much greater storage capacity for CO_2 than for O_2 in the tissues (6,7), particularly in the lungs (3), which tends to buffer rapid transients in CO_2.

The redistribution of blood volume shown in these experiments can be reproduced artificially without exercise by applying LBNP. Metabolic processes are presumably not affected by this manipulation, so that any fluctuations in pulmonary gas exchange and ventilation are attributable to shifts in blood volume alone. In previous studies in this laboratory (to be published), it was noted that application of -40 mmHg LBNP in the supine position, which caused an increase in LV of about 2% as seen at the onset of S and after U, was associated with a significant reduction in $\dot{V}O_2PC$, a marked rise in $PETO_2$, and a lesser fall in $PETCO_2$. However, these changes were not as abrupt as in the exercise study and were sustained for more than two min. During this time the O_2 requirements of the legs were being in part defrayed from O_2 stores in the pooled blood, thus reducing $\dot{V}O_2PC$. On releasing LBNP after a 10-min exposure, LV dropped by 1.5% during the first 10 s and returned to the original volume after one min. The rate of $\dot{V}O_2PC$ was double the control value at 10 s, but returned to baseline during the following two min. The peak in $\dot{V}O_2PC$ was followed shortly by a drop in $PETO_2$ of 10 mmHg and a rise in $PETCO_2$ of 3 mmHg. These fluctuations in gas exchange were much more drastic than the ones at the onset of LBNP. They signify the sudden arrival of markedly reduced blood in the lungs and are analogous to the course of events at the onset of U and the end of S in the exercise experiments. The LBNP studies demonstrate clearly what can only be inferred from the exercise tests, namely that the return of pooled blood to the lungs produces a rapid, transient increase in apparent gas exchange, whereas the redistribution of blood volume into the peripheral vascular bed is associated with a more gradual reduction in $\dot{V}O_2PC$ as some O_2B is being utilized in the periphery.

Although relatively small, the described changes in O_2B at the beginning and the end of exercise will enter into the estimation of the O_2 deficit and the fast component of the O_2 debt. For instance, Ceretelli et al. (2), who recently studied the on- and off-transients of exercise $\dot{V}O_2$ in the sitting and supine positions, noticed that $\dot{V}O_2$ rose more rapidly during upright work (half-time 36 s) than supine (half-time 49 s) and that the O_2 deficit was correspondingly smaller in the former (0.73 ℓ) than in the latter (1.30 ℓ). Differences in the utilization of O_2B with posture probably contributed to these discrepancies.

The ventilatory response at the beginning of exercise followed a similar pattern in both postures. However, in U, the rate of increase of $\dot{V}I$ (Fig. 3) was faster with a half-time only one-third of that in S. Apparently, the sudden influx of venous blood to the lungs, which must have been greater in U than in S, augmented the initial respiratory drive. More than 30 years ago, Mills (22) demonstrated that hyperpnea can be produced in man by the sudden release of blood previously sequestered by cuffs applied to the limbs. The mean time lag was only 2 s, a much faster response than could be attributed to systemic arterial chemoreceptors. He demonstrated by peripheral injections of cyanide that the mean response time of the latter was about 12 s and concluded that hyperpnea is induced via pressoreceptors in the pulmonary vasculature when more blood arrives in the lungs. Wasserman *et al.* (30) reported rapid hyperpnea following a sudden increase in cardiac output induced with isoproteronol or cardiac pacing in dogs, which was demonstrable even after removal of the carotid sinus bodies and resection of the carotid sinus nerve. They also provided evidence suggesting intrapulmonary receptors that are responsive to the CO_2 level in the blood reaching the lungs by intravenous loading in dogs (31). This produced a rapid increase in ventilation sufficient to maintain $PaCO_2$ close to control levels despite a fourfold increase in CO_2 output. The response was not affected by resection of the carotid bodies nor of the vagus nerve (33). Other investigators have disputed the presence of pulmonary chemoreflexes sensitive to CO_2, notably Gonzales and Fordyce (12). They injected $NaHCO_3$, HCl, and KCN in cold saline solutions and measured the transit time of the thermal transient to the aorta and carotid bifurcation. No responses were observed before the bolus had reached the ascending aorta and usually the carotid bifurcation. Other experiments by Levine (17) on animals after spinal cord transection (L-2) and carotid denervation even suggest the presence of humoral agents other than arterial CO_2, pH, or O_2 saturation that stimulate $\dot{V}I$. Furthermore, Hildebrandt and Winn (14) observed the response time of $\dot{V}E$ and $PETCO_2$ during exercise when occlusion cuffs on the legs were suddenly released. CO_2 started to rise at 10 s, and $\dot{V}E$ increased 15 s to 20 s after release of the cuffs, which is compatible with known chemoreceptor function as sole mediators of the response. Another interesting hypothesis has been advocated by Filley and Heineken (9). They believe that alterations in the alveolar-mixed venous gradients (A-$\bar{v}$) for O_2 and CO_2 cause a chemical disequilibrium in the pulmonary microvasculature that could account for the early hyperpnea of exercise via intrapulmonary chemoreceptors. Indirect evidence for such a mechanism was obtained by manipulating the A-$\bar{v}$ gradient with appropriate inspired gas mixtures and recording the immediate ventilatory response (8).

Whether it was the augmented venous return and cardiac output at the start of U in our experiments that caused the faster rise in $\dot{V}I$ or whether it was due to the high PCO_2 and low PO_2 in the pooled venous blood returning to the lungs cannot be decided from our data. However, it is clear that the initial response in both postures was faster than could be explained by known chemoreceptors and could be due either to intrapulmonary sensors or peripheral proprioceptive reflexes. But the marked difference between U and S that appeared before the systemic chemoreceptors could have been activated speaks more in favor of pulmonary receptors being involved.

SUMMARY

Breath-by-breath measurements of ventilation, gas exchange, and leg volume were made on six subjects during mild exercise and recovery in the upright and supine postures. The study demonstrated that marked differences related to posture in leg volume, ventilation, and O_2 transfer at the pulmonary capillary level were evident during early exercise and recovery. Increasing venous return to the heart and lungs, inferred from leg volume measurements, and increased leg blood flow during the onset of upright and end of supine exercise could account for these differences. The replenishing of venous O_2 stores, depleted prior to exercise in the upright posture, accounted for the faster rise in pulmonary O_2 transfer and ventilation seen during the first 40 s of upright exercise and a loss in lung O_2 stores during this time. These changes are quite similar to those seen after the release of LBNP. When exercise was terminated, O_2 transfer fell off more rapidly while upright because venous O_2 stores were again being depleted similar to the onset of LBNP, with a drop in ventilation compared to the supine recovery. $\dot{V}O_2 PC$ in early supine recovery was transiently elevated above the exercise baseline, reflecting an increased venous return to the heart and lungs. In analyzing the time courses of ventilation and gas exchange at the beginning and the end of exercise, the redistribution of blood volume and consequent changes in O_2 stores, which are posture dependent, should be taken into account in order to identify actual metabolic events and thus avoid errors in estimating the O_2 deficit and the early phase of the O_2 debt.

This work was carried out in compliance with the Recommendations of the Declaration of Helsinki. It was supported by the National Aeronautics and Space Administration under Contract NAS9-14920 with the L. B. Johnson Space Center, Houston, Texas.

REFERENCES

1. Auchincloss, J.H., Jr., R. Gilbert, and G.H. Baule. Effect of ventilation on oxygen transfer during early exercise. *J. Appl. Physiol.* 21:810-818, 1966.
2. Ceretelli, P., D. Shindell, D.P. Pendergast, P.E. diPrampero, and D.W. Rennie. Oxygen uptake transients at the onset and offset of arm and leg work. *Respir. Physiol.* 30:81-97, 1977.
3. Dubois, A.B., W.O. Fenn, and A.G. Britt. CO_2 dissociation curve of lung tissue. *J. Appl. Physiol.* 5:13-16, 1952.
4. Edwards, R.H.T., D.M. Denison, G. Jones, C.T.M. Davies, and E.J.M. Campbell. Changes in mixed venous gas tensions at start of exercise in man. *J. Appl. Physiol.* 32:165-169, 1972.
5. Ekelund, L.G. and A. Holmgren. Central hemodynamics during exercise. In: *Physiology of muscular exercise.* Amer. Heart Assoc. Monograph No. 15, 33-43, 1967.
6. Farhi, L.E., M.S. Nesarajah, A.J. Olszowka, L.A. Metildi, and A.K. Ellis. Cardiac output determination by simple one-step rebreathing technique. *Respir. Physiol.* 28:141-159, 1976.

7. Farrell, E.J. and J.H. Seigel. Estimation of blood gas contents from expired air under normal and pathological conditions. *Respir. Physiol.* 26:303-325, 1976.
8. Filley, G.F. The hyperpnea of exercise in man. *Fed. Proc.* 35:633, 1976.
9. Filley, G.F. and F.G. Heineken. A blood gas disequilibrium theory. *Br. J. Dis. Chest* 70: 223-245, 1976.
10. Foux, A., R. Seliktar, and A. Valero. Effects of lower body negative pressure (LBNP) on the distribution of body fluids. *J. Appl. Physiol.* 41:719-726, 1976.
11. Gauer, O.H. and H.L. Thron. Postural changes in the circulation. In: *Handbook of physiology. Circulation.* Washington, D.C.: Am. Physiol. Soc., 1965, sec. 2, vol. III, chap. 67, pp. 2409-2439.
12. Gonzales, F. and W.E. Fordyce. Evidence against a CO_2 mediated pulmonary chemoreflex in dogs. *Fed. Proc.* 35:553, 1976.
13. Guyton, A.C., B.H. Douglas, J.B. Langston, and T.Q. Richardson. Instantaneous increase in mean circulatory pressure and cardiac output at onset of muscular activity. *Circ. Res.* 11:431-441, 1962.
14. Hildebrandt, J.R. and R.K. Winn. Human cardiorespiratory responses to impulses of work during circulatory occlusion. *Physiologist* 19:225, 1976.
15. Holling, H.E., H.C. Boland, and E. Russ. Investigation of arterial obstruction using a mercury-in-rubber strain gauge. *Am. Heart J.* 62:194-205, 1961.
16. Krogh, A. and J. Lindhard. The regulation of respiration and circulation during the initial stages of muscular work. *J. Physiol. (London)* 47:112-136, 1913.
17. Levine, S. Ventilatory response to exercise of denervated extremities: observations in peripheral chemodenervated dogs. *Physiologist* 19:268, 1976.
18. Linnarsson, D. Dynamics of pulmonary gas exchange and heart rate changes at start and end of exercise. *Acta Physiol. Scand.* Suppl. 415:1-68, 1974.
19. Loeppky, J.A. and U.C. Luft. Fluctuations in O_2 stores and gas exchange with passive changes in posture. *J. Appl. Physiol.* 39:47-53, 1975.
20. Loeppky, J.A. and U.C. Luft. O_2 stores and ventilation in response to lower body negative pressure. *Physiologist* 19:275, 1976.
21. Luft, U.C., D. Cardus, T.P.K. Lim, E.C. Anderson, and J.L. Howarth. Physical performance in relation to body size and composition. *Ann. N.Y. Acad. Sci.* 110:795-808, 1963.
22. Mills, J.N. Hyperpnea in man produced by sudden release of occluded blood. *J. Physiol. (London)* 103:244-252, 1944.
23. Montgomery, L.D., P.J. Kirk, P.A. Payne, R.L. Gerber, S.D. Newton, and B. A. Williams. Cardiovascular responses of men and women to lower body negative pressure. *Aviat. Space Environ. Med.* 48:138-145, 1977.
24. Musgrave, F.S., F.W. Zechman, and R.C. Mains. Changes in total leg volume during lower body negative pressure. *Aerosp. Med.* 40:602-606, 1969.
25. Raynaud, J., H. Bernal, J.P. Bourdarias, P. David, and J. Durand. Oxygen delivery and oxygen return to the lungs at onset of exercise in man. *J. Appl. Physiol.* 35:259-262, 1973.
26. Reeves, J.T., R.F. Grover, G.F. Filley, and S.G. Blount, Jr. Circulatory changes in man during mild supine exercise. *J. Appl. Physiol.* 16:279-282, 1961.
27. Reeves, J.T., R.F. Grover, S.G. Blount, Jr., and G.F. Filley. Cardiac output response to treadmill walking. *J. Appl. Physiol.* 16:283-288, 1961.
28. Sjöstrand, T. The regulation of the blood distribution in man. *Acta Physiol. Scand.* 26: 312-327, 1952.
29. Tønnesen, K.H. Blood-flow through muscle during rhythmic contraction measured by [133]Xenon. *Scand. J. Clin. Lab. Invest.* 16:646-654, 1964.
30. Wasserman, K., B.J. Whipp, and J. Castagna. Cardiodynamic hyperpnea: hyperpnea secondary to cardiac output increase. *J. Appl. Physiol.* 36:457-464, 1974.

31. Wasserman, K., B.J. Whipp, R. Casaburi, D.J. Huntsman, J. Castagna, and R. Lugliani. Regulation of arterial P_{CO_2} during intravenous CO_2 loading. *J. Appl. Physiol.* 38:651-656, 1975.
32. Waterfield, R.L. The effect of posture on the volume of the leg. *J. Physiol. (London)* 72: 121-131, 1931.
33. Whipp, B.J., D.J. Huntsman, and K. Wasserman. Evidence for a CO_2-mediated pulmonary chemoreflex in dog. *Physiologist* 18:447, 1975.
34. Wolthuis, R.A., S.A. Bergman, and A.E. Nicogossian. Physiological effects of locally applied reduced pressure in man. *Physiol. Rev.* 54:566-595, 1974.

IV

Cold Stress
Chairperson

Hisato Yoshimura

Hyogo College of Medicine
Nishinomiya, Japan

COLD STRESS: A SELECTIVE REVIEW

E. R. Buskirk

Laboratory for Human Performance Research
Intercollege Research Programs
The Pennsylvania State University
University Park, Pennsylvania

INTRODUCTION

Man can experience cold in a variety of ways and some of these ways can prove traumatic without appropriate thermal protection. The study of man in cold has been accomplished by studying him during chamber, natural, whole body, as well as body part, segment, or organ exposures. Air, gas mixture, and water environments have been utilized as have hypo and hyperbaric conditions. The results of accidental exposure have been documented, as have the results of numerous treatments and therapies.

The literature on cold exposure is far too vast to cover comprehensively in the time available. Thus, only a few problems have been selected for discussion. Not too long ago, the Josiah Macy conferences on cold injury and subsequent publication of the presented papers provided a unique arrangement for assimilating the results of studies on man in the cold. It should be mentioned that the gentleman honored by this symposium, Steven Horvath, was a prominent participant in these excellent conferences; and his interest in the physiology of man in the cold has continued, as evidenced by his many subsequent publications, many of which are intentionally cited in this brief review of certain aspects of human applied physiology and cold exposure.

TEMPERATURE REGULATION MODELS

An important aspect of experimental design is the appropriate use of models that attempt to depict the regulated system. With respect to the physiological changes that occur with acute cold exposure, the recent contribution of Gordon, Roemer, and Horvath (13) is noteworthy. Their control system is based on a feedback signal involving cerebral core temperature and feed-forward signals arising from skin receptors for temperature and heat flux. The feed-forward concept is somewhat unique among biological mechanisms, virtually all of which are assumed to involve feedback. According to the authors, a heat-flux-based control system overcomes the necessity for building in a separate time delay because peripheral heat flux has a natural time delay. The concept of a heat flux system implies thermal receptor sites at various depths within and perhaps just beneath the skin.

In order to more readily conceptualize Gordon, Roemer, and Horvath's model, a colleague of mine (J. Loomis) and I took the liberty of redrawing it, taking out about half of the detail. Nevertheless, even the abbreviated model (Fig. 1) provides a useful description of variables not previously considered to any major extent, i.e., thermophysical properties of bone, muscle, and other tissues as well as regional blood flow. The calculated data utilizing the model were fitted to direct experimental results obtained from the cold air exposures performed by Raven *et al.* (32,33) and substantial agreement was achieved. The authors concluded that skin heat flux may provide an important input for temperature regulation.

The paper by Gordon *et al.* (13) points out an interesting problem in relation to cold exposure. The time course and extent of vasoconstriction with cold exposure, in terms of specific tissue involvement, is not well known. Thus, in the preparation of their model, it was assumed that 38% of the constriction occurred in the arms, 2% in the hands, 58% in the legs, and 2% in the feet. Obviously, such arbitrary assignments only reflect the meagerness of our knowledge. To quote these investigators, "The skin constriction signal weighting factors were obtained by a laborious trial and error procedure which involved adjusting them to obtain skin surface temperatures that were the same as those obtained from experimental data. This was necessary due to the paucity of skin blood flow data in the literature." These statements clearly indicate an area for future research with due recognition of the methodological difficulty such experimentation will encounter. The partitioning of blood flow by region and by layers within regions poses many technical problems, particularly if non-invasive techniques are employed.

It is not my intention to discuss temperature regulation models, but several other models (in their design) include the important models of Stolwijk and Hardy (36), Wissler (39), and Mitchell and Myers (23). A review of models of temperature regulation prepared by Hardy (14) provides a comprehensive background for interpreting the unique features and potential of the different types of model systems. Similarly, Bligh contributed an interesting model review that involved consideration of neural transmitters and inhibitors at the recent international symposium on temperature regulation held in Lille, France, July 1977.

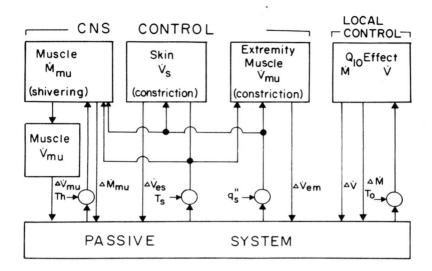

Fig. 1. Block diagram of control and passive system for human temperature regulation during transient cold exposure. A variety of nonlinear elements are involved, and the importance of heat flux through the skin is emphasized as an important variable sensed. $\dot{M}$ is Metabolic Rate (kcal·h^{-1}). $\dot{M}_{mu}$ is Metabolic Rate, Muscle (kcal·h^{-1}). $\Delta\dot{M}$ is Change in Metabolic Rate (kcal·h^{-1}). $\Delta\dot{M}_{mu}$ is Change in Metabolic Rate, Muscle (kcal·h^{-1}). q_s'' is Heat Flux, Skin (kcal·m^{-2}·h^{-1}). T_h is Temperature, Hypothalamic; Heat Core (°C). T_0 is Temperature, Set Value (°C). T_s is Temperature, Skin (°C). $\Delta\dot{V}_{em}$ is Change in Flow Rate, Efferent Muscle (ℓ·h^{-1}·ℓ^{-1}). $\Delta\dot{V}_{es}$ is Change in Flow Rate, Efferent Skin Constriction (ℓ·h^{-1}·ℓ^{-1}). $\dot{V}_{mu}$ is Blood Flow Rate, Muscle (ℓ·h^{-1}·ℓ^{-1}). $\dot{V}_s$ is Blood Flow Rate, Skin (ℓ·h^{-1}·ℓ^{-1}). Adapted from Gordon, Roemer, and Horvath, 1976

One further reference is in order, for Bullard and Rapp (4) presented a model for water immersion that treated heat flow as analogous to current flow. The heat flow from core to skin is modified by two parallel resistors: one a variable resistor associated with peripheral vasoconstriction and the other a fixed resistor associated with the insulation provided by subcutaneous fat. Thus, the interested reader can consult these references for extension of the model work in which he may be interested or to design an experiment to check certain model features.

ADAPTATION TO COLD

A variety of mechanisms are activated by moderate body cooling, i.e., cooling in the range above hypothermia. Some of these mechanisms are listed in Table 1. The circulation, endocrine glands, CNS, and muscular system are all importantly involved. Certainly, work on animals has clearly demonstrated acclimation, acclimatization, and adaptation to cold. Nevertheless, if one lumps all such observed changes under the label "adaptation," clear demonstrations of cold adaptation in man are scarce. Some selected examples of adaptive responses are shown in Table 2 that involve a shift in threshold body temperatures for shivering or a gain (increased slope) in the metabolic response to lowered body temperature.

LeBlanc (22) prepared a list of observations pointing out the specificity of cold adaptation in relation to improved cold tolerance (Table 3). LeBlanc feels that the important environmental modifiers linked with cross adaptation are hypoxia or altitude exposure and physical conditioning or training. There are many adaptive changes possible as a result of chronic or repeated cold exposure and these changes may be either morphological or regulatory. A listing of each type appears in Table 4.

Table 1. Mechanisms Activated by Moderate Body Cooling

Stimulation of cutaneous thermal receptors
Cutaneous vasoconstriction
Increased secretion of epinephrine and norepinephrine
TSH secretion increased slightly

Semiconscious increase in motor activity
Shivering
Piloerection

Body curling and extremity protection
Increased clothing
Seek warm environment

Table 2. Examples of Acclimative or Adaptive Responses to Cold Exposure

Hypothermic (Insulative) - Shift in Threshold (Critical Temperature) for Shivering

Korean pearl divers	Hong, 1963, 1973
Repeated cold air exposure to $5^{\circ}C$ 5 to 7 times within 2 weeks	Brück, 1971
Four weeks of cold air exposure, $15^{\circ}C$, 8 hours/day, 32 days	Davis, 1961
Soldiers living in Arctic	LeBlanc, 1956

Metabolic - Shift in Gain of Metabolic Resonse to Lowered Body Temperature

Neonates	Brück, 1961
Norwegians, 6 weeks in mountains	Scholander, *et al.*, 1958

Table 3. Specificity of Cold Adaptation

Tolerance ≈ Type of Exposure ≈ Exposure Specificity

Modes: Physiological]
          Psychological] – Habituation
          Behavioral

Alleviated Cold Discomfort – CNS Effect, Sympathetic

Pain, Pressor, Shivering

Face or Extremity, Whole Body

Modifiers: Hypoxia, Physical training

[Adapted from LeBlanc (22)]

Table 4. Possible Cold Adaptive Changes

Morphological
     Body composition
     Body size and shape
     Tissue changes, e.g., capillarization

Regulatory
     Thresholds or set-points
     Control precision
     Rate of change
     Change in gain
     Change in excitation or excretion pattern
     Tolerable range

A problem in interpreting whether or not a morphological or regulatory change is adaptive hinges on the rigorousness of the experimental design. Careful control of complicating variables becomes mandatory (25). The many investigations of the physiological effects of cold exposure have revealed that various treatments, conditions, and gross biological differences affect results. Those found to influence the resting metabolic response to cold as measured by oxygen consumption appear in Table 5. The importance of these factors was reviewed previously by Buskirk, *et al.* (7) and Buskirk (5), and only a brief amplification of the effects of repeated cold exposure and physical conditioning will be treated in the next section.

A newborn infant increases heat production markedly with very slight reduction in mean skin temperature from a value considerably higher than that of an adult [Brück (3)]. Presumably, this marked response is associated with the presence of brown fat and its extraordinary metabolic capability. A plot of the relationship

Table 5. Influence of Various Treatments, Conditions, and Gross Biological Differences on the Resting Metabolic Response to Cold in Man

Repeated natural exposure	Age	Alcohol
Repeated chamber exposure	Sex	Season
Ethnic background	CO_2	Thyroid hormone
Physical conditioning	O_2	Diurnal variation
Body composition	Food	Hyperbaria
Prior exposure to heat		

found by Brück between mean skin temperature and heat production in the new-born (neonate) as compared with an adult is shown in Fig. 2. The lower critical mean skin temperature for initiation of the metabolic response in the adult is considerably below that in the newborn.

REPEATED COLD EXPOSURE, PHYSICAL CONDITIONING, AND OBESITY

The effects of physical conditioning on the physiological responses to cold and the tolerance of cold have been studied by several investigators, but not in an exhaustive way. The studies have been largely descriptive and confined to non-invasive measurements of metabolism and heat exchange. The results are also quite mixed.

Adams and Heberling, whose work was summarized by LeBlanc (22), physically conditioned a group of five young men for eight weeks and exposed them under semi-nude conditions to 10°C air for one hour. An average increase of 25% in heat production was found following conditioning. This work was supported by the experiments of Scholander *et al.* (34), who studied a group of six young Norwegian men who bivouaced and worked hard in the mountains for six weeks. They were studied during the night hours while nude in a chamber maintained at 20°C and also in a 2 Clo sleeping bag at 3°C. Under both experimental conditions, heat production was increased approximately 15% by the end of the bivouac.

Keatinge (18) exercised a group of 12 men for 17 days and compared their responses when they were exposed to 6°C air while wearing only pants, shoes, and socks. There was either no change or a reduction in heat production following the physical conditioning regimen. Eagan (12) was unable to demonstrate that the level of physical fitness had any effect on the resistance to finger cooling.

Brück (3) has stated that "in man a hypothermic type (tolerance adaptation) is found which is based on an adaptive deviation of the threshold for cold defense reactions." In addition, Brück postulated a positive cross adaptation between physical and cold stress so that marathon training produced a further lowering of the body temperature threshold for an elevation in heat production from that initially brought about by repeated cold exposure. In Brück's study, each of the subjects

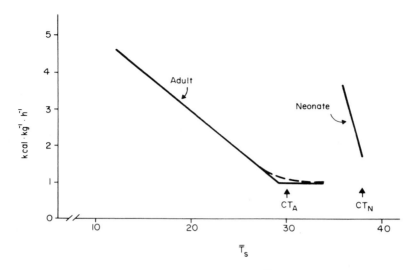

Fig. 2. Relationship between mean skin temperature ($\bar{T}_s$) and heat production in adults and newborn infants. [Redrawn from Brück (3), originally published in 1961.] The critical mean skin temperature (CT) for initiation of the metabolic response is indicated by an arrow.

(13 men and two women) were positioned supine and exposed in a chamber that was gradually cooled from 28°C to 5°C at a rate of 0.36°C to 0.55°C $\cdot$ min^{-1}. The cold exposures were repeated five to seven times within 14 days. Although the results on all 15 subjects were not consistent, nine of them showed a lowering of the body temperature threshold and a diminution of heat production. The lowered body temperature threshold involved both lowered core and skin temperatures.

Brück (3) postulated the possible mechanisms for the shift of thresholds for the elicitation of cold defense reactions to be as follows:

(1) Alteration in the function of cutaneous cold receptors
(2) Functional alteration of internal thermosensors
(3) Change in some nonthermosensitive central neurons – determinants of the thermal threshold curves.

The last possibility was favored as an explanation for the thermal threshold shift with repeated cold exposure and marathon training.

Our results (6,20) in young men who were physically conditioned, and in various groups of young men who differed in level of physical condition, indicated a relatively reduced heat production with conditioning (when exposed semi-nude for two hours to 10°C air) in conjunction with reduced skin and core temperatures and perhaps a reduced threshold. These results tend to agree with those of Keatinge (18) and Brück (3). The studies involved a physical conditioning regimen that lasted nine weeks and produced a significant increase in aerobic capacity and reduction in the time to run one mile. Interestingly, the young men displayed an increased temperature gradient from the underlying muscle surface to the skin surface for both the forearm and underarm after physical conditioning. The increased temperature gradient resulted from a lower skin temperature and a small elevation in muscle surface temperature. The implication is one of reduced blood flow in the skin and perhaps slightly increased flow in the underlying muscle.

The difference in the metabolic threshold among the groups that varied in level of physical fitness is shown in Fig. 3. The relation of relatively "steady state" mean heat production in relation to skin temperature for cold air exposures is shown. The lean unfit (LUF) subjects had a critical mean skin temperature ($\overline{T}_S$) for the metabolic response between 30°C and 28°C, whereas the lean fit (LF) subjects had a critical $\overline{T}_S$ of about 26°C. The large lean (LL) subjects had a $\overline{T}_S$ of about 25°C and these subjects were not "fit." Among the obese subjects, the critical $\overline{T}_S$ was also about 25°C. Although some of the obese subjects were quite active (OA), none could be regarded as fit as the lean fit subjects. The well-known decreased metabolic response to cold with an increase in body fatness is shown in Fig. 4 for these same subjects. In addition, the effects of large body size is also depicted, and even though these large men were not fat, their metabolic response to a two-hour cold air exposure is low. In contrast, the leaner subjects had a far greater metabolic response, with the lean unfit subjects tending to show the greatest response.

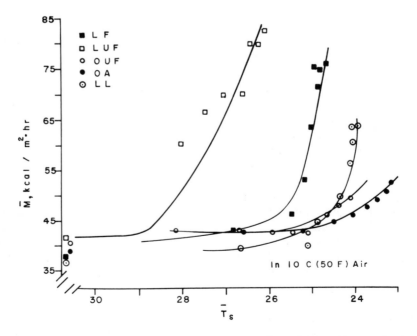

Fig. 3. Relationship between mean heat production ($\overline{M}$) and mean skin temperature ($\overline{T}_S$) for each group exposed to cool air 10°C (50°F). [From Buskirk and Kollias (6).]

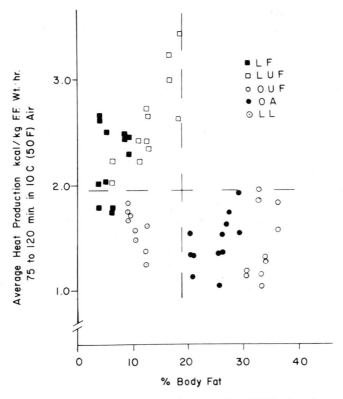

Fig. 4. Average heat production by groups between 75 and 120 min. of exposure to 10°C (50°F) air in relation to percent body fat content. Heat production is expressed per kilogram of fat-free body weight. [From Buskirk and Kollias (6).]

Similar experiments were conducted in cold water and a plot of calculated total body insulation in relation to the mean metabolic response for the various subjects is shown in Fig. 5. A curvilinear relationship is clearly shown, with the obese subjects showing the greatest total body insulation and the least metabolic response, and the lean subjects the greatest metabolic response. The large lean subjects gave intermediate results, presumably because of their large body size which provides a surface area to mass ratio that is less for them than for the smaller lean subjects.

A combined plot of the data of Buskirk and Kollias (6) for water exposures of from 30 min to one hour in 15°C or 20°C water, and those of Craig and Dvorak (9) in 24°C or 26°C water, shows that there is interaction between core and central drives for the metabolic response to cold in lean subjects (Fig. 6). Obtaining meaningful data of this type on obese subjects is most difficult, for their core temperatures do not decrease, or decrease only slightly with such exposures, even though their skin temperatures drop appreciably. In the absence of a reduced core temperature, the metabolic response in the obese is virtually absent, indicating a failure of the peripheral drive to operate without some lowering of the core temperature.

CARDIOVASCULAR RESPONSES TO COLD

That the fundamental relationships hold for cold exposure that were found during exercise for cardiac output, arteriovenous oxygen difference, and oxygen uptake was demonstrated by Raven *et al.* (32). Cardiac output increased directly with oxygen uptake during the two-hour exposures to 5°C air. In contrast to leg exercise, the increase in cardiac output was of lesser magnitude and was brought about largely by an increase in stroke volume and not by an increase in heart rate. Thus, both the increase in stroke volume and an enlarged arteriovenous oxygen difference accounted for the increase in oxygen uptake. The relative contributions of the respective variables are reported in Table 6. A rise in blood pressure indicated considerable peripheral vasoconstriction and probably an increased central blood volume which facilitated an increased stroke volume. Total peripheral resistance was reduced.

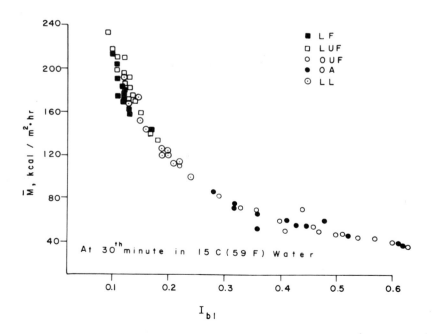

Fig. 5. Mean heat production ($\bar{M}$) at 30 min of exposure to 15°C (59°F) water in relation to total body insulation (I_{bl}). [From Buskirk and Kollias (6).]

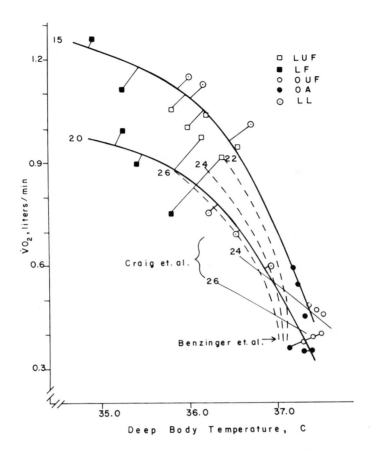

Fig. 6.   Comparison of the relationship of oxygen consumption ($\dot{V}O_2$) with deep body temperature in the study of Buskirk and Kollias (6), Benzinger *et al.* (2), and Craig and Dvorak (9).   Water temperatures in °C are indicated for the respective curves. Connecting lines are utilized to indicate which data points belong to which curve.

Table 6.   *Results of Near Nude Exposures of Young Men to 5°C Air for Two Hours (N = 11)   [From Raven et al. (32)]*

Variables	Control[a]	One Hour	% of Control	Two Hours	% of Control
$\dot{V}O_2$ (ml·min^{-1})	260	580	227	710	273
$\dot{Q}$ (ℓ·min^{-1})	4.93	8.53	173	9.66	196
HR (beats·min^{-1})	64	69	108	70	109
SV (ml)	77	124	161	138	179
C(a$-\bar{v}$)O$_2$ (ml·ℓ^{-1})	54	70	130	75	139
MBP(mmHg)	88	103	117	105	119

[a]Control at 28°C

In an extension of this work, Raven *et al.* (33) found that the increase in stroke volume was not linearly related to a decrease in ambient temperature, but by a triphasic curve that showed an increase in stroke volume over the ambient temperature range 28°C to 20°C, no change in stroke volume between 20°C and 10°C, and an increase in stroke volume from 10°C to 5°C. The environmental exposures were of two hours' duration and the two semi-nude young men involved were exposed to a series of ambient temperatures: 28, 25, 20, 15, 10, and 5°C. An explanation for the triphasic nature of the stroke volume—ambient temperature curve was not readily apparent, although the statement was made that the "cardiac output response was an integration of the depressed heart rate response and the increasing stroke output at these temperatures."

In regard to the central cardiovascular adjustments that result from cold exposure, Raven *et al.* (33) speculated that the increase in blood pressure mediated by norepinephrine release from the sympathetic nervous system provides baroreceptor and reflex vagal stimulation. The chronotropic response of the heart was decreased, and the increased oxygen demands associated with the muscular contractions of shivering were met by the increased stroke volume, cardiac output, and arteriovenous oxygen difference.

Breathing Cold Air

An interesting feature of exercise in the cold is the thermal protection of the upper airways during the performance of exercise such as jogging or cross-country skiing under cold conditions. Only a limited amount of work has been done in this area, but what has been done indicates that at -10°C air temperature and high ventilatory rates, pharyngeal temperatures as low as 25°C can be measured. Extrapolation to much colder temperatures (such as -40°C) indicates that pharyngeal temperatures might drop below 20°C. Even so, extremely cold air is warmed to values well above temperatures associated with tissue freezing before reaching the trachea or bronchi, which precludes the possibility of tissue freezing in the airways. The discomfort of nose breathing during hard exercise in the cold is well known, and oral breathing is substituted. To the best of my knowledge, upper airway freezing is relatively unknown under conditions of hard exercise in the cold, but the appearance of cold-induced pain in the nares at temperatures above those required to freeze tissue serves as a protective mechanism (37,17).

When the dewpoint moves within the upper airways, condensation occurs and the nose drips with excess condensate. If there is danger of facial frostbite, or the cold pain threshold of the upper airways is exceeded, provision of thermal protection through use of a face mask is helpful. Claremont (8) has recommended a "sheik-like" veil of terry cloth draped across the face, bridging the nose and fastened to the head cover at ear level. This partial covering provides a wind screen and thermal protection, but does not impede breathing.

Thermal Protection

No attempt will be made to discuss thoroughly the problem of thermal protection and clothing, but because many people are now jogging or cross-country skiing during the winter months, a brief comment was deemed appropriate about the relationship between exercise intensity and clothing requirements. Obviously, considerable thermal protection is required by the resting man exposed to cold until, at temperatures of 10°C and below, he can readily utilize the maximal insulation that can be practically provided (Fig. 7). Walking at three times the resting metabolic rate (or at 3 Mets) reduces the insulation requirements considerably, and running at a metabolic intensity of 15 Mets reduces the requirements to quite low values. The respective interrelationships are shown in Fig. 7.

Cold, wet, windy weather poses special problems because of the effective reduction in clothing insulation. In subjects who rode a cycle ergometer into a 15-kph wind, Pugh (29) found about a four-fold reduction in insulation of wetted clothing of the type commonly worn for outdoor walks in winter. Interestingly, the clothing insulation reduction when wet and worn by the non-exercising resting man was two-fold, while that for the exercising man was four-fold. Movement of the clothing and augmented convective cooling presumably accounted for the difference. In Pugh's study, it was also found that at work rates of 900 kg • m • min^{-1} and over

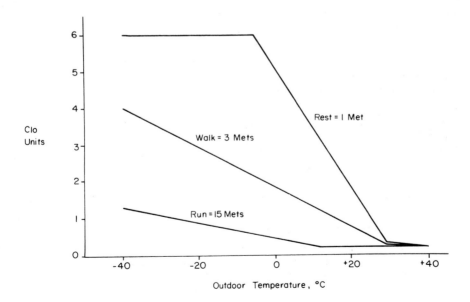

Fig. 7. Insulation requirements at different energy expenditures in the cold. Wind and rain decrease insulation. The Clo unit is defined as: $0 \cdot 18^{\circ}$C$\cdot$m^2 $\cdot$h$\cdot$kcal^{-1}.

there was no shivering or metabolic increase in response to cold, and core temperature was maintained. Shivering added to the metabolism at the lower work rates, but core temperature fell.

The dangers associated with excessive cooling are the debilitating features associated with hypothermia, i.e., progressive muscular fatigue, cessation of shivering, disorientation, etc. Thus, the winter outdoor recreationist is well advised to select his route and clothing carefully to minimize the consequences of cold injury.

Swimming in Cold Water

The thermal problems associated with diving as related to undersea activity have received considerable attention in the past few years, and this interest has also been reflected by the many investigations of man swimming at or near the surface in cold water. The invention of the swimming flume (1) has aided such research, and several of these water treadmills are now in service around the world. The heat conductance and specific heat of water are about 25 and 1000 times greater, respectively, than those in air. The heat transfer coefficient for cycling in air is of the order of $18W \cdot m^{-2} \cdot {}^{o}C^{-1}$ (26), and for swimming in water of $30^{o}C$ to $33^{o}C$, about $580W \cdot m^{-2} \cdot {}^{o}C^{-1}$ (24). Thus, a water environment can cool the body rapidly, and man's ability to protect himself against such cooling through peripheral vasoconstriction and increased heat production can be stressed beyond his capability. When man swims, the increased water turbulence and his movement through the water greatly increase convective cooling. The heat transfer coefficient data of Rapp (31) that were replotted by Webb (38) appear in Fig. 8. Although the conductive heat transfer coefficient remains relatively constant at about $11W \cdot m^{-2} \cdot {}^{o}C^{-1}$, the combined heat transfer coefficient that largely reflects the increase in convective heat transfer increases progressively with increased swimming velocity to about $400W \cdot m^{-2} \cdot {}^{o}C^{-1}$ at a velocity of one-half meter per second. Since the heat transfer coefficient during swimming in water is large, only small changes in skin blood flow are necessary to stabilize core temperature and heat balance in water that is not too cold. In colder water, heat loss can exceed the swimmer's capability to produce sufficient metabolic heat to preserve body core temperature.

In a recent experiment in Stockholm, Bergh et al. (1977, personal communication) found that 10 swimmers in the field of competitors swimming a 3.5-km race in $19^{o}C$ water had rectal temperatures at the end of the race of $35^{o}C$ and below. Five swimmers could not finish the race, but were helped out of the water with symptoms of hypothermia. The fall in core temperature was greatest in lean subjects who were slow swimmers. Thus, the reduction in core temperature was a function of both swimming speed (ability) and subcutaneous fatness — a fact previously emphasized by Pugh and Edholm (30), Craig and Dvorak (10), and Keatinge (19). Bergh and Ekblom (1977, personal communication) in a subsequent series of investigations found that reduced body temperature, i.e., below $36^{o}C$, was associated with significant reductions in aerobic capacity, maximal heart rate, and work time.

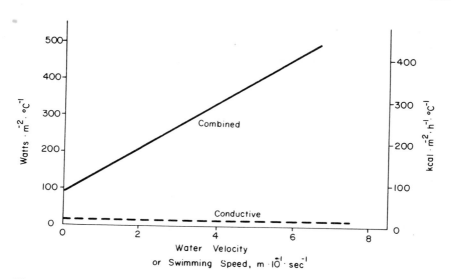

Fig. 8. Values for the conductive plus the combined convective and conductive heat transfer coefficient from skin to water with increased water velocity. [Data from Rapp (31).] A more comprehensive treatment has been presented by Webb (38).

Nadel et al. (24), in a partitional calorimetric analysis of heat loss during flume swimming, found that oxygen consumption was greater in $18°C$ than in $26°C$, and greater in $26°C$ than in $33°C$ water at any submaximal swimming speed. The increased metabolic cost of swimming was attributed to shivering. These results agree with those of Nielsen (27) on swimming man, but contrast to those of Pugh (29), who found no metabolic increment during cycling at work rates in excess of $900 \text{ kg} \cdot \text{m} \cdot \text{min}^{-1}$ in subjects exposed in wet clothing to wind and cold. Interestingly, Nadel et al. calculated the combined heat transfer coefficient to be $230\text{W} \cdot \text{m}^{-1} \cdot °\text{C}^{-1}$ for the subjects at rest in still water, $460\text{W} \cdot \text{m}^{-2} \cdot °\text{C}^{-1}$ when they were at rest in moving water, and $580\text{W} \cdot \text{m}^{-1} \cdot °\text{C}^{-1}$ while swimming regardless of the swimming speed. These heat transfer coefficient results are different from those plotted from the data of Rapp (31) and tend to confirm Webb's (38) statement that there is "considerable discrepancy between calculations, measurements taken on a physical model, and measurements on humans." Certainly the whole problem of accurately measuring heat storage, and hence heat flux, may account for some of the discrepancy.

The ratio of conductance with full cutaneous vasodilation compared with conductance in cold water with full vasoconstriction is of the order of 15:1, but this reduction is not uniform over the body for the head retains considerable cutaneous blood flow and hence warm temperatures (28). Loss of heat from the head, neck, and upper trunk is large and frequently underestimated in considerations of heat loss. The diving women in Japan and Korea wrap their head and neck in protective clothing, providing a practical solution to the problem.

The final paragraph in a paper by Bullard and Rapp (4) contains the still valid viewpoint: "As one looks at the literature in this field, it is disheartening to see the many small bits and pieces in which the information exists. Future impetus must be toward more complete studies in which wide ranges of water temperatures, varied metabolic rates, varied body compositions, and physical conditions are assimilated. More sophisticated physiological experimentation needs to be done on the regulation of shivering and conductance and the rate of countercurrent heat exchange.

SUMMARY

The literature on the physiology of cold exposure is vast and no attempt was made to prepare a thorough review. The temperature regulation model of Gordon *et al.* (13) is presented because it represents not only one of the latest attempts to model heat exchange in man exposed to cold, but also a sophisticated effort by the group hosting this honorary symposium. Some of the experiments associated with demonstration of an altered body temperature threshold for the metabolic response to cold were presented. It is conceivable that the temperature threshold does indeed shift to lower values with repeated cold exposure and improved physical condition. The increase in cardiac output found during two-hour exposures to cold air was largely accounted for by an increase in stroke volume. Breathing extremely cold air during jogging or cross-country skiing has little effect on the respiratory tract, pratically if some simple thermal protection is worn. Wetted clothing loses considerable insulation and places the wearer at risk for thermal injury. Swimming in cold water can lower core temperature drastically in lean, slow swimmers. In this brief compendium it should be readily apparent that Horvath and his colleagues have contributed measurably to our knowledge of man in the cold.

The experiments reported by the author and his collaborators in this manuscript were conducted following the recommendations for use of human subjects in research that appear in the Declaration of Helsinki and are endorsed by the American Physiological Society.

REFERENCES

1. Åstrand, P.O. and S. Englesson. A swimming flume. *J. Appl. Physiol.* 33:514, 1972.
2. Benzinger, T.H., C. Kitzinger, and A.W. Pratt. The human thermostat. In: *Temperature — its measurement and control in science and industry.* Vol. 3, Part 3. New York: Reinhold Publishing Corp., 1963, pp. 111-120.
3. Brück, K. Cold adaptation in man. In: *Regulation of depressed metabolism and thermogenesis.* Edited by L. Jansky and X.J. Musacchia. Springfield, Ill.: Charles C. Thomas Publisher, 1976, pp. 42-63.
4. Bullard, R.W. and G.M. Rapp. Problems of body heat loss in water immersion. *Aerosp. Med.* 41(11):1269-1277, 1970.

5. Buskirk, E.R. Variation in heat production during acute exposures of men and women to cold air or water. *Annals N.Y. Acad. Sci.* 134:733-742, 1966.

6. Buskirk, E.R. and J. Kollias. Total body metabolism in the cold. *N.J. Acad. Sci., The Bulletin, Special Symposium Issue,* pp. 17-25, March 1969.

7. Buskirk, E.R., R.H. Thompson, and G.D. Whedon. Metabolic response to cooling in the human: role of body composition and particularly body fat. In: *Temperature – its measurement and control in science and industry.* Vol. 3, Part 3. New York: Reinhold Publishing Corp., 1963, pp. 429-442.

8. Claremont, A.O. Taking winter in stride requires proper attire. *Phys. and Spts. Med.,* pp. 65-68, December 1976.

9. Craig, A.B., Jr. and M. Dvorak. Thermal regulation during water immersion. *J. Appl. Physiol.* 21:1577-1585, 1966.

10. Craig, A.B., Jr. and M. Dvorak. Thermal regulation of man exercising during water immersion. *J. Appl. Physiol.* 25:28-35, 1968.

11. Davis, T.R.A. Chamber cold acclimatization in man. *J. Appl. Physiol.* 16:1011-1015, 1961.

12. Eagan, C.J. Resistance to finger cooling related to physical fitness. *Nature (London)* 200:851-852, 1963.

13. Gordon, R.G., R. B. Roemer, and S.M. Horvath. A mathematical model of the human temperature regulatory system – transient cold exposure response. *IEEE Trans. Biomed. Eng.* 23(6):434-444, 1976.

14. Hardy, J.D. Models of temperature regulation – a review. In: *Essays on temperature regulation.* Edited by J. Bligh and R. Moore. New York: North Holland/Am. Elsevier, 1972, pp. 163-186.

15. Hong, S.K. Comparison of diving and nondiving women in Korea. *Fed. Proc.* 22:831-833, 1963.

16. Hong, S.K. Pattern of cold adaptation in women divers of Korea (Ama). *Fed. Proc.* 32:1614-1622, 1973.

17. Kaufman, W.C. Breathing cold air. (Letter to Editor) *Phys. and Spts. Med.* p. 7, May 1977.

18. Keatinge, W.R. The effect of repeated daily exposure to cold and of improved physical fitness on the metabolic and vascular responses to cold air. *J. Physiol.* 157:209-220, 1960.

19. Keatinge, W.R. *Survival in cold water: the physiology and treatment of immersion hypothermia and drowning.* Philadelphia: J.B. Lippincott Co., 1969.

20. Kollias, J., R. Boileau, and E.R. Buskirk. Effects of physical conditioning in man on thermal responses to cold air. *Int. J. Biometerol.* 16:389-402, 1972.

21. LeBlanc, J. Evidence and meaning of acclimatization to cold in man. *J. Appl. Physiol.* 9:395-398, 1956.

22. LeBlanc, J. *Man in the cold.* American Lectures and Environmental Series. Springfield, Ill.: Charles C. Thomas Publishing Co., 1975.

23. Mitchell, J.W. and C. E. Myers. An analytical model of the countercurrent heat exchange phenomena. *Biophys. J.* 8:897-911, 1968.

24. Nadel, E.R., I Holmer, U. Bergh, P.O. Åstrand, and J.A.J. Stolwijk. Energy exchanges of swimming man. *J. Appl. Physiol.* 36(4):465-471, 1974.

25. Nadel, E.R. and S.M. Horvath. Evaluation of method for investigating cold stress. Optimal evaluation of cold tolerance in man. *JIBP Synthesis* 1:89-117, 1975.

26. Nielsen, B. Thermoregulation in rest and exercise. *Acta Physiol. Scand. Suppl.* 323:7-74, 1969.

27. Nielsen, B. Metabolic reactions to cold during swimming at different speeds. *Arch. Sci. Physiol.* 27:207-211, 1973.

28. Nunneley, S.A., S.J. Troutman, and P. Webb. Head cooling in work and heat stress. *Aerosp. Med.* 42:64-68, 1971.

29. Pugh, L.G.C.E. Cold stress and muscular exercise, with special reference to accidental hypothermia. *Brit. Med. J.* 2:333-337, 1967.
30. Pugh, L.G.C.E. and O.G. Edholm. The physiology of channel swimmers. *Lancet* 269: 761-768, 1955.
31. Rapp, G.M. Convection coefficients of man in a forensic area of thermal physiology: heat transfer in underwater exercise. *J. Physiol. (Paris)* 63:392-396, 1971.
32. Raven, P.B., I. Niki, T.E. Dahms, and S.M. Horvath Compensatory cardiovascular responses during an environmental cold stress, 5°C. *J. Appl. Physiol.* 29(4):417-421, 1970.
33. Raven, P.B., J.E. Wilkerson, S.M. Horvath, and N.W. Bolduan. Thermal metabolic and cardiovascular responses to various degrees of cold stress. *Can. J. Physiol. Pharmacol.* 53:293-298, 1975.
34. Scholander, P.F., H.T. Hammel, K.L. Anderson, and Y. Løyning. Metabolic acclimation to cold in man. *J. Appl. Physiol.* 12:1-8, 1958.
35. Scholander, P.F., H.T. Hammel, J.S. Hart, D.H. LeMessurier, and J. Steen. Cold adaptation in Australian Aborigines. *J. Appl. Physiol.* 13:211-218, 1958.
36. Stolwijk, J.A.J. and J.D. Hardy. Temperature regulation in man – a theoretical study. *Pfluegers Arch. Ges. Physiol.* 291:129-162, 1966.
37. Webb, P. Air temperature in respiratory tracts of resting subjects in the cold. *J. Appl. Physiol.* 4:378-382, 1951.
38. Webb, P. Thermal stress in undersea activity. In: *Underwater physiology V.* Edited by C. J. Lambertson. Bethesda: Fed. Am. Soc. Exp. Biol., 1976, pp. 705, 724.
39. Wissler, E.H. A mathematical model of the human thermal system. *Bull. Math., Biophys.* 26:147-166, 1964.

AGE, SEX AND FITNESS,
AND THE RESPONSE TO LOCAL COOLING

J. LeBlanc
J. Côté
S. Dulac
F. Turcot

Department of Physiology
School of Medicine
Laval University
Quebec, Canada

The relationship of individual factors to cold stress is complex, because there are really two types of cold environment to which we are exposed. Prolonged exposure to moderate cold initially causes an increase in heat production through shivering. Subsequently, when adaptation is achieved, some species use non-shivering thermogenesis to increase heat production. In humans there is really no evidence for non-shivering thermogenesis, very likely because long exposure to moderate cold is not experienced by any group of humans (3). On the other hand, short exposures to severe cold are often experienced. Evidence for adaptation to this type of cold exposure is considerable; we have observed it in Eskimos, Gaspé fishermen, military personnel, mailmen, and medical students (3,4,5,6,7). The first thing that had to be decided was the type of testing to use. To replicate the natural conditions to which these groups are exposed, we measured cardiovascular responses to both hand and face stimulation by cold. The cold hand test consisted of placing the hand into cold water (5°C) for two minutes, during which time blood pressure and heart rate variations were measured. The cold face test also lasted two minutes, during which cold wind (0°C with 40-mph wind) was blown on the face of the subject. These two tests are about equally stressful and produce an activation of the sympathetic

nervous system. In addition, the face test causes bradycardia, which is due to a
vagal reflex action. In cold-adapted people, we have shown that while the activity
of the sympathetic system is decreased, that of the parasympathetic is enhanced (6).

   Another important consideration with relation to the response to cold is that of
individual variation. In a group of non-adapted subjects, the variations in the response to our tests were fairly large. Some subjects had a marked response to the
face test, but a normal response to the hand test; others responded normally to the
face test, but exhibited a high sensitivity to the hand test. In order to sort out these
responses and try to explain the variability observed, we examined the effect of age,
sex, and physical training on the responses to local cold stimulation.

   The importance of individual factors became evident when we evaluated the responses of a group of 27 male subjects aged 20 to 40 years. This group was composed of volunteers from the Department of Physical Education and included
students, professors, and staff. Fig. 1 shows a significant correlation between age
and the increase in systolic pressure after immersion of the hand for one minute into
cold water (8). Following this determination, $\dot{V}_{O_2}$ max was measured in all subjects,
revealing a negative correlation between the $\dot{V}_{O_2}$ max and the blood pressure response
to immersion of the hand into cold water for one minute (Fig. 2). In addition,
Fig. 3 shows a correlation between age and level of $\dot{V}_{O_2}$ max, the younger subjects
being more fit than the older. A partial correlation analysis was made on these

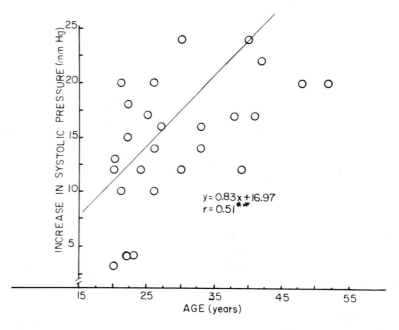

Fig. 1. Effect of age on the blood pressure response to a cold hand test in a group of subjects
from the Physical Education Department (8, reproduced with permission).

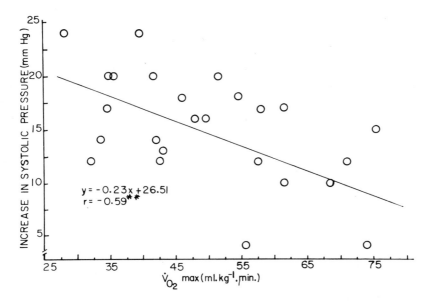

Fig. 2. Correlation between $\dot{V}_{O_2}$ max and the response to a cold hand test in the same subjects as for Fig. 1 (8, reproduced with permission).

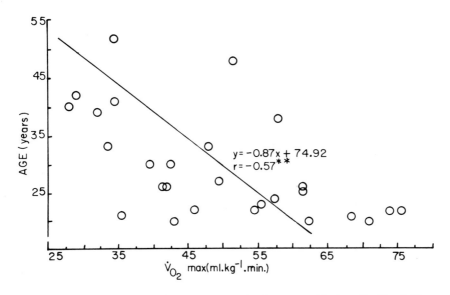

Fig. 3. Correlation between $\dot{V}_{O_2}$ max and age in the same subjects as for Fig. 1 (8, reproduced with permission).

results, which revealed that the level of training significantly affected the response to a cold hand test, whereas age had no real influence on this test. Fig. 4 shows responses of two groups of subjects to a cold pressor test (2 min). One group of 13 subjects all had a $\dot{V}O_2$max greater than 50 ml $\cdot$ kg^{-1} $\cdot$ min^{-1} (average 62.6 $\pm$ 1.7 ml $\cdot$ kg^{-1} $\cdot$ min^{-1}). The second group of 14 subjects all had a $\dot{V}O_2$max less than 50 ml $\cdot$ kg^{-1} $\cdot$ min^{-1} (average 38.7 $\pm$ 1.7 ml $\cdot$ kg^{-1} $\cdot$ min^{-1}). The decreased pressor response observed in the trained subjects is possibly related to a reduced cold vaso-constriction. Indeed, Adams and Heberling (1) have shown that the fall in skin temperature resulting from exposure to $10^{\circ}C$ is smaller in trained subjects. This was verified in the present study. Fig. 5 shows that the fall in cheek temperature when the face is exposed to a cold wind ($0^{\circ}C$ with 40-mph wind) is significantly reduced in the trained subjects. The lower blood pressure and higher skin tempera-ture would seem to indicate that the vasoconstriction caused by cold would be reduced in trained subjects. Similar findings were obtained on cold-adapted groups such as the Eskimos and the Gaspé fishermen (4,5,6). These changes suggest modi-fications in the response of the sympathetic nervous system to cold. It is not known whether or not this modification is central (due to a reduced overall activation of the sympathetic nervous system) or peripheral (due to some reduction in sensitivity of blood vessels to noradrenaline). A recent study would indirectly refute the second alternative (9). Indeed, Fig. 6 shows that the response to noradrenaline is at least as important in trained as it is in non-trained subjects.

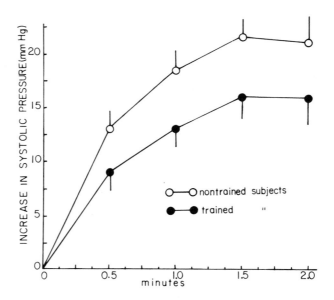

Fig. 4. Blood pressure response to a cold hand test in trained (average $\dot{V}_{O_2}$max: 62.6) and non-trained subjects (average $\dot{V}_{O_2}$max: 38.7) (8, reproduced with permission).

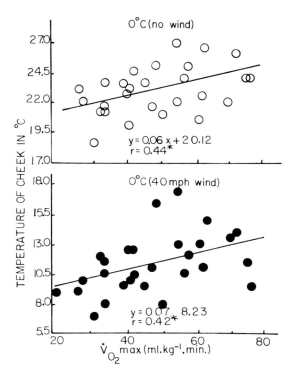

Fig. 5. Correlation between the fall in cheek skin temperature caused by cold air with and without wind and the $\dot{V}_{O_2}$ max of the subjects (8, reproduced with permission).

Another study of the response to cold stimulation was made on a younger population (33 subjects) with an average age of 14.5 (range 13 to 17 years). Four groups of subjects were used. Two groups (one composed of boys, the other of girls) were elite swimmers of the region of Quebec. The other two groups (one of boys, the other of girls) were non-trained subjects who served as controls. All subjects performed both the cold hand and the cold face tests. Results obtained with the cold hand test for systolic and diastolic pressure indicate significant differences between the trained and non-trained (Fig. 7). In terms of responses of the autonomic nervous system, these results suggest that the sympathetic is more active in trained subjects stimulated by cold. To our surprise, the results of this study on a young population were completely opposite to those obtained on an adult population. Since an effect related to a difference in protocol can be eliminated, there are two other factors that may be relevant: the age of the subjects and the type of training. The young subjects were short distance swimmers and their training was more of the resistance type. The average $\dot{V}_{O_2}$ max for these young trained subjects was 51.4 ± 1.9 compared to 35.1 ± 2.6 for the young non-trained subjects. In the older group, the

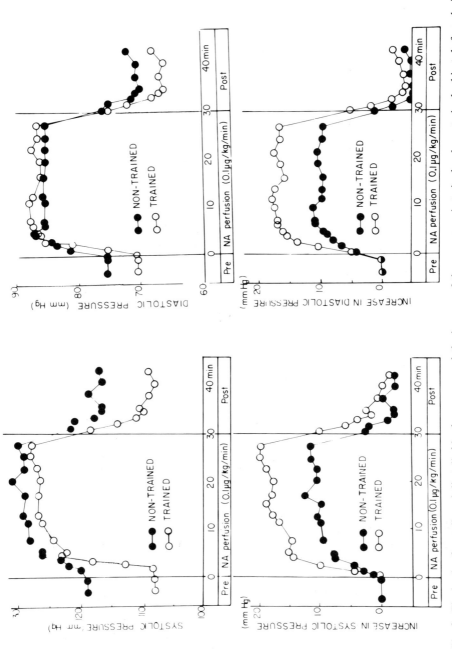

Fig. 6. The levels of systolic and diastolic blood pressures, and the increase of these pressures in trained and non-trained subjects before, during, and after noradrenaline perfusion (9).

trained subjects had a $\dot{V}O_2$ max of 62.6 ± 2.3 compared to 38.6 ± 1.7 in the non-trained subjects. The training in this case was of the endurance type; the subjects were marathon runners, cross-country skiers, and cyclists. Because of the difference in the type of sport activity between the younger and the older athletes, the influence of personality profile cannot be completely disregarded. The other factor that should be discussed in that respect is the age of the subjects. In Fig. 8 we have illustrated together (for comparison) the cold pressor responses of the trained and non-trained subjects of both age groups. The young non-trained group had a smaller increase in systolic pressure than the older non-trained group. On the other hand, training reduces the response in the older groups and increases it in the young group, so that the final result is a comparable response to cold in both these trained groups. We have shown that young Eskimos (6) have a very low response to cold stimulation, comparable to that of their parents. We expressed some surprise at the time these results were obtained, since the young Eskimos lived in heated houses and attended heated schools. The present results on the young swimmers suggest that the age factor may have been important in the young Eskimos as well.

In view of the differences observed between adults and subjects at the onset of puberty, we decided to investigate the response to cold stimulation in male subjects (ages 55 to 60 years) who also reported a decline in sexual activity. Figs. 9 and 10 compare the heart rate responses to a cold hand and a cold face test in subjects aged 20 to 40 years and subjects aged 53 to 60 years. The response to the cold hand test is significantly reduced in the older group of subjects, whereas that of the cold face test is more pronounced. Many functions and responses are affected by age.

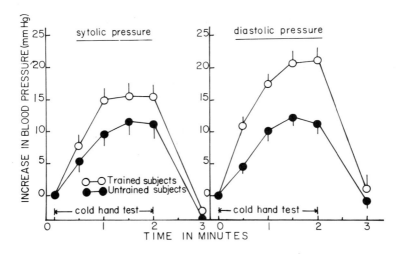

Fig. 7. Effect of a cold hand test on systolic and diastolic blood pressure increase in trained and non-trained young subjects (8, reproduced with permission).

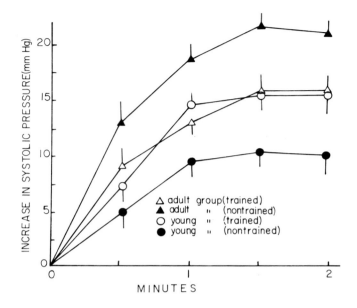

Fig. 8. Effect of a cold hand test on blood pressure of trained and non-trained subjects in groups of young and adult subjects (8, reproduced with permission).

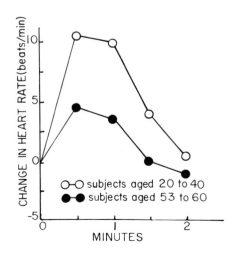

Fig. 9. Effect of a cold hand test on heart rate of young and older adults (8, reproduced with permission).

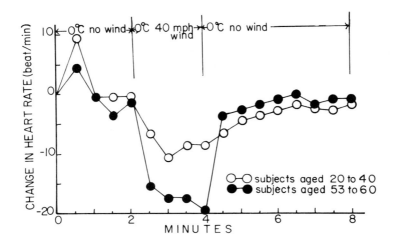

Fig. 10. Effect of a cold face test on heart rate of young and older adults (8, reproduced with permission).

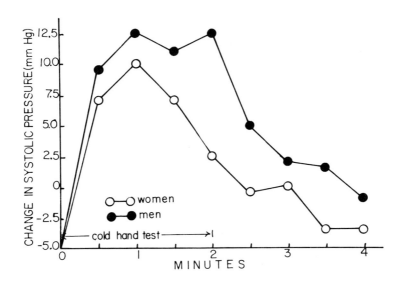

Fig. 11. Effect of a cold hand test on systolic blood pressure of adult men and women (8, reproduced with permission).

For instance, the increased adenyl cyclase activity, normally produced by noradrenaline, remains relatively unchanged in older animals (2). More studies are
needed before we can explain the effect of age as well as the effect of training on
the response to cold stimulation.

To examine the effects of sex differences, a study was performed with two
groups of adult subjects, one male and the other female. At the end of the cold
water test, the female group showed a significantly lower increase in systolic blood
pressure, compared to the male group (Figs. 11 and 12). The important individual
factors in response to cold are adaptation and training, whereas age and sex are
relatively less important. These factors, however, cannot explain all individual
variations. It is suggested that personality profiles or other inherited characteristics
could help explain the large individual variation in the response to stressful conditions.

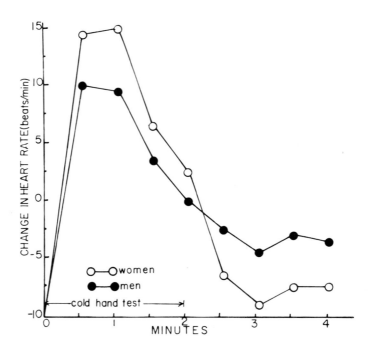

Fig. 12. Effect of a cold hand test on heart rate in adult men and women (8, reproduced
with permission).

REFERENCES

1. Adams, T. and E. Heberling. Effect of training on response to cold. *J. Appl. Physiol.* 13:226-230, 1958.
2. Ericsson, E. and L. Lundholm. Adrenergic beta-receptor activity and cyclic AMP metabolism: variation with age. *Mechanisms of Ageing and Development* 4:1-6, 1975.
3. LeBlanc, J. *Man in the cold.* Springfield, Ill.:Charles C. Thomas, 1975.
4. LeBlanc, J. Local adaptation to cold of Gaspé fishermen. *J. Appl. Physiol.* 17:950-954, 1962.
5. LeBlanc, J., J.A. Hildes, and O. Héroux. Tolerance of Gaspé fishermen to cold water. *J. Appl. Physiol.* 15:1031-1035, 1960.
6. LeBlanc, J., S. Dulac, J. Côté, and B. Girard. Autonomic nervous sytem and adaptation to cold in man. *J. Appl. Physiol.* 39:181-187, 1975.
7. LeBlanc, J. and P. Potvin. Studies on habituation to cold in man. *Can. J. Physiol. Biochem.* 44:287-294, 1966.
8. LeBlanc, J., J. Côté, S. Dulac, and F. Turcot. Physical fitness and response to cold. *J. Appl. Physiol.* (May 1978, in press).
9. LeBlanc, J., M. Boulay, S. Dulac, M. Jobin, A. Labrie, and S. Rousseau-Migneron. Metabolic and cardiovascular responses to norepinephrine in trained and non-trained human subjects. *J. Appl. Physiol.* 43:166-173, 1977.

SEASONAL VARIATION OF AEROBIC WORK CAPACITY IN AMBIENT AND CONSTANT TEMPERATURE

Hideji Matsui
Kiyoshi Shimaoka
Miharu Miyamura
Kando Kobayashi

Research Center of Health, Physical Fitness and Sports
Nagoya University
Nagoya, Japan

The maximum oxygen uptake is widely accepted as the best measure for evaluating an individual's aerobic work capacity, since performance capacity depends mainly on the ability to take up, transfer, and deliver oxygen to working muscles. In 1923 Hill and Lupton (8) reported a plateau in oxygen uptake during runs at increasing speeds, suggesting that maximum oxygen uptake had been reached and could not exceed this level. Subsequent experimental reports indicate, however, that maximum oxygen uptake may be affected by various factors such as type of exercise, age, sex, and training status of the individual (3,20). Conversely, environmental temperatures do not appear to substantially influence maximum oxygen uptake during acute exposure to hot and/or cold temperatures (2,14,18,21,25).

Most European and American investigators have failed to find seasonal variation in the basal metabolic rate in man, while Japanese and Korean investigators have unanimously reported the presence of a seasonal variation of basal metabolism (13, 22,26). There is, however, no available data concerning the effect of season on maximum oxygen uptake except that reported by Nagasawa and Watanabe (15). They observed no seasonal variation in maximum oxygen uptake, although performance of a 5-min run in the winter season was superior to that in summer. The present study was undertaken to (a) ascertain if seasonal variation of aerobic work

279

capacity does occur as indicated by changes in oxygen uptake, and (b) compare the cardiovascular responses during maximal exercise to cold (C), moderate (M), and hot (H) environmental temperatures.

METHODS

Nineteen healthy males, aged 18 to 36 years, from the university population, were used as subjects. These individuals were divided into two groups. The first group was used to determine the seasonal variation of maximum oxygen uptake over one year, and the second group was studied to compare the maximum oxygen uptake in three different environmental temperatures.

Seasonal variation of maximum oxygen uptake. The subjects consisted of six male students aged 18 to 29 years, and seven laboratory members aged 27 to 36 years. Their anthropometric data are shown in Table 1. The subjects performed maximal exercise twice each month for one year, once at ambient and once at a constant temperature. The maximum oxygen uptake was determined first at ambient room temperature (subject to external temperature changes – Fig. 1) and one week later at constant temperature and humidity ($18 \pm 0.5^\circ C$ db, $60 \pm 3\%$) in a climatic chamber.

All subjects were given a preliminary test on a motor-driven treadmill in order to accustom them to the procedures and to determine the initial work load (treadmill speed) for each subject. The experiments were always done between 1:00 p.m. and 5:00 p.m. at least one hour after the last meal. The exercise was carried out using stepwise incremental loading on a treadmill with a constant grade of 8.6%. The speed was increased by 10 m/min once every minute to exhaustion. The initial speed was

Table 1. Physical Data of Subjects

Subject	Age	Weight	Height
Y.H.	18	48.0	163.0
Y.Y.	19	54.0	172.0
K.M.	19	52.0	165.0
Y.I.	18	52.0	168.8
S.T.	18	60.5	166.0
T.B.	19	60.0	165.5
K.K.	27	69.0	175.5
K.S.	29	56.5	163.5
H.S.	32	70.0	167.0
N.F.	32	70.5	168.5
I.O.	33	57.5	167.0
M.M.	33	54.5	165.0
S.M.	36	56.0	172.0

chosen so that the subjects could run for 5 to 9 min before they were exhausted. In most cases, the initial speed was found to be 120 to 140 m/min.

Each individual was weighed prior to testing. After electrocardiographic leads had been placed, the subjects rested for 30 min to become acclimatized to the room temperature. They then warmed up for 4 min at a speed of 100 to 110 m/min on the treadmill at an 8.6% grade. This warm-up was followed by a 2-min rest period; thereafter a maximal exercise test was conducted.

Oxygen uptake during exercise was determined by the Douglas bag technique. Expired gas was collected into a Douglas bag every minute until exhaustion. The collected gas volume was measured with a wet-gasometer, and gas analysis was performed in duplicate by Scholander micro-gas analyzer. The diameter of the respiratory valve and connecting tube were 30 and 33 mm, respectively. Heart rate was recorded continuously on a multichannel recorder. The ambient room temperature was determined using an Assman thermometer.

Maximum oxygen uptake at different environmental temperatures Subjects were six male students whose age, height, and weight were 18.7 ± 0.5 yr, 172.0 ± 3.6 cm, and 56.6 ± 7.7 kg, respectively (mean $\pm$ S.D.). The room temperature of the climatic chamber was maintained at either $5 \pm 1.0°C$ db, $18 \pm 0.5°C$ db, or $35 \pm 2.0°C$ db. The relative humidity was maintained at $60 \pm 3\%$ at moderate and hot temperature, but it ranged from 80% to 90% at the cold temperature, owing to the limitation of the humidity control in the climatic chamber. Under these three conditions, the maximum oxygen uptake was determined for the six subjects using the same techniques as described above. Each subject performed only a single test each day, and one week elapsed between each of the three experimental tests. In this series of experiments, additional data were collected, including calf blood flow, hematocrit, hemoglobin concentration, blood water volume, lactate, glucose, non-esterified fatty acids (FFA).

Calf blood flow in recumbent subjects was measured by the use of venous occlusion plethysmography before and immediately after the maximal exercise. A mercury-in-rubber strain-gauge, under a tension of 25 g, was wound lightly around the calf of the left limb at the point of maximal girth. Plaster adhesive was used to prevent slipping along the limb during exercise. In addition, a 24-cm wide pneumatic cuff was placed around the ankle. To measure calf blood flow, the ankle cuff was inflated to 220 to 240 mmHg 30 s prior to venous occlusion to exclude circulation to the foot. The venous pressure cuff was inflated to 90 to 100 mmHg during blood flow determinations. The calf volume changes were recorded four times at rest, and continuously for 5 min, at 1-min intervals, beginning immediately after exercise. Calf blood flow was calculated as described by Whitney (24). A 6-ml blood sample was drawn from the antecubital vein at rest prior to exercise and again immediately after exercise. Hemoglobin concentration was determined by a cyanmethemoglobin method. Plasma lactate and glucose concentrations were estimated by Hohorst's (9) and Froesch and Renold's (6) methods, respectively. FFA in the blood was ascertained by a colorimetric micro-method described by Laurell and Tibbling (12).

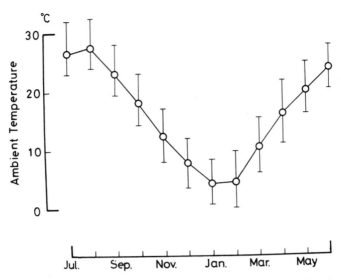

Fig. 1. Monthly change of mean ambient temperature at the Nagoya City from July 1976 to June 1977.

RESULTS

Fig. 1 indicates the monthly change of ambient temperature around the City of Nagoya in 1976–77. Although the mean ambient temperature in that summer and winter were about 25°C db and 3°C db, the temperature and humidity of the experimental room throughout the year varied between 8.6°C db and 32.2°C db, and 40.0% and 92.0%, respectively.

Individual values for maximum oxygen uptake ($\dot{V}O_2$ max) and maximum oxygen uptake per kg of body weight ($\dot{V}O_2$ max/W) in 13 subjects at ambient and constant room temperature are presented in Tables 2 and 3. $\dot{V}O_2$ max ranged from 2.36 to 4.8 ℓ/min at ambient temperature, and from 2.29 to 4.25 ℓ/min at constant temperature. The annual mean and standard deviation of $\dot{V}O_2$ max for each individual subject ranged from 2.52 ± 0.07 ℓ/min to 4.06 ± 0.11 ℓ/min at ambient temperature and from 2.57 ± 0.13 ℓ/min to 4.10 ± 0.10 ℓ/min at constant temperature. The mean and standard deviations of $\dot{V}O_2$ max, $\dot{V}O_2$ max/W, maximal pulmonary ventilation ($\dot{V}$Emax) and maximal heart rate (HRmax) are plotted in Fig. 2. It was found that HRmax was higher in the summer season (Fig. 2) than in winter, particularly at ambient temperature. Although there is no significant seasonal variation in mean $\dot{V}O_2$ max, it was observed that the mean $\dot{V}O_2$ max was slightly higher at constant room than at ambient temperature. It was almost the same in September, April, and May. In order to determine the peak $\dot{V}O_2$ max of each subject, the maximum oxygen uptake was expressed by a vector; each arrow in Fig. 3 shows the vector of

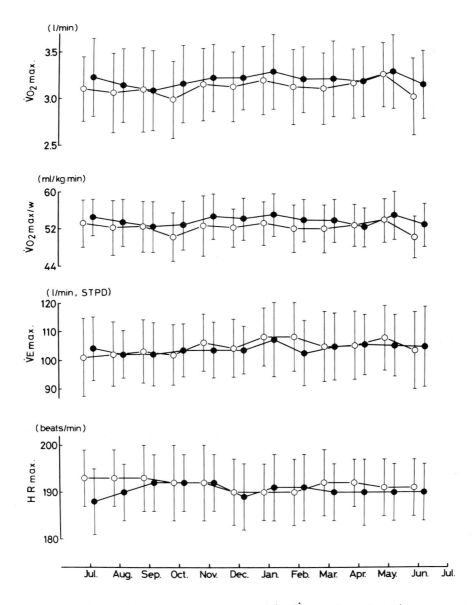

Fig. 2.   Change of mean maximum oxygen uptake ($\dot{V}_{O_2}$max), mean maximum oxygen uptake per kg of body weight ($\dot{V}_{O_2}$max/W), mean maximal pulmonary ventilation ($\dot{V}_E$max), and mean maximal heart rate (HR_{max}) during maximal exercise at both ambient (○) and constant (●) temperature. Ordinate lines represent the standard deviations.

Table 2. Individual Values of Maximum Oxygen Uptake (ℓ/Min, Upper Value in the Column) and Maximum Oxygen Uptake Per Kilogram of Body Weight ($Ml \cdot Kg^{-1} \cdot Min^{-1}$, Lower Value in the Column) as Determined at Ambient Temperature

Subject	July	Aug.	Sept.	Oct.	Nov.	Dec.	Jan.	Feb.	March	April	May	June
Y.H.	2.64	2.45	2.41	2.36	2.68	2.65	2.72	2.74	2.63	2.60	2.78	2.39
	55.1	52.2	50.1	49.1	55.2	52.9	53.9	53.2	51.6	50.5	53.9	46.3
Y.Y	3.19	3.10	3.07	3.02	3.12	3.07	3.19	3.05	3.04	3.25	3.21	2.92
	59.0	57.9	56.8	57.0	56.8	56.9	59.0	55.9	56.8	60.7	59.4	54.1
K.M.	3.12	3.01	3.11	2.99	3.08	2.85	3.06	2.82	2.95	2.93	3.29	2.98
	60.0	58.4	59.2	55.8	55.1	55.3	57.7	54.2	55.2	55.4	59.8	54.1
Y.I.	2.67	2.60	2.44	2.47	2.62	2.78	2.70	2.77	2.84	2.82	3.08	2.65
	51.4	48.5	46.5	46.5	49.1	50.5	49.0	50.9	50.3	49.9	54.5	47.3
S.T.	2.96	2.78	2.72	2.79	2.98	3.07	2.99	2.93	2.60	3.02	3.14	2.73
	48.9	46.0	45.4	45.7	48.0	50.3	48.6	48.9	43.0	49.9	51.8	46.3
T.B.	3.39	3.14	3.34	3.23	3.43	3.16	3.46	3.30	3.59	3.35	3.56	3.31
	55.2	51.6	53.9	51.2	54.5	50.5	54.6	52.8	56.1	51.9	54.8	50.6
K.K.	3.84	4.14	4.10	3.98	4.17	4.18	4.00	4.17	3.88	4.13	4.09	4.08
	55.7	60.0	59.4	57.7	59.5	59.3	57.6	58.8	55.1	57.8	56.7	56.7
K.S.	3.37	3.43	3.58	3.22	3.30	3.22	3.34	3.22	3.39	3.31	3.32	3.27
	59.7	60.2	61.1	55.5	56.9	55.5	58.1	56.4	59.5	58.1	58.2	56.7
H.S.	3.37	3.32	3.34	3.37	3.37	3.39	3.48	3.39	3.22	3.38	3.33	3.20
	48.1	46.7	46.7	46.9	46.3	46.8	47.9	46.4	44.7	46.6	45.4	44.3
N.F.	3.04	2.85	3.27	2.91	3.08	3.01	3.22	2.96	3.02	3.21	3.33	3.11
	43.1	40.7	46.7	41.5	44.3	44.0	45.4	41.9	44.4	46.9	48.0	45.7
I.O.	3.09	3.12	3.13	2.95	3.10	3.17	3.07	3.18	3.19	3.02	3.11	2.98
	53.8	53.3	53.1	50.0	53.0	53.3	51.3	53.0	53.2	51.2	52.2	49.2
M.M.	2.51	2.54	2.51	2.39	2.61	2.59	2.55	2.41	2.50	2.62	2.55	2.42
	46.0	46.5	45.7	42.7	46.7	46.3	45.5	43.0	44.6	46.7	45.5	42.8
S.M.	3.08	3.17	3.04	2.89	3.18	3.14	3.39	3.32	3.14	3.14	3.27	3.15
	55.0	55.6	54.8	49.8	54.3	53.6	57.4	56.7	54.6	55.1	56.9	54.3

Table 2. Individual Values of Maximum Oxygen Uptake (ℓ/Min, Upper Value in the Column) and Maximum Oxygen Uptake Per Kilogram of Body Weight (Ml $\cdot$ Kg^{-1} $\cdot$ Min^{-1}, Lower Value in the Column) as Determined at Constant Temperature

Subject	July	Aug.	Sept.	Oct.	Nov.	Dec.	Jan.	Feb.	March	April	May	June
Y.H.	–	2.55	2.29	2.47	2.62	2.81	2.65	2.65	2.53	2.52	2.71	2.50
	–	52.2	48.2	50.4	54.0	56.3	53.1	53.0	49.5	49.4	52.1	50.0
Y.Y	3.13	2.91	3.17	3.12	3.18	3.18	3.31	3.24	3.29	–	3.30	3.13
	58.4	55.4	58.7	57.8	59.9	58.8	61.9	59.4	60.9	–	61.0	57.4
K.M.	3.14	3.12	3.07	–	3.17	3.09	3.02	2.96	3.01	3.16	3.28	3.20
	59.3	60.6	58.5	–	57.6	58.3	57.5	56.9	56.8	58.5	60.8	58.2
Y.I.	2.70	–	2.47	2.65	2.78	2.92	2.90	2.83	3.00	2.76	2.94	2.99
	50.9	–	47.5	50.0	52.0	53.1	52.8	52.0	53.6	48.9	53.4	53.8
S.T.	3.06	2.96	2.86	2.89	2.92	3.11	3.00	3.12	3.03	3.03	3.24	2.93
	49.7	49.3	47.7	47.8	47.9	50.6	49.6	52.0	50.5	50.1	54.0	49.7
T.B.	3.40	3.19	3.26	3.40	3.33	3.51	3.45	3.36	3.31	3.45	3.45	3.39
	56.6	51.0	52.6	54.3	54.6	55.7	54.8	54.1	52.1	53.5	53.5	52.1
K.K.	4.25	4.00	4.07	4.05	4.15	4.07	4.24	3.94	4.20	4.02	4.22	4.01
	61.6	59.3	60.3	58.8	60.6	58.1	61.0	55.9	59.5	56.2	59.4	56.6
K.S.	3.40	3.37	3.32	3.48	3.55	3.44	3.38	3.42	3.45	3.37	3.39	3.36
	60.2	58.6	57.7	61.0	61.8	59.2	59.2	61.1	60.0	57.6	59.5	57.8
H.S.	3.52	3.40	3.29	3.42	3.37	3.33	3.60	3.43	3.48	3.35	3.58	3.27
	50.3	48.0	46.9	47.5	46.5	46.6	50.8	47.0	48.3	46.2	49.1	45.5
N.F.	3.12	3.29	3.20	3.04	3.25	3.07	3.34	3.30	2.95	3.27	3.15	3.18
	44.6	46.7	45.0	44.0	47.1	45.2	47.4	47.8	43.7	47.4	46.0	46.4
I.O.	3.12	3.06	3.21	3.17	3.16	3.13	3.34	3.20	3.17	3.17	3.24	3.07
	53.9	52.7	54.8	54.1	54.0	53.0	55.6	54.2	53.3	53.7	54.5	51.6
M.M.	2.59	2.45	2.47	2.66	2.83	2.64	2.69	2.54	2.62	2.67	2.53	2.52
	47.4	44.6	44.9	47.5	50.9	47.1	48.0	45.0	46.8	46.9	45.1	45.0
S.M.	3.20	3.26	3.11	3.29	3.32	3.28	3.47	3.34	3.43	3.17	3.50	3.31
	57.2	58.1	56.1	57.2	57.2	56.5	59.4	57.6	59.1	55.1	60.8	58.1

each subject obtained at both ambient and constant temperature. There are no vectors for four subjects at constant temperature because their $\dot{V}O_2$max could not be determined for health reasons. It is shown in this figure that, in 9 of 13 subjects at ambient temperature and in all subjects at constant temperature, the direction of the $\dot{V}O_2$max vector is located in the cold season (i.e. $\dot{V}O_2$max tended to be higher in winter, especially at constant room temperature).

Table 4 summarizes the average values and standard deviations for treadmill endurance time (T), maximum oxygen uptake ($\dot{V}O_2$max), maximum oxygen uptake per kg of body weight ($\dot{V}O_2$max/W), maximal pulmonary ventilation ($\dot{V}E$max) and maximal heart rate (HRmax) during maximal exercise at cold, moderate, and hot temperatures in the climatic chamber. Calf blood flow (CBF), hematocrit (Hct), hemoglobin concentration (Hb), blood water volume (WV), and plasma glucose (Glu), lactate (LA), and non-esterified fatty acid (FFA) obtained before and immediately after exercise are also listed in Table 4.

There are no statistically significant differences in $\dot{V}O_2$max, $\dot{V}O_2$max/W, treadmill endurance time, and $\dot{V}E$max at the three different temperatures. However, HRmax was significantly higher in the hot temperature than in the cold ($p < 0.001$. Although not statistically significant, a slightly higher calf blood flow at rest was observed in the hot temperature. No significant differences in resting or post-exercise measurements of CBF, Hct, Hb, WV, glucose, and FFA were found between the hot and cold temperatures. However, post-exercise lactate concentrations were significantly higher following the cold environmental condition ($p < 0.05$).

DISCUSSION

In competitive sports, performance capacity may be determined objectively as a function of distance or time, or it may be judged subjectively as in gymnastics and dancing. The individual's performance is considered to to be the combined result of various factors, and it is impossible to explain all factors by a single formula as described by Åstrand and Rodahl (3). It is interesting, however, that the individual maximal performance capacity changes from time to time when performing the same task.

As stated previously, maximum oxygen uptake is the best measure to evaluate individual aerobic work capacity. However, maximum oxygen uptake may vary, depending on the measurement technique employed. For the measurement of $\dot{V}O_2$max, two types of ergometers are generally used to assess aerobic power, the treadmill and the bicycle. In the present study, $\dot{V}O_2$max was determined using maximal treadmill exercise because $\dot{V}O_2$max measured on a bicycle ergometer is approximately 6% to 11% less (3,20). It was found that in all subjects, the maximal heart rate and respiratory quotient determined at $\dot{V}O_2$max exceeded 180 bpm and 1.14, respectively (4,11). We concluded, therefore, that each of the subjects had reached his maximum oxygen uptake.

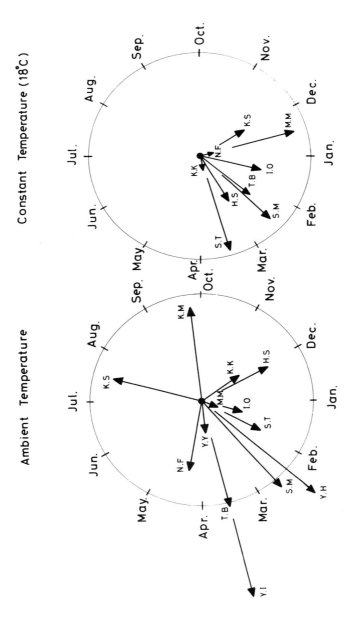

Fig. 3. Maximum oxygen uptake of each subject as expressed by vector that was obtained at ambient (left) and constant (right) temperature.

It has been observed that there is a very small day-to-day variability in $\dot{V}_{O_2}$max and that it is either unaffected or reduced only a few percent by acute exposure to either cold or hot temperature (20). Although Pirnay *et al.* (17) observed a 25% reduction of $\dot{V}_{O_2}$max in hot environment (Ta = 46°C, Twb = 35°C), we found no differences in $\dot{V}_{O_2}$max between cold, moderate, or hot room temperature (Table 4). These results are in agreement with the data of Mostordi *et al.* (24), Saltin *et al.* (21), Rowell *et al.* (18), Adams *et al.* (2), and Williams *et al.* (8), and suggest that, at least under our experimental conditions, the acute change of room temperature does not have an effect on $\dot{V}_{O_2}$max. However, the heart rate during maximal exercise was significantly higher under the hot condition than under the cold condition (Fig. 2 and Table 4). These results are supported by the data of Pandolf *et al.* (16), Strydom (23), Consolazio *et al.* (5), and Williams *et al.* (25), but not by Rowell *et al.* (19).

As described above, Yoshimura *et al.* (26) and Lee *et al.* (13) have observed a seasonal variation in basal metabolism, increasing in winter and decreasing in summer. They suggested that the seasonal change of basal metabolism is due to an adaptive change of metabolism to change of climate. However, Nagasawa and Watanabe (15) have reported no seasonal variation in $\dot{V}_{O_2}$max and $\dot{V}_{O_2}$max/W. In the present study, the mean $\dot{V}_{O_2}$max was higher at a constant room temperature than at ambient temperature, but was almost the same in September, April, and May. The difference in $\dot{V}_{O_2}$max between ambient and constant temperature may be due to the various factors such as cardiac output, muscle blood flow, and adaptability for environmental temperature. The lack of a difference in $\dot{V}_{O_2}$max in April, May, and September may have reflected the fact that the difference between the ambient temperature and the conditions in the climatic chamber were insignificant. However, the higher $\dot{V}_{O_2}$max at constant temperature at other times may also be a result of the order of measurement. We determined $\dot{V}_{O_2}$max first at ambient temperature and one week later at constant temperature, as described in the procedure. The possibility of a training effect needs further investigation.

Some authors (1,7) have reported that an increased level of physical fitness is accompanied by physiological adjustments very similar to those attributed to adaptation to cold. Hong *et al.* (10) found the $\dot{V}_{O_2}$max was increased in Ama divers during the winter season when the daily work, although more intense, is shorter in duration due to the low ambient temperature. Furthermore, it is well known that performance in marathon running is much better under cool ambient conditions. Another finding in this study is that the $\dot{V}_{O_2}$max of each subject (except K.S. and K.M.) tends to peak in the winter season (Fig. 3), although a clear seasonal cycle is not apparent. We are presently unable to explain, on a physiological basis, why the $\dot{V}_{O_2}$max of these two subjects was higher in the hot season. However, in the case of subject K.S., his treadmill endurance time was always 7 min, at both ambient and constant temperature, throughout the year.

Sasaki (22) found that the seasonal variation of basal metabolism decreased progressively with age, and that the rate of energy production from carbohydrate

Table 4. Average value and standard deviations of
physiological parameters in maximal treadmill exercise
at cold, moderate, and hot temperature

		Room Temperature		
		Low (5°C db)	Moderate (18°C db)	Hot (35°C db)
T		6'56" ± 55"	7'08" ± 58"	7'04" ± 51"
$\dot{V}_{O_2}$max		3.21 ± 0.48	3.21 ± 0.46	3.15 ± 0.38
$\dot{V}_{O_2}$max/kg		57.2 ± 4.4	56.6 ± 4.3	55.9 ± 4.2
$\dot{V}_E$max		98.8 ± 12.1	99.3 ± 10.0	100.1 ± 8.6
HRmax		195 ± 7	203 ± 6	210 ± 10
CBF	R	3.42 ± 0.83	3.80 ± 0.75	4.97 ± 0.87
	E	27.5 ± 4.47	28.3 ± 1.38	26.7 ± 3.12
Hct	R	43.2 ± 1.8	43.1 ± 1.4	42.3 ± 1.2
	E	49.3 ± 1.4	49.8 ± 2.1	48.8 ± 2.2
Hb	R	14.3 ± 0.5	14.3 ± 0.6	14.1 ± 0.3
	E	16.1 ± 0.5	16.1 ± 0.5	16.2 ± 0.4
WV	R	81.4 ± 0.5	80.4 ± 0.3	81.1 ± 0.5
	E	78.9 ± 0.5	78.9 ± 0.7	78.6 ± 0.5
Glu	R	104 ± 32.2	85 ± 7.2	107 ± 13.5
	E	79 ± 5.0	94 ± 11.1	83 ± 7.9
LA	R	0.96 ± 0.24	0.95 ± 0.24	1.08 ± 0.21
	E	11.0 ± 0.15	9.70 ± 1.17	9.14 ± 2.79
FFA	R	0.28 ± 0.05	0.37 ± 0.12	0.27 ± 0.05
	E	0.30 ± 0.06	0.37 ± 0.05	0.30 ± 0.06

Values are mean ± SD. T = Treadmill endurance time (min & s).
$\dot{V}_{O_2}$max = maximum oxygen uptake (ℓ/min). $\dot{V}_{O_2}$max/W = maximum
oxygen uptake per kg of body weight (ml·kg^{-1}·min^{-1}). $\dot{V}_E$max = maximal pulmonary ventilation (ℓ/min, STPD). HRmax = maximal heart rate
(bpm). CBF = calf blood flow (ml·min^{-1}·dℓ$^{-1}$). Hct = hematocrit (%).
Hb = hemoglobin concentration (g/dl). WV = blood water volume (%).
Glu = glucose concentration (mg %). LA = lactic acid concentration
(mM/ℓ). FFA = non-esterified fatty acid concentration (mEq/ℓ). R =
rest. E = immediately after exercise.

sources decreased year after year, although there was no change in total energy production. In recent years, more Japanese spend their time in temperature-controlled
rooms during both summer and winter. It is conceivable that a change in dietary or
living habits may be related to the seasonal variation of maximum oxygen uptake.
Although our results do not demonstrate any specific effect of acute exposure to
cold on maximum oxygen uptake, it is possible that a higher maximum oxygen uptake in winter, as seen in the present study, may result from a chronic adaptation to
cold. This possibility needs further investigation, however.

SUMMARY

The present study was undertaken (a) to ascertain seasonal variation of aerobic work capacity as determined by monthly measurements of maximum oxygen uptake, both at ambient and constant temperature, and (b) to compare the cardiovascular responses during maximal exercise in three different temperatures (5, 18, and 35°C db). We found that the maximum oxygen uptake tends to be higher in winter than in summer, although the maximum oxygen uptake was the same at cold, hot, or moderate temperature. These results suggest that a higher maximum oxygen uptake in winter may be due to chronic adaptation to cold.

REFERENCES

1. Adams, T. and E.J. Heberling. Human physiological responses to a standardized cold stress as modified by physical fitness. *J. Appl. Physiol.* 13: 226-230, 1958.
2. Adams, W.C., R.H. Fox, A.J. Fry, and I.C. MacDonald. Thermoregulation during marathon running in cool, moderate, and hot environments. *J. Appl. Physiol.* 38: 1030-1037, 1975.
3. Astrand, P.-O. and K. Rodahl. *Textbook of work physiology.* San Francisco: McGraw-Hill, 1970.
4. Balke, B., G.P. Grillo, E.B. Konecci, and U.C. Luft. Work capacity after blood donation. *J. Appl. Physiol.* 7:231-238, 1954.
5. Consolazio, C.F., L.R.O. Matoush, R.A. Nelson, J.B. Torres, and G.J. Issac. Environmental temperature and energy expenditures. *J. Appl. Physiol.* 18: 65-68, 1963.
6. Froesch, E.R. and A.E. Renold. Specific enzymatic determination of glucose in blood and urine using glucose oxydase. *Diabetes* 5: 1-6, 1956.
7. Heberling, E.J. and T. Adams. Relation of changing levels of physical fitness to human cold acclimatization. *J. Appl. Physiol.* 16: 226-230, 1961.
8. Hill, A.V. and H. Lupton. Muscular exercise, lactic acid, and the supply and utilization of oxygen. *Q. J. Med.* 16: 135-171, 1923.
9. Hohorst, H. In: *Methoden der Enzymatischen Analyse.* Edited by H.U. Bergmeyer. Weinheim: Verlag-chemie, 1962, p. 226.
10. Hong, S.K., P.K. Kim, H.K. Pak, J.K. Kim, M.J. Yoo, and D.W. Rennie. Maximal aerobic power of Korean women divers. *Fed. Proc.* 28: 1284-1288, 1969.
11. Issekutz, B., N.C. Birkhead, and K. Rodahl. Use of respiratory quotients in assessment of aerobic work capacity. *J. Appl. Physiol.* 17: 47-50, 1952.
12. Laurell, S. and G. Tibbling. Colorimetric micro-determination of free fatty acids in plasma. *Clin. Chim. Acta* 16: 57-62, 1967.
13. Lee, K.Y., S.H. Cheen, S. Hong, and Y.H. Sung. Seasonal variations in the basal metabolic rate of Korean airmen volunteers (Korean). *Korean J. Physiol.* 6:23-30, 1972.
14. Mostadi, R., R. Kubica, A. Veicsteinas, and R. Margaria. The effect of increased body temperature due to exercise on the heart rate and on the maximal aerobic power. *Europ. J. Appl. Physiol.* 33: 237-245, 1974.
15. Nagasawa, H. and Y. Watanabe. Seasonal variation in basal oxygen intake, oxygen requirement for exercise and maximum oxygen intake (Japanese). *Sci. Report Facul. Educ., Gifu Univ. (Natural Science).* 5(3): 261-276, 1974.
16. Pandolf, K.B., E. Cafarelli, B.J. Noble, and K.F. Mets. Hyperthermia: Effect on exercise prescription. *Arch. Phys. Rehabil.* 56: 524-526, 1975.

17. Pirnay, F., R. Deroanne, and J. M. Petit. Maximal oxygen consumption in a hot environment. *J. Appl. Physiol.* 28: 642-645, 1970.
18. Rowell, L.B., J.R. Blakmon, R.H. Martin, J.A. Mazzarella, and R.A. Bruce. Hepatic clearance of indocyanine green in man under thermal and exercise stresses. *J. Appl. Physiol.* 20: 384-394, 1965.
19. Rowell, L.B., H.J. Marx, R.A. Bruce, R.D. Conn, and F. Kusumi. Reductions in cardiac output, central blood volume, and stroke volume with thermal stress in normal men during exercise. *J. Clin. Invest.* 45: 1801-1816, 1966.
20. Rowell, L.B. Human cardiovascular adjustments to exercise and thermal stress. *Physiol. Rev.* 54: 75-159, 1974.
21. Saltin, B., A.B. Gagge, U. Bergh, and J.A.J. Stolwijk. Body temperatures and sweating during exhaustive exercise. *J. Appl. Physiol.* 32: 635-643, 1972.
22. Sasaki, T. Relation of basal metabolism to changes in food composition and body composition. *Fed. Proc.* 25: 1165-1168, 1966.
23. Strydom, N.B., C.H. Wyndham, C.G. Williams, J.F. Morrison, G.A.G. Bredell, M.J. Von Rahden, and J. Peter. Energy requirements of acclimatized subjects in humid heat. *Fed. Proc.* 25: 1366-1371, 1966.
24. Whitney, R.J. The measurement of volume changes in human limbs. *J. Physiol.* 121: 1-27, 1953.
25. Williams, C.G., G.A.G. Bredell, C.H. Wyndham, N.B. Strydom, J.F. Morrison, J. Peter, P.W. Fleming, and J.S. Ward. Circulatory and metabolic reactions to work in heat. *J. Appl. Physiol.* 17: 625-638, 1962.
26. Yoshimura, M., K. Yukiyoshi, T. Yoshioka, and H. Takeda. Climatic adaptation of basal metabolism. *Fed. Proc.* 25: 1169-1174, 1966.

REVIEW OF STUDIES ON COLD ADAPTATION
WITH SPECIAL REFERENCE TO THOSE IN JAPAN

Hisato Yoshimura

Department of Physiology
Hyogo College of Medicine
Mukogawa-cho 1-1
Nishinomiya, Hyogo
Japan

A large number of studies have been carried out on cold adaptation of various species of animals, but there are few studies on humans. About 15 years ago, Scholander and his colleagues performed a series of investigations on cold tolerance of primitive men and circumpolar people. The method they used was the so-called eight-hour cold tolerance test. In this test, thermal and metabolic responses were observed in men exposed to moderate cold during the night. They pointed out that there are different patterns in human adaptation to cold, which could be classified into two groups, insulating adaptation and metabolic adaptation. Following Scholander's work, the Lange-Anderson group in Oslo reported another different pattern of adaptation to the cold. The methods used were similar to Scholander's experiments. Anderson experimented with nomadic Lapps and used members of the research team as controls. During cold exposure at night, the skin temperatures of the limbs fell remarkably in the case of the controls and the rectal temperature was maintained constant. The oxygen consumption increased about 50% in the controls, who slept very poorly, while the nomadic Lapps slept well and their metabolic rate remained almost at the same level as before sleeping. The fall in the skin temperature of nomadic Lapps was not remarkable and the skin was kept warmer than the control. On the other hand, the rectal temperature of the nomadic Lapps showed a slight fall during the cold exposure. This indicates that the pattern of responses of the nomadic Lapps was similar to that of a conforming type of adaptation, as seen in poikilo-thermic animals. The cause and mechanism of the differentiation of these three types of pattern of responses to the cold in cold-adapted subjects are not yet clarified. The proposal of IBP in 1964 in Paris had considerable influence on studies of cold adaptability of Japanese people on the basis of either ethnic or individual physiology.

Among various highlights of these studies of environmental physiology in Japan were those of Dr. Itoh's group in Hokkaido (3). They studied the mechanism of cold adaptation in the Ainu, who have been living for several hundred years in the cold regions of Japan. First, he compared the rise of heat production between the Ainu and proper Japanese by immersing the subjects in cold water of 23 to 25°C for 30 min (Fig. 1). The rise of heat production was significantly lower in the Ainu. Then they exposed Japanese and Ainu subjects clad in light clothes to cold in a climatic chamber at 10°C for 90 min. The oxygen consumption and the urinary excretion of vanilmandelic acid, a catecholamine metabolite, and FFA in plasma were measured. The results suggest that the increase in heat production in cold exposure is possibly associated with mobilization of FFA from depot fat, and that this lipolysis was stimulated by catecholamine released in cold exposure. In this instance, we should recall that the heat production of the Ainu is not so sensitive as that of the control Japanese. In the experiment, the plasma FFA increased more in the Japanese control than in the Ainu, while the ketone body concentration in plasma was higher in the latter than in the former, as seen in Fig. 1. In an attempt to clarify the role of catecholamines in cold adaptation of the Ainu, norepinephrine was administered to the Ainu subjects and the proper Japanese controls under basal conditions in the morning. The results are shown in Fig. 2.

It is demonstrated that the calorigenic action of norepinephrine is very strong for the Ainu compared with the control Japanese. Plasma level of FFA also markedly increased in the Ainu, while it was confirmed that the increase of ketone bodies is more remarkable in the Ainu compared with the control. Itoh suggested that the turnover rate of plasma FFA increased markedly in the Ainu. The plasma FFA mobilized in the Ainu may be rapidly converted into ketone bodies in the liver and oxidized to produce heat. The results are quite different from those of the Eskimo who are living in a warm micro-climate devised by their long experience in the cold, i.e. by cultural adaptation. The Ainu, whose living habits are similar to those of ancient Japanese, had very poor protection against winter cold and tolerated it mainly by physiological adaptation. A long, continued repetition of cold acclimatization from generation to generation may have enabled them to possess a specific trait for fat metabolism. They are very sensitive to norepinephrine, and the FFA mobilized from depot fat can be readily utilized to produce heat. Thus, the Ainu appear to be equipped with adaptive mechanisms characterized by norepinephrine-sensitive thermogenesis. Itoh's group further confirmed this observation, using rats which had been reared at 5°C for more than ten generations. Davis suggested that cold acclimatization in man may be accompanied by the development of non-shivering thermogenesis. In rats, Robert Smith's and Carlson's groups clarified that

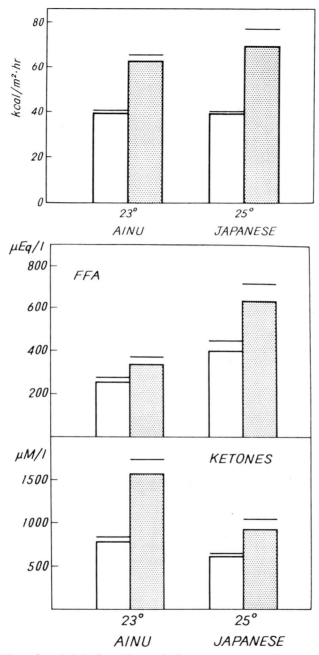

Fig. 1. Effect of a whole body cold water bath on energy metabolism, plasma levels of free fatty acids (FFA), and total ketone bodies in the Ainu and proper Japanese. (From Ref. 3, reproduced by permission)]

non-shivering thermogenesis is effected by the stimulation of brown adipose tissue (BAT) by norepinephrine. In man, however, insufficient BAT is present to explain the elevated metabolic adaptation to the cold. Recently, the Hokkaido groups have been studying skeletal muscle metabolism of the rat adapted to the cold with special reference to non-shivering thermogenesis. Several groups in Japan are studying the role of CNS in heat and cold adaptation. Dr. Miura (4) is studying a chronologic change of comfort zone of man in winter. According to him, the comfort zone in winter was 16 to 18°C for male and female Japanese in 1947, while it rose from 20 to 22°C in 1960 and attained 24 to 25°C in 1970. It is thus suggested that the capacity for adaptability (2) to cooling of room temperature seems to be reduced. On the other hand, Dr. Momiyama (5) studied a chronological change of seasonal variation of mortality in Japan. She concluded that the summer peak of mortality of Japanese, which was very high before the war, disappeared recently, and the winter peak also seems to be reduced at present. She explained this as being due to the development of industrialized control of living climate in Japan. Thus, the advance of cultural or industralized control of living climate seems to affect our lives under environmental stress beneficially on the one hand, while it may act unfavorably in terms of resistance against natural environmental stress.

These are outlines of present Japanese studies in the field of environmental physiology. Many problems remain to be solved in this field. Many geophysicists predict the imminent approach of a glacial epoch. Survival of *Homo sapiens* may depend on their cold adaptability through this crucial period. As was suggested by a Japanese physiologist, our adaptability to the cold tends to be reduced by being habituated to a comfortable climate produced artifically. We must keep our cold adaptability at a strong level similar to that which enabled *Homo sapiens* to tolerate the cold environment after the Wurm IV glacial epoch, about 25,000 years ago. For this purpose, our future studies on cold adaptation should clarify the mechanism of cold adaptability in more detail and devise methods to strengthen this ability. At this stage, I would cite the words of Rene DuBos (1):

> Any activity, however trivial, that deals with submicroscopic particles or subcellular chemical phenomena is labeled as fundamental, whereas efforts to formulate, investigate, or teach the phenomena of life as experienced by the whole organism are considered scientifically unsophisticated. Yet, the study of these phenomena demands great ingenuity, originality and initiative precisely because they are so complex and so neglected. Organismic and especially environmental medicine constitute today virgin territories to be studied further. They will remain undeveloped unless a systematic effort is made to give them academic recognition and to provide adequate facilities for their exploration.

The work of Itoh's group was the outcome of systematic efforts made for IBP where worldwide cooperation was obtained.

Cooperative studies in this area between Japanese and American investigators have been mutually beneficial. Many of us shall retire before long. I hope that young environmental physiologists in both countries will inherit and maintain the

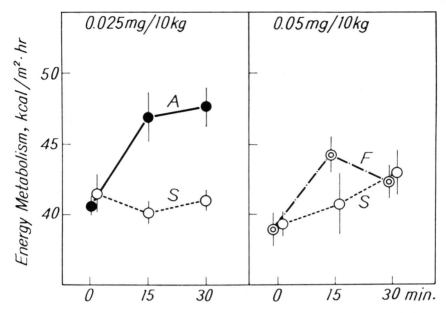

Fig. 2-a. Effect of norepinephrine on energy metabolism in Asahikawa Ainu (A), Sapporo students (S), and Huren farmers (F). (From Ref. 3, reproduced by permission.)

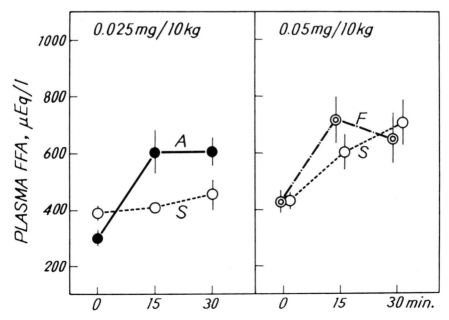

Fig. 2-b. Effect of norepinephrine on plasma FFA concentration in Asahikawa Ainu (A), Sapporo students (S), and Huren farmers (F). (From Ref. 3, reproduced by permission.)

friendship that we initiated and expand this cooperation internationally so as to advance environmental physiology and to help man to tolerate the coming cold epoch. In closing, I wish to present my sincere thanks to Dr. Horvath and his colleagues for their long continued friendship and cooperation both in research and training of the Japanese young generation.

REFERENCES

1. DuBos, R. *Man adapting.* New Haven: Yale Univ. Press, 1965, p. 446.
2. Hensel, H. and G. Hildebrandt. Organ systems in adaptation-nervous system. In: *Adaptation to the environment.* Washington, D.C.: Am. Physiol. Soc., 1964, pp. 55-72.
3. Itoh, S. *Physiology of cold-adapted men.* Sapporo: Hokkaido Univ., School of Medicine (in Japanese), 1974.
4. Miura, T. *Physiology of cold-adapted man.* Tokyo: Chuo-Koronsha, 1977.
5. Momiyama, M. *Seasonality in human mortality.* Tokyo: Univ. of Tokyo Press, 1977.

BODY FAT AND COOLING RATES
IN RELATION TO AGE

W. R. Keatinge

Department of Physiology
The London Hospital Medical College
University of London
London, England

Although evidence has existed for half a century that body fat can influence heat loss in man, it has only become clear recently that surface fat thickness is usually the only important individual factor determining heat loss when normal people are exposed to a cold environment. This is the case particularly during immersion in cold water, when external insulation is virtually removed and the rate of heat loss to the body surface is determined by the internal insulation of the tissues. There are, of course, other tissues than fat interposed between the body core and the skin surface, but for the most part high blood flow in these tissues keeps their insulation low. Muscle is a deep tissue that can provide considerable insulation, but even this becomes small when muscle flow is increased by voluntary exercise or by shivering. All deep tissues are therefore relatively ineffective in providing insulation in cold surroundings. On the other hand, vasoconstriction in the skin and subcutaneous tissues enables the skin and subcutaneous fat to act virtually as an inert layer of insulation during moderate cold exposure. A number of experiments have accordingly shown a close correlation between the reciprocal of people's surface fat thickness and their body insulation in water at $15^{\circ}C$ - $20^{\circ}C$ (2,6,10). The most surprising feature of the second of these studies was the small amount that variations in skin blood flow contributed to individual variations in heat loss. Peripheral blood flow did vary between individuals in water at $15^{\circ}C$, and contributed to individual variations in heat loss that were not due to fat, but peripheral flow was always small and its influence was much less than that of fat.

A great many of the differences in body heat loss by groups of individuals of different age and sex can also be explained by differences in subcutaneous fat between these groups. For example, men are generally thinner than women, and children are generally thinner than adults; men cooled faster than women, and children faster than adults in cold water (11). More important, the differences between the groups as well as between the individuals were largely accounted for by the differences in subcutaneous fat thickness. The only other important factor, particularly in the case of children, was the higher surface area/mass ratio of small individuals. This led to faster cooling by children than adults of given fat thickness, but when this factor was allowed for as well as fat, no systematic differences in cool-rate remained between the groups. Kollias et al. (8) similarly found that women's generally smaller size (and consequent high surface area/mass ratio) was sufficient to account for the fact that they cooled faster than men of comparable fat thickness. The surface pattern of heat loss is generally similar in men and women (5), but it is of interest that the distribution of surface fat, as well as its quantity, differs between men and women; the fat thickness on the limbs is considerably greater in relation to trunk fat in women than in men (11). This should place women at a relative advantage during swimming when limb muscles are active, whereas at rest in cold water the advantage may be less. Whether it does so has not been fully established, but in our study the women's extra limb fat was at least associated with a lower overall cooling rate during exercise than was seen in men with similar trunk fat and total body size.

Fat is not only important in limiting body heat loss during cold immersion. Apart from providing insulation, fat provides bouyancy, and this also has major survival value in water near $0^{o}C$. The high viscosity of water at these low temperatures leads to rapid exhaustion during the vigorous swimming that thin people require to keep their heads above water (7). Thin people collapsed, sank and required immediate rescue after as little as 90 sec swimming in water at $4.7^{o}C$, while fatter and generally older people could swim for 10 min or more. The reason seemed to be that the thin people without bouyancy had to swim harder to stay afloat and so tired more rapidly. This is in practice a frequent cause of death when young, fit, thin people overturn a boat in cold water and try to swim ashore without an artificial bouyancy aid.

The insulation provided by fat can enable fat men to maintain body temperature in water as cold at $12^{o}C$ (1); but in colder water, blood flow returns to the skin, particularly of the extremitities, in the reaction known as cold vasodilation. This reaction represents cold paralysis of the blood vessels, and no practical way has been found to prevent it other than by providing external insulation by clothing. It leads to blood flow bypassing the insulation of fat and causing rapid heat loss during prolonged immersions in water near $0^{o}C$, however fat the individual. Very few other factors can break down the intense cutaneous vasoconstriction produced by cold exposure. The most important one is hypoglycaemia brought on by alcohol following a period of physical exercise. The alcohol blocks gluconeogeneis; and at a time when exercise has used up the body's carbohydrate reserves, this causes a large fall in blood glucose which in turn almost eliminates both the normal vasoconstrictor and metabolic response to cold, leading to a rapid fall in deep body temperature (4).

In a more neutral thermal environment, fat thickness becomes less important in determining heat loss, cutaneous blood flow being higher, so that small adjustments in it and in metabolic rate have a larger effect on total heat balance. In the older age groups, Wagner et al. (12) found that metabolic rate is usually lower than in young people; this is sometimes not fully compensated for by vasoconstriction and leads to reduced deep as well as surface temperature. There was some indication of a tendency to reduced vasoconstrictor response to cold in spite of the low peripheral temperature, since blood flow recordings during cold exposure generally showed higher values in the old than the young subjects. With respect to temperature changes, these results were confirmed by Collins et al. (3), though greater individual variation was observed. It is notable that no illness or other ill-effect was generally observed to result from the change. On the other hand, a few old people have gross defects in vasoconstriction and in metabolic response to cold (9), and these people are likely to develop severe hypothermia under even moderate cold stress. This small group of old people presumably have undergone a selective degenerative loss of central thermoregulatory neurons, although the precise deficiency has not been revealed by simple postmortem examinations.

SUMMARY

In a warm or neutral environment, many factors affect body temperature, including differences in vasoconstrictor tone and sweat rates. However, experiments during cold stress, particularly cold immersion, show a close linear correlation between body cooling rates and reciprocal subcutaneous fat thickness. It appears that in these conditions normal people undergo almost complete vasoconstriction in the skin, and that insulation provided by subcutaneous fat then largely determines heat loss. Men in general are thinner and cool faster than women in the cold; children are thinner and cool faster than adults, though children's higher surface area/mass ratio also contributes to their rapid cooling. The only major exceptions seem to be conditions such as alcohol-induced hypoglycaemia and occasional degenerative conditions in old age, when breakdown of the vasoconstrictor mechanism leads to rapid body cooling even with a thick layer of subcutaneous fat.

REFERENCES

1. Cannon, P. and W.R. Keatinge. The metabolic rate and heat loss of fat and thin men in heat balance in cold and warm water. *J. Physiol.* 154: 329-344, 1960.
2. Carlson, L.D., A.C.L. Hsieh, F. Fullington, and R.W. Elsner. Immersion in cold water and total body insulation. *J. Aviat. Med.* 29: 145-152, 1958.
3. Collins, K.J., C. Dore, A.N. Exton-Smith, R.H. Fox, I.C. MacDonald, and P.M. Woodward. Accidental hypothermia and impaired temperature homeostasis in the elderly. *Brit. Med. J.* 1: 353-356, 1977.
4. Haight, J.S.J. and W.R. Keatinge. Failure of thermoregulation in the cold during hypoglycaemia induced by exercise and ethanol. *J. Physiol.* 229: 87-97, 1973.
5. Hayward, J.S., M. Collis, and J.D. Eckerson. Thermographic evaluation of relative heat loss of man during cold water immersion. *Aeros. Med.* 44: 708-711, 1973.
6. Keatinge, W.R. The effects of subcutaneous fat and of previous exposure to cold on the body temperature, peripheral blood flow and metabolic rate of men in cold water. *J. Physiol.* 153: 166-178, 1960.
7. Keatinge, W.R., C. Prys-Roberts, K.E. Cooper, A.J. Honour, and J.S.J. Haight. Sudden failure of swimming in cold water. *Brit. Med. J.* 1: 480-483, 1969.
8. Kollias, J., L. Barlett, V. Bergsteinova, J.S. Skinner, E.R. Buskirk, and W. C. Nicholas. Metabolic and thermal response of women during cooling in water. *J. Appl. Physiol.* 36: 577-580, 1974.
9. MacMillan, A.L., J.L. Corbett, R.H. Johnson, A. Crampton-Smith, J.M.K. Spalding, and L. Wollner. Temperature regulation in survivors of accidental hypothermia of the elderly. *Lancet* 2: 165-169, 1967.
10. Pugh, L.G.C. and O.G. Edholm. The physiology of channel swimmers. *Lancet* 2: 761-768, 1955.
11. Sloan, R.E.G. and W.R. Keatinge. Cooling rates of young people swimming in cold water. *J. Appl. Physiol.* 35: 371-375, 1973.
12. Wagner, J A., S. Robinson, and R.P. Marino. Age and temperature regulation of humans in neutral and cold environments. *J. Appl. Physiol.* 37: 562-565, 1974.

THE EFFECT OF ETHANOL CONSUMPTION
ON HUMAN HEAT EXCHANGE

L. Kuehn
S. Livingstone
R. Limmer
B. Weatherson

Defence and Civil Institute of Environmental Medicine
Downsview, Ontario, Canada

INTRODUCTION

It has long been an admonition of survival experts and others (3,8,16) that consumption of ethanol-containing beverages in cold environments provokes a predisposition to accelerated heat loss from the human body, principally due to enhanced vasodilation of peripheral blood vessels. Although this long-standing belief has been widely circulated, there is little factual evidence in the literature to substantiate it.

In 1940, Horton *et al.* (11) reported that cheek and hand temperatures increased after ingestion of 15 cc of 95% pure ethanol in one subject, and increases in finger temperature and toe temperature occurred in two female subjects who were administered 30 cc of 95% pure ethanol. Although results were graphically portrayed in this publication, there was little discussion of the experimental methodology to permit assessment of the experiment protocol involved in obtaining these results.

Keatinge and Evans (12) examined this purported increase in heat loss by examining the fall in body temperature of men immersed in cold water. Ten nearly nude subjects were involved in a series of experiments in which changes in rectal temperature, skin temperature, heart rate, metabolism, and finger heat flow were assessed

303

before and during immersion in water of 15°C for a period of 30 min. Those subjects who imbibed 75 ml of absolute ethanol 45 min prior to the immerson showed no difference in rate of fall of rectal temperature during immerson compared to that for the control situation. However, alcohol usually reduced metabolic rate and the initial increase in heart rate during immersion as well as the subjective sensations of discomfort and cold in the water. It was implied by conjecture that the cold stress may have provoked a peripheral vasoconstriction sufficiently intense to overcome the vasodilator effect of the alcohol.

In 1963, Anderson *et al.* (2) examined the effect of 1 g/kg of alcohol on heat balance and sleep in men exposed nude to both 20°C and 15°C air environments. They found that alcohol did not alter metabolic rate or rectal or skin temperature changes from control values during nights of sleep. They also found that the subjects slept better and that they were more relaxed after consuming alcohol. It was suggested that the cold exposure may have been severe enough to override any vasodilative effects of the alcohol.

That ethanol is responsible for peripheral vasodilation has been well substantiated (1,7), particularly by Gillespie (9), who found that a dose of Scotch whiskey of 2 ml/kg body weight was required to produce sustained vasodilation in healthy subjects. This action was restricted only to skin blood flow and was not evident in muscle blood flow observed at the forearms and calves by venous occlusion plethysmography via mercury strain gauges. Further experiments by Gillespie indicated that alcohol acts as a vasodilator only when it has been partly metabolized, and that it has a direct vasoconstrictor effect when infused intra-arterially.

More recent work by Martin *et al.* (14) has indicated that consumption of ethanol produces no change in mean skin, rectal, and aural temperatures in subjects immersed in water at 13.6°C. Recently, Hobson and Collis (10) reported that consumption of 1.87 ml of 70-proof whiskey per kilogram of body weight resulted in deceleration of cooling of three nude subjects in water of 7.5°C. These workers also found no evidence that the vasodilatory properties associated with ethanol accelerate the cooling rates of humans in cold water.

Several speculations can then be immediately raised. One is that if the vasodilative effects of ethanol are only manifested in the extreme regions of the periphery, such as the face, hands, and feet, then any increase in heat loss in these ares may have only a relatively small effect on body core temperature or mean skin temperature. Another is that the cold stress experienced in intensely adverse conditions, such as cold water immersion may provoke such extensive vasoconstriction that the vasodilative effects of ethanol are overwhelmed. A corollary of this hypothesis is that the vasodilative effects of ethanol on enhanced heat loss will only be observed in conditions of relatively mild cold stress.

Because of this dearth of evidence for accentuated heat loss, we decided to perform a number of experiments to resolve the issue.

METHODOLOGY

In the first series of experiments, infrared thermographic techniques (utilizing the AGA Thermovision 680 system) were used to display the surface temperatures of intoxicated subjects in various air environments of mild or moderate cold stress. The hypothesis was that accelerated or increased heat loss should be manifested by a rise in body surface temperature.

Experiment I-1. Four sober young male Caucasian subjects dressed only in shorts sat in a controlled temperature room set at a temperature of 25°C and a relative humidity of 40%, a condition in which mild vasoconstrictor tone was assumed to exist in the subjects. Their frontal surface temperatures were recorded photographically from the infrared thermographic apparatus in which temperatures were displayed in a color-coded fashion on a color television screen (13). Such measurements were obtained from all four subjects in a standing pose every 10 to 15 min during the experimental period.

After 1.5 hr of pre-experimental exposure to this environment, two of the four subjects were each given a mixture of orange juice and 50 cc of absolute ethanol to drink; the other two subjects were administered only an equivalent total volume of orange juice. Such a dose of ethanol was considered to be approximately equivalent to five bar whiskey drinks. Thermographic photographs were then taken over a 2-hr period. On the succeeding day, the subjects' roles were reversed and the experiment was repeated. The entire experiment was repeated a week later in experimental conditions maintained at 30°C air temperature and 40% relative humidity, conditions chosen to abolish any vasoconstrictor tone so that any vasodilator effect could be noted.

Experiment I-2. Concern was expressed that the reported vasodilative effect of ethanol may not be due to ethanol itself, but to impurities inherent in commercially available spirits and beverages. To investigate this possibility, four sober young male Caucasian subjects were again tested in the manner described above at the two environmental conditions of 25°C (40% relative humidity) and 30°C (40% relative humidity). In this instance the two experimental subjects were administered 125 cc cognac brandy, equivalent in ethanol content to 50 cc of absolute ethanol. The two control subjects were administered an equivalent total volume of water. The roles of the subjects were again reversed for a second day of testing at each environmental condition, and infrared thermographic techniques were used to record frontal surface skin temperatures in standing poses.

Experiment I-3. Although substantial testing had verified the use of infrared thermographic techniques for monitoring differences in body surface temperatures as small as 0.1°C, it was decided to check this technique by testing the procedure on one subject, using skin thermistors (Yellow Springs Instruments) to record temperatures at the left forefinger, left forearm, and the left side of the neck; and such measurements were taken every 10 to 15 min as described earlier. During these

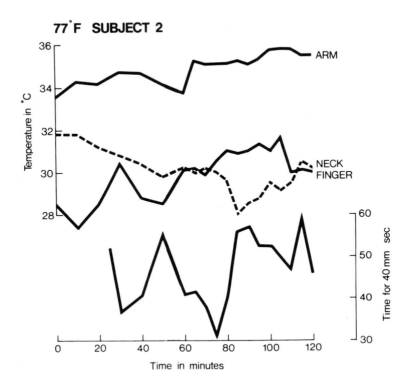

Fig. 1. The effect of drinking 125 cc cognac brandy on skin temperatures and finger blood flow of a nearly nude subject in a conditioned room set at 25°C air temperature and 40% relative humidity. Ethanol consumption denoted by the stippled area. Temperatures were measured with thermistors. The plethysmograph readings were calculated as the amount of time required for the polygraph pen to move 40 mm across the graphic reading paper. Greater blood flow is indicated by shorter times.

sessions in which the subject imbibed 125 cc cognac brandy at environmental temperatures of 25°C and 30°C on two different days, the subject's finger blood flow was measured with a plethysmograph to observe changes in vasodilation consequent to drinking the brandy.

Experiment I-4. Since the measurements so far pertained to nearly nude young men in temperate air environments, whereas the anecdotal information on ethanol-induced vasodilation pertained to cold exposure of clothed individuals, three sober young male Caucasian subjects were tested in a cold environmental room at a temperature of -23°C, dressed in the Canadian Forces Arctic clothing ensemble (thermal insulation about 4 clo). As before, two of the subjects were given a mixture of orange juice and 50 cc absolute ethanol to consume, 1.5 hr after exposure to this environment, and the other subject was given an equivalent total volume of orange juice for control purposes. Surface skin temperature measurements were obtained by skin thermistors at the forefinger, forearm, cheek, and foot sites. The experiment was repeated on the succeeding day with the subjects' roles reversed, two now serving as control subjects and only one consuming the ethanol-containing beverage.

Experiment II. In the next series of experiments, the convective heat loss and metabolic heat production of eight young male Caucasian subjects in cold water of 25°C temperature was determined, each with and without an ethanol consumption of 30 cc per 70 kg body weight, in a human water calorimeter (4). Calorimetry is considered the proper technique for direct measurement of heat loss from human subjects, and the technique employed has been determined to be accurate to within 3% of actual body heat loss.

In these experiments the subjects were maintained in a nearly nude condition, wearing only swimming shorts, in a laboratory with controlled conditions of 20°C and 65% relative humidity for a pre-experimental period of 15 to 30 min. The ethanol dose was administered with orange juice precisely 10 min prior to immersion. They were then immersed in a supine position into the 25°C temperature water bath of the calorimeter via a special calorimeter subject rack in the same room. The calorimeter was then closed, leaving only the subject's head protruding. Rectal and skin temperatures were obtained from the subject with YSI thermistor probes, and the subject's metabolic rate was determined with the conventional Douglas bag technique. The period of immersion lasted one hour, and the subject was free to terminate the experiment whenever he chose.

RESULTS

Experiment I-1. The results for the first series of experiments in which infrared imaging techniques were used to capture the skin temperature responses in four young subjects after imbibing 50 cc pure ethanol showed no significant skin temperature increases or shifts of temperature isotherms over any of the subjects' frontal surfaces for the 2-hr period. This was the case whether the room was at 25°C or 30°C. These experimental observations are at variance with the hypothesis that ethanol of this dosage induces greater heat loss from subjects in thermoneutral or cool environments.

Experiment I-2. The results were similarly negative for the experiments in which the pure ethanol doses were replaced by brandy in an attempt to ascertain whether the purported accelerated heat loss was due to impurities or higher alcohols in the intoxicating beverage.

Experiment I-3. Figs. 1 and 2 show the effect of ethanol ingestion on the three nearly nude subjects who were monitored with skin thermistors and plethysmograph in temperate conditions of 25°C and 30°C. No significant increases were apparent in the temperatures of the forearm, neck, or fingers due to the ethanol dose, despite the fact that the plethysmograph indicated an increase in vasodilation soon after imbibing of the ethanol, as reported earlier by Gillespie (9).

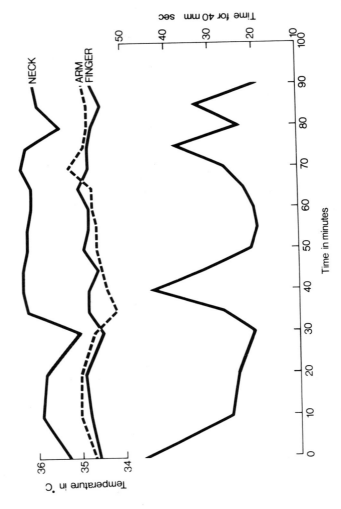

Fig. 2. The effect of consumption of 125 cc cognac brandy on skin temperatures and finger blood flow of a nearly nude subject in a conditioned room set at 30° C air temperature and 40% relative humidity. Ethanol consumption is denoted by the stippled area. Temperatures and plethysmograph readings as in Fig. 1.

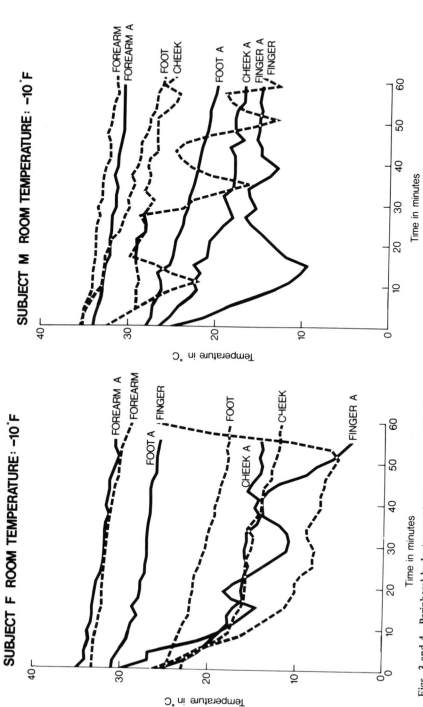

Figs. 3 and 4. Peripheral body temperatures as measured by thermistors in two clothed subjects (4 clo of clothing insulation) in a conditioned room set at -23°C air temperature with (A) and without the influence of imbibing 50 cc pure ethanol. Ethanol consumption is denoted by the stippled area.

Experiment I-4. Such negative results were also observed in the case of the three clothed subjects in which thermistors were used to monitor the temperature of the cheek, forearm, finger, and foot in cold environments with and without ethanol consumption. Figs. 3, 4, and 5 show that no differences were observed between the two conditions that could be attributed to a greater increase of heat loss at the body periphery due to enhanced vasodilation. Sometimes different temperature values were observed at the same site on the two different days; however, the relative rates of temperature decline were similar, indicating no enhanced heat loss.

Experiment II. With regard to the second series of experiments involving calorimetric assessment of convective heat loss and metabolic heat production, results presented in Table 1 were similarly negative. During the 1-hr immersion at 25°C, the rectal temperature decrease, convective heat loss, and metabolic rates were not significantly different for each subject for the two conditions (ethanol or no ethanol).

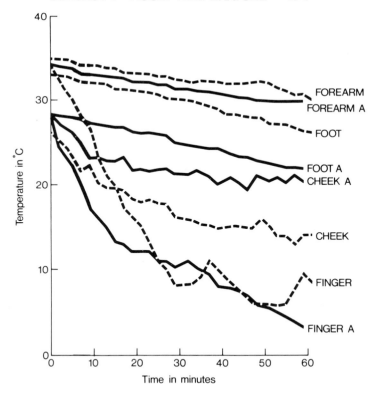

Fig. 5. Peripheral body temperatures as measured by thermistors in a clothed subject (4 clo of clothing insulation) in a conditioned room set at -23°C air temperature with (A) and without the influence of imbibing 50 cc pure ethanol. Ethanol consumption is denoted by the stippled area.

Table 1. Calorimetric Heat Loss under the Influence of Ethanol

Subject	Age yr	Weight kg	Surface Area m²	Test	Rectal Temperature Final, °C	Rectal Temperature Drop, °C	Metabolic Rate, kcal	Convective Heat Loss, kcal
1	25	64.0	1.74	control	37.0	0.6	109.1	192.3
				ethanol	37.1	0.5	94.2	180.1
2	25	69.0	1.86	control	37.0	0.4	84.0	200.0
				ethanol	37.0	0.6	81.3	210.5
3	24	80.0	1.95	control	36.9	-0.1	81.6	172.2
				ethanol	37.0	0.1	112.2	191.7
4	22	88.6	2.21	control	37.3	0.2	134.7	224.4
				ethanol	37.6	0.0	170.7	249.4
5	38	95.5	2.25	control	37.3	0.1	151.8	248.1
				ethanol	37.2	0.1	114.9	251.6
6	25	75.0	1.98	control	36.9	0.2	257.1	272.1
				ethanol	36.9	0.4	304.9	264.5
7	21	69.0	1.73	control	36.3	0.9	255.6	273.4
				ethanol	36.3	1.1	315.9	247.7
8	22	77.3	1.98	control	36.4	0.7	107.7	262.4
				ethanol	36.4	1.2	119.4	284.7
Average (± Std Dev)	25 ± 5	77.3± 10.6	1.96± 0.19	control	36.9± 0.4	0.4± 0.3	147.7± 71.0	230.8± 39.0
				ethanol	36.9± 0.4	0.5± 0.4	164.2± 94.0	235.0± 36.8

Even if each subject is compared only to himself, the metabolic rate under the influence of ethanol is 2.10 ± 0.22 that in the control condition and the convective heat loss under the influence of ethanol, as measured by the calorimeter, is 1.02 ± 0.08 that in the control condition.

DISCUSSION

The foregoing results have shown that alcohol has little effect on heat loss from persons at thermoneutral, mildly cool, or cold air environments. Despite plethysmographic evidence of alcohol-induced vasodilation in the extreme periphery, no evidence of increased skin temperature or heat loss was observed. This result is in agreement with the observation that ingestion of niacin, an effective peripheral vasodilator (6) does not increase skin temperature, although flushing has been observed (Livingstone, unpublished observations).

It is possible that the dose of alcohol used in these experiments (approximately 1 ml pure ethanol/kg body weight) was not sufficient to induce extensive vasodilation and a concomitant increased heat loss, but since the results were all negative with this rather large dose, it seems unlikely that such an effect would suddenly appear with larger, excessive doses. Furthermore, Gillespie (9) has reported sustained vasodilation in the skin with even smaller doses of alcohol. The fact that these results are at variance with those reported by Horton *et al.* (11) is still not explained.

It is interesting that nearly all the subjects reported feelings of warmth at their face, chest, and shoulders after imbibing the ethanol, despite the fact that infrared imaging, standard thermistor techniques, and calorimetry showed no change in body temperature or heat loss. The sensation of warmth may be due to direct stimulation of the heat receptors in the skin as is the case for intravenous injection of calcium salts (15) and not due to effects of ethanol-induced vasodilation.

Flushing of the face and upper torso was noted in several subjects, indicative of the vasodilative effects of the ingested ethanol. Early in the investigation, such flushing was postulated to cause an increased insensible heat loss from the subjects, due to enhanced evaporation of sweat in the affected regions; this possibility was one of the reasons that recourse was made to the use of calorimetry in the second series of experiments. However, it is important to note that no conclusion can be drawn from skin color about either skin temperature or skin blood flow. Color depends on the amount of blood present in the subpapillary venous plexus, while skin temperature depends mostly on the rate of blood flow (15).

SUMMARY

Concern has often been expressed regarding the advisability of consumption of alcoholic beverages by persons exposed to cold environments, partially because of the presumed accentuated heat loss due to ethanol-induced vasodilation of the body "shell." This paper describes the experimental investigation of heat loss as a function of ethanol ingestion in young male Caucasian subjects in air and water environments. In the first series of experiments, infrared thermographic techniques were used to display the surface temperatures of subjects after ethanol ingestion in various air environments. Whether the subjects were dressed only in shorts at air temperatures of 25°C and 30°C or in Arctic clothing at air temperatures of -23°C, it was found that individual consumption of 50 cc pure ethanol or 125 cc cognac brandy (equivalent to 50 cc pure ethanol) did not result in any significant changes compared to experiments in which the same subjects drank equal volumes of water or orange juice. This finding was substantiated by use of skin thermistors in further experiments. A second series of experiments was conducted in which the convective heat loss and metabolic heat production of eight subjects in cold water at 25°C were determined, each with and without an ethanol dose of 30 cc per 70 kg body weight in a human water calorimeter. No significant differences were observed in any subject regarding rate of heat loss, skin temperatures, rectal temperature, or metabolic rate as a consequence of the ethanol consumption. It is concluded that the hypothesis of increased heat loss via ethanol-induced vasodilation is not supported by this experimental evidence.

REFERENCES

1. Abramson, D., H. Zazeela, and N. Schkloven. The vasodilating action of various therapeutic procedures which are used in the treatment of peripheral vascular disease. *Am. Heart J.* 21: 756-766, 1941.
2. Anderson, K.L., B. Hellstrom, and F. Vogt Lorentzen. Combined effect of cold and alcohol on heat balance in man. *J. Appl. Physiol.* 18: 975-982, 1963.
3. Anonymous. Don't walk and drink. *Lancet* 1: 816, 1973.
4. Craig, A. and M. Dvorak. Heat exchanges between men and the water environment. *Proceedings of the Fifth Underwater Symposium.* Bethesda: FASEB, 1976.
5. Cushny, A. *Pharmacology and therapeutics.* Philadelphia: Lea & Febiger, 1910, p. 138.
6. Darby, W., K. McNutt, and E. Todhunter. Niacin. *Nutr. Rev.* 33: 289-297, 1975.
7. Fewings, J., M. Hanna, J. Walsh, and R. Whelan. The effects of ethyl alcohol on the blood vessels of the hand and forearm in man. *Br. J. Pharmacol. Chemother.* 27: 93-106, 1966.
8. Gaddum, J. *Pharmacology.* London: Oxford University Press, 1953, pp. 110-117.
9. Gillespie, J. Vasodilator properties of alcohol. *Brit. Med. J.* 2: 274-277, 1967.
10. Hobson, G. and M. Collis. The effect of alcohol upon cooling rates of humans immersed in 7.5°C water. *Can. J. Physiol. Pharmacol.* 55: 744-746, 1977.
11. Horton, B., C. Sheard, and G. Roth. Vasomotor regulation of temperatures of extremities in health and disease. *J. Med. Soc. N. J.* 37: 311-323, 1940.
12. Keatinge, W. and M. Evans. Effect of food, alcohol and hycosine on body-temperature and reflex responses of men immersed in cold water. *Lancet* 2: 176-178, 1960.
13. Krog, J. Methods in circulatory research. In: *The physiology of cold weather survival.* Edited by A. Borg and J. Veghte. AGARD Report No. 620. June 1974.
14. Martin, S., K. Cooper, and R. Diewald. The effect of cold water immersion and alcohol ingestion on body temperature responses. *Can. Physiol.* 8: 51, 1977.
15. Rothman, S. *Physiology and biochemistry of the skin.* Chicago: University of Chicago Press, 1954.
16. Victor, M. and R. Adams. Alcohol. In: *Principles of internal medicine.* Edited by T. Harrison. London: Mc-Graw-Hill, 1958, pp. 747-755.

V

Altitude
Chairperson

David B. Dill

Desert Research Institute
Boulder City, Nevada

SOME HIGH POINTS IN HIGH ALTITUDE PHYSIOLOGY

Ralph H. Kellogg

Department of Physiology
University of California
San Francisco, California
and
Laboratoire de Physiologie Respiratoire
Centre National de la Recherche Scientifique
Strasbourg, France

It is a pleasure to have the opportunity to pay tribute to Professor Steven M. Horvath and simultaneously to share with you my interest in the history of high altitude physiology as a background for the contemporary science to follow. So relax and come back with me to earlier centuries.

I would like to begin before the beginning, so to speak. It is usually said, and rightly so, that the first clear description of mountain sickness is that of the Jesuit missionary, Father Joseph de Acosta (ca. 1539-1600), writing in 1590 upon his return from Peru about his experiences in crossing a high Andean pass (1,2). But for over three centuries, many authors reviewing the history of high altitude physiology have qualified this by saying that, long before de Acosta, Aristotle (384-322 B.C.) described high altitude symptoms on Mt. Olympus (2,910 m). One can cite as an early example of this the statement by Robert Boyle (1627-1691), writing in 1660 (11). The odd thing about this and all other references to this supposedly Aristotelian statement is that none of the authors cite where in Aristotle's works this statement is located. I presume subsequent authors have merely copied earlier authors, as John F. Fulton (1899-1960) in 1948 (15) obviously copied Boyle, without checking the original.

I became curious about this some time ago and tried to locate the passage in Aristotle's works, even paging through every volume of the Oxford English translation of his complete works (38) when indexes and classics scholars were unable to help me. I can therefore say with great assurance that there is no such statement anywhere in the works of Aristotle known today. It was Dr. Charles Webster of Corpus Christi College, Oxford (and now head of the Wellcome Unit for the History of Medicine at Oxford), who kindly directed me to the passage that has obviously by mis-attribution at some time in the past given rise to these statements. It appears to be a passage in *De Genesi Contra Manichi* (5) by Saint Augustine (354-430 A.D.) that is the real basis for the statements. It is clear from that passage that St. Augustine was referring to effects of low humidity rather than hypobaric hypoxia. Thus, de Acosta's position as the first to describe true hypoxic mountain sickness (19) is secure, and the first description known to me of the effects of low humidity on a mountain top is perhaps seven and a half centuries younger than supposed.

What are some of the other high points in the historical development of our knowledge of high altitude physiology, first discoveries of matters now so well accepted that we take them for granted without thinking of the discoverer or his evidence?

After Evangelista Torricelli (1608-1647) and his associates had made the first mercurial barometer (32) about 1643, it was the French philosopher Blaise Pascal (1623-1662) who demonstrated that barometric pressure declined with increasing altitude (37.) To demonstrate this, he persuaded his brother-in-law, Périer, to take a barometer to the top of the Puy-de-Dôme (1,465 m) near Clermont-Ferrand, where Pascal had been born. The ruins of a temple of Mercury there bear witness that this summit has been frequented by man at least since Roman times. (Today it also bears a temple to a modern deity in the form of a television transmitter.)

Of course, decrease of total pressure with altitude, shown by Pascal, does not necessarily mean decrease in partial pressure of oxygen, although that seems to have long been assumed implicitly. What was missing was the demonstration that the percentage of oxygen was the same at different altitudes. For the first clear demonstration of this, I would cite the work of the Englishman, James Glaisher (1809-1903), meterologist of the Royal Greenwich Observatory. To examine this question, he made various balloon ascensions, most notably on September 5, 1862, when he was taken up from Wolverhampton by a balloonist named Coxwell nearly to the altitude of Mt. Everest, establishing a new world record. The dramatic account of their ascent (in which Glaisher collapsed in the basket and Coxwell also became partly paralyzed but managed to save them by grasping with his teeth the cord that vented hydrogen to bring them down) was written up with dramatic illustrations in Glaisher's book. This volume was first published in French (16), perhaps because of his collaboration with French balloonists, followed by a second edition in English (17), which was later translated into German (18).

The role of oxygen in high altitude effects could not be known until the discovery of oxygen and the demonstration of its role in respiration. The latter can be dated from 1777, when Lavoisier presented his magnificent mémoire on oxygen consumption and carbon dioxide production to the Académie Royale des Sciences in Paris, which was published three or four years later (30). But even then, it was not obvious that lack of oxygen was the essential cause of mountain sickness. Far from it.

The first book devoted exclusively to mountain sickness that I know of, *Die Berg-krankheit...* (31), published in 1854 by a Zurich physician, Conrad Meyer-Ahrens, lists many theories. One of the most popular of these was that expressed by Albrecht von Haller (1708-1777) in his encyclopedic treatise on physiology in 1761. This "mechanical" theory was that the effects of altitude or decompression were due purely to the effects of the lowered barometric pressure *per se*. The idea was that the disturbance was primarily circulatory, arising from failure of the air pressure to support adequately the outside walls of the superficial blood vessels, thereby causing blood volume redistribution from deeper vessels, superficial vaso-dilation and pulmonary congestion, and even superficial hemorrhages such as nose-bleeds, along with difficulty in drawing the air into the lungs (20). Haller obviously had not grasped the significance of "Pascal's Law," that pressure is immediately transmitted equally in all directions throughout a fluid such as the body, and thus a reduction in atmospheric pressure on the body surface results in identical reduction inside the body, with no differential pressures resulting to affect the distribution of blood.

Nevertheless, this erroneous idea of Haller's was revived as one of the proposed explanations for "caisson disease," which became a problem when the use of com-pressed air in caissons became popular for building bridge foundations in the 19th century (12,13). The compressed air was supposed to collapse the superficial ves-sels and drive the blood into the interior of the body, from which it rushed out, producing symptoms, when the superficial vessels were no longer compressed by the air pressure.

I think the last revival of this theory occurred with Hugo Kronecker (1839-1914), Professor of Physiology in Berne, was asked to predict the physiological problems that might be encountered in connection with the proposed cog railway up the Jungfrau (4,166 m) that was subsequently constructed only as far as the Jungfraujoch (3,454 m). Kronecker recognized that ascent without exertion on a cog railway would be very different physiologically from climbing a mountain on foot, so he arranged to have seven subjects of diverse ages and both sexes carried up the Breithorn (3,750 m) on September 13, 1894. He submitted his reports two months later (27), then expanded his discussion in a book a decade later (28). He was one of a group of physiologists (25,26) who did not believe that the mountain sickness ordinarily seen was due to hypoxia, contrary to the observations and con-clusions of Paul Bert (1833-1886) (6-10). On the contrary, he thought the syn-drome that developed was one of pulmonary edema, fluid leaking out of the pul-monary capillaries because of the partial vacuum outside. As late as 1911, the year of the famous international physiological expedition to Pike's Peak (14) led by John S. Haldane (1860-1936), who already understood and accepted the hypoxic etiology of mountain sickness, Kronecker maintained this "mechanical" theory and even used it to explain acclimatization (29).

In the absence of clear evidence, the theory that mountain sickness was due to hypoxia was perhaps most strongly defended in the 1860's by Denis Jourdanet, a French physician who had spent many years in Mexico and had become interested in the medical effects of the different altitudes upon the populations there (21-24). He was the first to coin the word, "anoxemia," drawing a parallel between the symptoms of anemia and of high altitude, both of which he thought represented a lack of oxygen in the blood, although for different reasons (23). It was Jourdanet who interested Paul Bert in investigating high altitude effects in his physiological laboratory at the Sorbonne and provided Bert with the financial support to have a decompression chamber built and to publish his most famous book.

Of course this book, *La Pression Barométrique...* (9), published in Paris in 1878, is now very well known, although increasingly scarce and expensive in the rare-book market. It was even translated into English by Professor and Mrs. Hitchcock of Ohio State University in 1943 (10). Somewhat less well known, but comparably rare and expensive, is the shorter research monograph that he published twice in 1874, first in the *Annales des Sciences Naturelles* (7) and then later in the same year (despite a misprinted date) in the *Bibliothèque de l'École des Hautes Études* (8). But these publications are actually summaries of several years' work on various respiratory subjects relating to high as well as low pressures. I wonder how many people are familiar with the original research reports on which they are based, and particularly with the first of these 13 notes, in which he reported his demonstration of the importance of the partial pressure of oxygen in a communication to the Académie des Sciences of Paris on July 17, 1871 (6).

The reason that date is so remarkable is related to contemporary French historical events. In the summer of 1870, Napoleon III unwisely declared war on Prussia, whose army very quickly captured the French Army (including Napoleon himself). There was then a grass roots effort on the part of patriotic Frenchmen to stem the German invasion, in which Paul Bert was heavily involved. Returning to his native Auxerre, 167 km southeast of Paris, he was made Secretary-General to the new Republican prefect of the Department of Yonne, with whom he organized local militia and the destruction of bridges and creation of other obstacles to block the German advance. When this failed, he joined the Republican leader, Léon Gambetta, in Bordeaux. There he was appointed to be the new prefect in Lille for the Departement du Nord, which he reached only by going around the German lines by way of England. Thus, he was not quietly working away in his Sorbonne laboratory during the long siege of Paris, but on the contrary was thoroughly preoccupied elsewhere, at least until the armistice was signed after the fall of Paris on January 28, 1871. Thus, it was within less than six months of these major disruptions in his life that Paul Bert was able to present his fundamental observations showing that in sparrows, guinea pigs, and frogs, the product of barometric pressure and percentage oxygen (and hence the partial pressure of oxygen) at which each species died of hypoxia was a constant, regardless of the absolute values of the barometric pressure or of the oxygen percentage looked at individually. As we have already seen from Hugo Kronecker's views, his experimental evidence was not universally accepted for several decades, but that is another story (25,26).

Because of the interest of Dr. Horvath in metabolic as well as other aspects of physiology, I thought I would look up for this paper the first discovery that the metabolic rate is practically independent of the environmental oxygen pressure over a very wide range. This is a fundamental concept that we all take for granted when we say that a fall in alveolar carbon dioxide pressure in resting man, on ascending to high altitude, must mean an increase in alveolar ventilation. Unfortunately, I was quickly reminded of the fact that this is a very messy field historically, remaining controversial for more than a century after Lavoisier's early measurements of oxygen consumption. It would be misleading to cite just a single paper and date.

But I see that there is a paper on today's program about lactic acid. Therefore, I shall pinpoint for you the demonstration that lactic acid is produced as a result of hypoxia. This dates back to 1891, when one of the first Japanese biochemists, Trasaburo Araki, working in Hoppe-Seyler's laboratory, pointed out the unifying fact of hypoxia as the causative agent in his experiments and those of others (3). He himself produced his hypoxia both by letting his animals rebreathe air in the presence of a CO_2 absorber and by poisoning them with carbon monoxide, an agent that Dr. Horvath was interested in many years ago. Araki's priority was promptly challenged by Albert J. Dastre (1844-1917), one of Claude Bernard's students, but Araki defended himself against this attack (4).

Since there is also a paper on sleep hypoxia at altitude on the program, I might cite one more first, the original description of periodic breathing of the Cheyne-Stokes type during sleep at high altitude. This was first described by the Italian physiologist, Angelo Mosso (1846-1910), Professor of Physiology in Turin, who did experiments over many years on Monte Rosa (36). Many of his ideas were wrong, including his failure to believe Paul Bert about the hypoxic etiology of mountain sickness (25,26,36), but he was right about periodic breathing. He first reported his observation to the *Reale Accademia dei Lincei* on January 4, 1885 (33), shortly after he had started work at high altitude, and then published the complete report in both French (34) and German (35) in the following year.

I hope these samplings of high points or "firsts" (including a false one) in the history of high altitude physiology will provide some long-term perspective for the papers to follow and perhaps stimulate further interest in the history of our field. But now it is time to get on to more current work.

ACKNOWLEDGMENTS

I would like to thank the Director, Professor Pierre Dejours, and the Laboratoire de Physiologie Respiratoire, CNRS, Strasbourg, where this was written while I was on sabbatical leave from the University of California, San Francisco. Most of the material was collected previously with support from USPHS Grant HL-13841 from the National Heart and Lung Institute of the National Institutes of Health.

REFERENCES

1. Acosta, I. de. *Historia natvral y moral de las Indias...* Seville: Iuan de Leon, 1590, Book 3, Chap. 9.
2. Acosta, I. *The naturall and morall historie of the East and West Indies...* London: Val: Sims for E. Blount and W. Aspley, 1604, Book 3, Chapter 9.
3. Araki, T. Ueber die Bildung von Milchsäure und Glycose im Organismus bei Sauerstoffmangel. *Hoppe-Seyler's Zeitschr. Physiol. Chem.* 15:335-370, 1891.
4. Araki, T. Ueber Bildung von Glycose und Milchsäure bei Sauerstoffmangel. Entgegnung. *Hoppe-Seyler's Zeitschr. Physiol. Chem.* 16:201-204, 1892.
5. Augustine, St. De Genesi Contra Manichi, Chap. 15, Sec. 24. *Patrologia Latina* Vol. 34, Col. 184.
6. Bert, P. Recherches expérimentales sur l'influence que les changements dans la pression barométrique exercent sur les phénomènes de la vie. *C.R. Acad. Sci. Paris* 73:213-216, 1871.
7. Bert, P. Recherches expérimentales sur l'influence que les modifications dans la pression barométrique exercent sur les phénomènes de la vie. *Ann. Sci. Nat., Paris, Sér. 5* 20 (Art. 1):1-167, 1874.
8. Bert, P. Recherches expérimentales sur l'influence que les modifications dans la pression barométrique exercent sur les phénomènes de la vie. *Bibliothèque de l'École des Hautes Études: Sect. Sci. Nat.* 10 (Art. 2):1-167, 1874.
9. Bert, P. *La pression barométrique: recherches de physiologie expérimentale.* Paris: G. Masson, 1878.
10. Bert, P. *Barometric pressure: researches in experimental physiology.* Columbus, Ohio: College Book Co., 1943.
11. Boyle, R. *New experiments physico-mechanicall, touching the spring of the air, and its effects...* Oxford: H: Hall for Tho: Robinson, 1660, pp. 357-358.
12. Bucquoy. *De l'air comprimé.* Thèse. Strasbourg, 1861. [Cited by (10), pp. 373-375 and 457-460.]
13. Clark, E.A. Effects of increased atmospheric pressure upon the human body. With a report of thirty-five cases brought to City Hospital from the caisson of the St. Louis and Illinois bridge. *Med. Arch., St. Louis* 5:1-30, 1870-71.
14. Douglas, C.G., J.S. Haldane, Y. Henderson, and E.C. Schneider. Physiological observations made on Pike's Peak, with special reference to adaptation to low barometric pressure. *Phil. Trans. Roy. Soc. London, Series B* 203:185-318, 1913.
15. Fulton, J.F. *Aviation medicine in its preventive aspects, an historical survey.* London, New York, Toronto: Geoffrey Cumberlege, Oxford Univ. Press, 1948, p. 4.
16. Glaisher, J., C. Flammarion, W. de Fonvielle, and G. Tissandier. *Voyages Aériens.* Paris: L. Hachette, 1870.
17. Glaisher, J., C. Flammarion, W. de Fonvielle, and G. Tissandier. *Travels in the air.* London: Bentley; Philadelphia: J.B. Lippincott, 1871.
18. Glaisher, J., C. Flammarion, W. v. Fonvielle, and G. Tissandier. *Luftreisen.* Leipzig: Friedrich Brandstetter, 1884.
19. Günther, S. Die ältesten Beobachtungen über die Bergkrankheit der Kordilleren. *Arch. Geschichte Naturwiss. Technik. Leipzig* 6:122-131, 1913.
20. Haller, A.v. *Elementa physiologiae corporis humani, vol. 3.* Lausanne: sumptibus Sigismundi d'Arnay, 1761, Lib. 8, Sec. 3, § 7, p. 196.
21. Jourdanet, D. *Les altitudes de l'Amérique tropicale comparées au niveau des mers au point de vue de la constitution médicale.* Paris:J.-B. Baillière, 1861.
22. Jourdanet, D. *Du Mexique au point de vue de son influence sur la vie de l'homme.* Paris: J.-B. Baillière, 1861.

23. Jourdanet, D. *De l'anémie des altitudes et de l'anémie en général dans ses rapports avec la pression de l'atmosphère.* Paris: J.-B. Baillière, 1863.

24. Jourdanet, D. *Le Mexique et l'Amérique tropicale: climats, hygiène et maladies.* Paris: J.-B. Baillière, 1864.

25. Kellogg, R.H. Altitude acclimatization, a historical introduction emphasizing the regulation of breathing. *Physiologist* 11:37-57, 1968.

26. Kellogg, R.H. Balloons and mountains. *Oxford Med. Sch. Gaz.* 23 (No. 3): 15-18, 1971.

27. Kronecker, H. *Ueber die Bergkrankheit mit Bezug auf die Jungfraubahn.* Gutachten über die Frage: "Ob und unter welchen Bedingungen sowohl der Bau als der Betrieb einer Eisenbahn auf die Jungfrau ohne ausnahmsweise Gefährdung von Menschenleben (Gesundheit) möglich sei." Bern, November 21, 1894. (In Schweiz. Landesbibliothek, Bern.)

28. Kronecker, H. *Die Bergkrankheit, mit Unterstützung der "Elizabeth Thompson-Stiftung."* Berlin and Wien: Urban and Schwarzenberg, 1903.

29. Kronecker, H. Das Wesen der Bergkrankheit und ein seltener Fall derselben. *Biol. Centralbl.* 31:771-777, 1911.

30. Lavoisier. Expériences sur la respiration des animaux, et sur les changemens qui arrivent à l'air en passant par leur poumon. *Hist. Acad. Roy. Sci., Paris: Mém. Math. Phys.* (1777): 185-194, 1780. (*Oeuvres de Lavoisier.* Paris 2: 174-183, 1862.)

31. Meyer-Ahrens, C. *Die Bergkrankheit, oder der Einfluss des Ersteigens grosser Höhen auf den thierischen Organismus.* Leipzig: F.A. Brockhaus, 1854.

32. Middleton, W.E.K. *The history of the barometer.* Baltimore: Johns Hopkins Press, 1964.

33. Mosso, A. Sulla respirazione di lusso e la respirazione periodica. *Atti della Reale Accad. dei Lincei, Rendiconti* Anno 282, Serie 4, Vol. 1, Classe di scienze fisiche, matematiche e naturali, Seduta del 4 gennaio 1885, pp. 45-46.

34. Mosso, A. La respiration périodique et la respiration superflue ou de luxe. *Arch. Ital. Biol.* 7-48-127, 1886.

35. Mosso, A. Periodische Athmung und Luxusathmung. *Arch. (Anat.) Physiol.* 1886 Suppl.-Bd.: 37-116, 1886.

36. Mosso, A. *L'Uomo sulle Alpi. Studii fatti sul Monte Rosa...* Milano: Fratelli Treves, 3rd Ed., 1909.

37. Pascal. *Traitez de l'équilibre des Liqvevrs, et de la pesantevr de la masse de l'air...* Paris: Gvillavme Desprez, 1663, pp. 52-55 and 164-195.

38. Smith, J.A. and W.D. Ross, Editors. *The works of Aristotle translated into English.* Oxford: Clarendon Press, 1908-1931, 11 Volumes.

ADAPTATION TO HIGH ALTITUDE

Robert F. Grover *

Cardiovascular Pulmonary Research Laboratory
Division of Cardiology, Department of Medicine
University of Colorado Medical Center
Denver, Colorado

"Variety's the very spice of life."

To paraphrase this well-known quotation of William Cowper (5), "variability's the very spice of adaptation to stress." Certainly this is true of human adaptation to high altitude. To illustrate, let us examine one of the major and most significant aspects of adaptation to altitude, namely the decrease in aerobic working capacity, a phenomenon which has been documented by numerous investigators over the years. In 1969, Buskirk (4) collected all available published reports, from which he summarized the general relationship between altitude and maximum oxygen uptake ($\dot{V}O_2$ max): from sea level to about 1,500 m, there is no consistent alteration in $\dot{V}O_2$ max; however, with further increase in altitude, there is a linear decrease in $\dot{V}O_2$ max at the rate of 10% per 1,000 m. Thus, at 3,000-m altitude, i.e., 1,500 m above the threshold of 1,500 m, $\dot{V}O_2$ max will be 15% less than at sea level. This reduction of 15% in $\dot{V}O_2$ max is an average value obtained from eight different studies conducted around 3,000-m altitude and reported group mean reductions in $\dot{V}O_2$ max, which varied from 7% to 20%. Even this does not reflect the full extent of variability. For example, when Balke (3) studied six men at 3,000 m, individuals

---

*Dr. Grover is the recipient of Senior Investigator Award 010-745 from the Colorado Heart Association.

were found to have values of $\dot{V}O_2$ max which were from as little as 7% to as much as 22% less than when they returned to sea level.

Clearly, there is marked individual variability in this measure of adaptation to high altitude. Such variability is usually considered anathema to the physiologist, for he believes that he will be unable to see the forest for the trees. However, to the extent that this variability is biological, reducing it to a single mean value may well result in discarding valuable information. Stephen Weinstein recognized this ten years ago when he expressed the opinion that, "In these differences between people who show a great decrement and those who do not, we are very likely to find great leverage in working with the effects of altitude." (26).

The physiological importance of measuring maximum oxygen uptake is that it relates to the integrated function of the body's various oxygen transport mechanisms. Hence, as pointed out by Åstrand et al. (2) and again by Saltin et al. (18), a reduction in $\dot{V}O_2$ max results from a reduction in maximum oxygen transport. Therefore, if we wish to understand the physiological basis for the individual variability in the reduction of $\dot{V}O_2$ max at high altitude, we should endeavor to identify the source of variability in the adaptation of the various oxygen transport mechanisms. Furthermore, that mechanism that shows the greatest variability in response may well be the mechanism most important in the overall adaptive process.

Oxygen transported by the systemic circulation has two major components: the quantity of oxygen in each unit volume of arterial blood (termed arterial oxygen content, CaO_2), and the number of unit volumes of blood pumped by the heart per minute (the cardiac output). The former (CaO_2) is a function of the oxygen-carrying capacity of the blood as determined by the hemoglobin concentration, and the extent to which that hemoglobin is saturated with oxygen (SaO_2). In turn, this is influenced by pulmonary ventilation. Cardiac output is determined by the volume of blood ejected by the left ventricle per beat (termed stroke volume) and the number of beats per minute (heart rate). Therefore, in seeking the sources of variability in oxygen transport, we shall examine the nature of adaptation in ventilation, blood oxygenation, and cardiac output.

As man ascends from sea level to high altitude, he is exposed to a progressive decrease in atmospheric pressure (6) and an associated decrease in inspired oxygen pressure. Consequently, alveolar and hence arterial oxygen pressure (PaO_2) falls, stimulating ventilation. However, the ventilatory response to acute hypoxia varies markedly among individuals (12). Furthermore, this is not the only mechanism that operates to alter ventilation during chronic hypoxia. Therefore, it is not surprising that the net increase in ventilation at high altitude is greater in some individuals than in others. This is illustrated by the data of Alexander et al. (1), who studied eight young men, first at sea level and then again after ten days' adaptation to 3,100-m altitude. The decrease in barometric pressure from 760 to 530 Torr reduced inspired oxygen pressure 50 Torr. Had there been no change in ventilation, PaO_2 would also have been lowered by 50 Torr. However, a variable increase in alveolar ventilation did occur, which lowered arterial CO_2 tension by as much as 14 Torr in some individuals, but as little as 3 Torr in others (Fig. 1). Consequently,

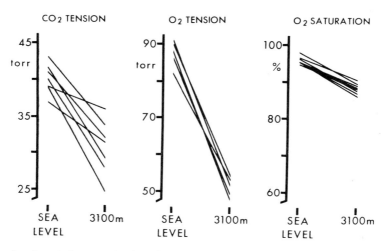

Fig. 1. Ascent from sea level to 3,100-m altitude produces a fall in arterial O_2 tension, which stimulates ventilation. The resulting increase in alveolar ventilation is greater in some individuals than in others. Nevertheless, the fall in O_2 saturation is consistently small (1).

the decrease in PaO_2 varied from 39 to 48 Torr. The resulting absolute values in PaO_2 during exercise ranged from 47 to 56 Torr. In spite of this large and variable reduction in PaO_2, the resulting effect on saturation was small, since the hemoglobin-oxygen dissociation curve is relatively flat in this range of PaO_2. Consequently, in these eight men at 3,100 m, the decrease in SaO_2 varied only from -6% to -9%. Obviously, at higher altitudes with lower values of PaO_2, saturation would be more variable. However, at 3,100 m, the marked individual variability in the response of ventilation had only a minor influence on SaO_2, and hence made little contribution to the variability in oxygen transport.

As stated above, the two determinants of arterial O_2 content are hemoglobin saturation (SaO_2) and hemoglobin concentration which is normally reflected by the hematocrit. After several days at high altitude, an increase in hematocrit is observed in most individuals. This results from fluid shifts and a decrease in plasma volume (7). Hence, the increase in hematocrit indicates a constant total red cell mass contained in a smaller total plasma volume; total blood volume is therefore reduced also. Limited data indicate that rise in hematocrit is of equal magnitude in young men and young women (9,10), i.e., no sex difference. Also, the hemoconcentration response may not be attenuated with increasing age, but the time of onset appears to be delayed (13) (Fig. 2). In the study by Alexander *et al.* (1), all eight young men demonstrated an increase in hematocrit, with some individual variability. Overall, the associated increase in hemoglobin concentration offset the small decrease in saturation, with the result that CaO_2 showed no consistent change

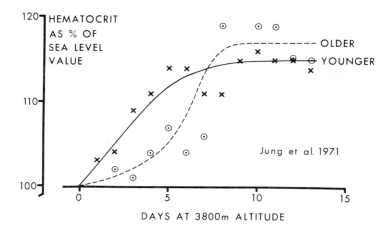

Fig. 2. During the first days following ascent to high altitude, hemoconcentration occurs. The resulting increase in hematocrit is delayed in older men, but is ultimately as great as in younger men (13).

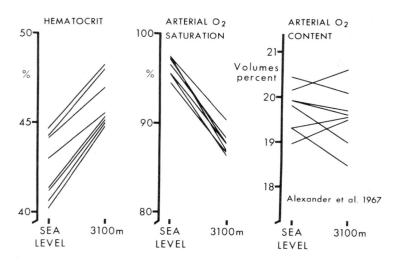

Fig. 3. Adaptation to high altitude includes an increase in hematocrit. The associated rise in hemoglobin concentration increases the O_2 carrying capacity of the blood. This offsets the decrease in O_2 saturation, and as a consequence, arterial O_2 content is not less than at sea level (1).

following ten days' adaptation to high altitude (Fig. 3). From this we must conclude that the variability in oxygen transport does not result from variability in hemoconcentration, ventilation, or blood oxygenation.

In addition to arterial O_2 content, the other major determinant of systemic oxygen transport is cardiac output, which has two components: stroke volume and heart rate. Upon arrival at high altitude, heart rate and cardiac output are greater than at sea level during submaximal exercise, while maximum heart rate and cardiac output remain unaltered (21). However, after several days of adaptation to high altitude, stroke volume decreases. Concurrently, heart rate also decreases. Thus, Alexander et al. (1) found that after ten days' adaptation to 3,100-m altitude, heart rate during several levels of submaximal exercise was essentially the same as it had been at sea level. Hence, with a decrease in stroke volume and no change in heart rate, cardiac output was less than at sea level during submaximal exercise (Fig. 4). Vogel et al. (24) showed that maximal cardiac output is also reduced. It is this decrease in cardiac output that reduces systemic oxygen transport and accounts for the decrease in $\dot{V}O_2$ max at high altitude. Furthermore, the individual variability in the decrease of $\dot{V}O_2$ max results largely from the variability in reduction of stroke volume.

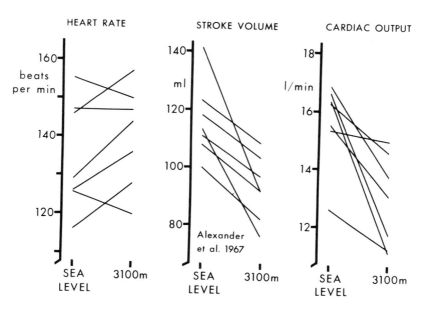

Fig. 4. Following ten days' adaptation to 3,100-m altitude, individual heart rates during submaximal exercise show no consistent change from sea level values. However, all persons showed a decrease in stroke volume that was greater in some individuals than in others. Consequently, the decrease in cardiac output shows marked individual variability (1).

Racial variability is also observed in response to change in altitude. The decrease in $\dot{V}_{O_2}$ max and associated changes in oxygen transport described above have been observed consistently by all investigators taking sea level residents to high altitude. In contrast, the Quechua Indians native to high altitude in the Andes do not demonstrate these changes. Vogel et al. (23) observed that young men native to 4,300 m in Peru have virtually the same $\dot{V}_{O_2}$ max at both 4,300 m and at sea level, whereas in sea level natives, $\dot{V}_{O_2}$ max was reduced 30% following ascent to 4,300 m (24). Furthermore, the Quechua Indians had a $\dot{V}_{O_2}$ max at 4,300 m equal to that of the lowlanders at sea level. These remarkable observations appear to be explained by the fact that in the high altitude natives, their cardiac index is no less at 4,300 m than at sea level (Fig. 5). By maintaining cardiac output, they also preserve oxygen transport and hence suffer no decrease in $\dot{V}_{O_2}$ max at high altitude. How they accomplish this, we do not know. This acclimatization to high altitude results from something more than simply birth and lifelong exposure to the hypoxic environment. When Hartley et al. (11) studied men native to 3,100 m in Colorado, they found that cardiac output was low by sea level standards — comparable to the reduced cardiac output that developed in the sea level residents transported to 3,100 m. Furthermore, when the high altitude natives of Colorado were taken to sea level, they demonstrated an increase in stroke volume (11) and in $\dot{V}_{O_2}$ max (8), changes that are the reciprocal of those seen with ascent to altitude. Clearly, then, there is an important aspect of variability between high altitude natives in North America and in South America.

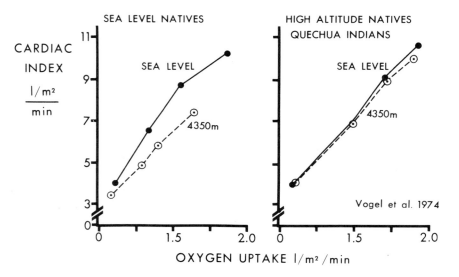

Fig. 5. When sea level natives are taken to high altitude, cardiac index decreases, which reduces O_2 transport and maximum O_2 uptake (24). These changes are not observed in Quechua Indians native to the high Andes (23). Hence, this phenomenon shows racial variability.

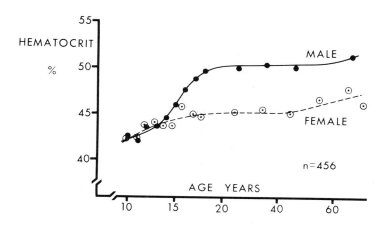

Fig. 6. Age and sex variations in hematocrit among long-term residents at 3,100-m altitude (17,22).

Among people native to 3,100 m in Colorado, age and sex account for significant variability in the hematologic response to chronic hypoxia. At all ages and in both sexes, values of hematocrit are higher at 3,100-m altitude than at sea level. Prior to puberty, boys and girls have the same hematocrit (42%). However, with sexual maturation, males develop hematocrits that are significantly higher than in females (22). Thus, in healthy adults, the mean values for hematocrit are 50% in males and 45% in females (17) (Fig. 6). These increases in hematocrit reflect a larger total body red cell mass which, as demonstrated by Weil *et al.*, increases in response to the decrease in arterial O_2 saturation (not O_2 tension). Since the O_2 saturation is decreased from 96% at sea level to 89% at 3,100 m, only a modest increase in red cell mass and hematocrit would be expected.

Leadville residents at 3,100-m altitude have arterial blood oxygenation which is precariously located on the "shoulder" of the hemoglobin-oxygen dissociation curve at PaO_2 57 Torr and SaO_2 89% (25). Consequently, any variation in PaO_2 produces proportional changes in SaO_2 and consequently hematocrit. Okin *et al.* (17) found the normal range in hematocrit for adults at 3,100 m to be 46%-58% for men and 40%-53% for women. A significant number of men living at 3,100 m develop excessive polycythemia with hematocrits greater than 60%, and as would be anticipated, they are also abnormally hypoxemic. Because of associated mental symptoms, this condition resembles chronic mountain sickness, or Monge's disease (16), and has age and sex characteristics; it is observed almost exclusively in older men who have lived at high altitude for over 30 years.

Recently, Kryger *et al.* (15) studied 17 men living at 3,100 m who had hematocrits over 60%. The mean SaO_2 was 84% when these men were awake, but fell markedly during sleep. They had only minimal CO_2 retention and the majority were clinically free of lung disease. Although their hypoxemia was not obviously due to generalized alveolar hypoventilation (and therefore probably due to poor matching of regional perfusion to ventilation), they did respond dramatically to chronic stimulation of ventilation with medroxyprogesterone acetate (Provera) (14). Ironically then, this maladaptation to high altitude in older males responded to treatment with a female sex hormone derivative. Once again, we see the roles of age and sex in the variable responses to high altitude.

One other maladaptation to high altitude that varies markedly with age is high altitude pulmonary edema (HAPE). This is a relatively rare but serious condition which develops in otherwise healthy young individuals within one to three days following rapid ascent to altitudes above 2,800 m. The largest experience has been reported by the army of India (20), when they moved vast numbers of young men from low elevations to high areas in the Himalaya. Paradoxically, high-altitude residents are also susceptible when they re-enter their usual high altitude environment following a sojourn at low altitude. Scoggin et al. (19) reviewed the records of St. Vincent's Hospital in Leadville and found 39 admissions for HAPE in a six-year period. All but three were less than 21 years old and equally divided between males and females. Thus, the occurrence of HAPE shows marked variability with age, being primarily an illness of children and young adults. Since we do not know the pathophysiology of HAPE, we cannot explain this prevalence in the younger age group.

Now that we have examined the individual variability in adaptation to high altitude, we may reflect on the significance of this "adaptation." Clearly the process of adaptation does not counteract the fundamental hypoxic stress. In response to the decrease in atmospheric oxygen tension, the fall in alveolar PO_2 is lessened only slightly by hyperventilation. While the resulting small decrease in oxygen saturation is offset by hemoconcentration to restore arterial O_2 content, systemic oxygen transport is not corrected since cardiac output is reduced. Consequently, $\dot{V}O_2$ max is never restored to its pre-altitude value. Hence, "adaptation" does not restore the aerobic working capacity which the individual had at sea level. In addition to this physical handicap, the "adapted" individual must also tolerate impairments of central nervous system function, including a slowing of decision-making, errors in judgment, errors in measuring and computation, sleep disturbances, difficulty in learning new material, and impaired vision in dim light. Adaptation is therefore the process of adjustment to the hypoxic stress of the high altitude environment. Presumably with these adjustments, the hypoxic stress is more tolerable, but physical and mental function remain impaired. As with adjustment to any stress, age and sex contribute to the variable responses among individuals.

In summary, when man ascends to high altitude, one of the most significant consequences is a reduction in his aerobic working capacity. Individual variability in this response reflects variability in adaptation of the various oxygen transport mechanisms. Analysis of these mechanisms reveals that at altitudes up to 4,300 m, variabilities in hyperventilation, hemoconcentration, and blood oxygenation are not the major limiting factors in oxygen transport. Rather, it is the variable decrease in cardiac stroke volume (and hence in cardiac output) that reduces oxygen transport and limits oxygen uptake. Variability in response to high altitude also includes maladaptation. Soon after ascent, pulmonary edema may occur; this is most prevalent in children and young adults of both sexes. With prolonged residence at high altitude, excessive polycythemia may develop; this is observed almost exclusively in older males. Virtually all of the normal adaptations as well as the maladaptations to high altitude are reversible by descent to sea level.

REFERENCES

1. Alexander, J.K., L.H. Hartley, M. Modelski, and R.F.Grover. Reduction of stroke volume during exercise in man following ascent to 3,100 m altitude. *J. Appl. Physiol.* 23:849-858, 1967.
2. Åstrand, P.O., T.E. Cuddy, B. Saltin, and J. Stenberg. Cardiac output during submaximal and maximal work. *J. Appl. Physiol.* 19:268-274, 1964.
3. Balke, B. Work capacity and its limiting factors at high altitude. In: *The physiological effects of high altitude.* Edited by W. H. Weihe. New York: Macmillan, 1964, pp. 233-240.
4. Buskirk, E.R. Decrease in physical work capacity at high altitude. In: *Biomedicine problems of high terrestrial elevations.* Edited by A. H. Hegnauer. Natick, Mass.: U.S. Army Res. Inst. Environ. Med., 1969, pp. 204-222.
5. Cowper, W. The task. Book II: The timepiece, line 606, 1785. In: *Familiar quotations,* 13th edition. Edited by J. Bartlett. Boston: Little, Brown, and Co., 1955, p. 364.
6. Dill, D.B. and D.S. Evans. Report barometric pressure! *J. Appl. Physiol.* 29:914-916, 1970.
7. Dill, D.B., S.M. Horvath, T.E. Dahms, R.E. Parker, and J.R. Lynch. Hemoconcentration at altitude. *J. Appl. Physiol.* 27: 514-518, 1969.
8. Grover, R.F., J.T. Reeves, E.B. Grover, and J.E. Leathers. Muscular exercise in young men native to 3,100 m altitude. *J. Appl. Physiol.* 22: 555-564, 1967.
9. Hannon, J.P., J.L. Shields, and C.W. Harris. Effect of altitude acclimatization on blood composition of women. *J. Appl. Physiol.* 26: 540-547, 1969.
10. Hansen, J.E., J.A. Vogel, G.P. Stelter, and C.F. Consolazio. Oxygen uptake in man during exhaustive work at sea level and high altitude. *J. Appl. Physiol.* 23:511-522, 1967.
11. Hartley, L.H., J.K. Alexander, M. Modelski, and R.F. Grover. Subnormal cardiac output at rest and during exercise in residents at 3,100 m altitude. *J. Appl. Physiol.* 23:839-848, 1967.
12. Hirshman, C.A., R. E. McCullough, and J.V. Weil. Normal values for hypoxic and hypercapnic ventilatory drives in man. *J. Appl. Physiol.* 38: 1095-1098, 1975.
13. Jung, R.C., D.B. Dill, R. Horton, and S.M. Horvath. Effects of age on plasma aldosterone levels and hemoconcentration at altitude. *J. Appl. Physiol.* 31: 593-597, 1971.
14. Kryger, M., R.E. McCullough, D.D. Collins, C.H. Scoggin, J.V. Weil, and R.F. Grover. Treatment of excessive polycythemia of high altitude with respiratory stimulant drugs. *Amer. Rev. Resp. Dis.* (In press).
15. Kryger, M., R.E. McCullough, R.C. Doekel, D.D. Collins, J.V. Weil, and R.F. Grover. Excessive polycythemia of high altitude. *J. Appl. Physiol.* (In press).
16. Monge-M, C. and C. Monge-C. Chronic mountain sickness. In: *High altitude diseases.* Springfield, Ill.: Charles C. Thomas, 1966, pp. 32-50.
17. Okin, J.T., A. Treger, H.R. Overy, J.V. Weil, and R.F. Grover. Hematologic response to medium altitude. *Rocky Mountain Med. J.* 63:44-47, 1966.

18. Saltin, B., R.F. Grover, C.G.Blomqvist, L.H. Hartley, and R. L. Johnson, Jr. Maximal oxygen uptake and cardiac output after two weeks at 4,300 m. *J. Appl. Physiol.* 25: 400-409, 1968.
19. Scoggin, C.H., T.M. Hyers, J.T. Reeves, and R.F. Grover. High altitude pulmonary edema in the children and young adults of Leadville, Colorado. *New Eng. J. Med.* 297: 1269-1272, 1977.
20. Singh, I. and S.B. Roy. High altitude pulmonary edema. In: *Biomedicine problems of high terrestrial elevations.* Edited by A.H. Hegnauer. Natick, Mass.: U.S. Army Res. Inst. Environ. Med., 1969, pp. 108-120.
21. Stenberg, J., B Ekblom, and R. Messin. Hemodynamic response to work at simulated altitude, 4,000 m. *J. Appl. Physiol.* 21:1589-1594, 1966.
22. Treger, A., D.B. Shaw, and R.F. Grover. Secondary polycythemia in adolescents at high altitude. *J. Lab. Clin. Med.* 66: 304-314, 1965.
23. Vogel, J.A., L.H. Hartley, and J.C. Cruz. Cardiac output during exercise in altitude natives at sea level and high altitude. *J. Appl. Physiol.* 36: 173-176, 1974.
24. Vogel, J.A., L.H. Hartley, J.C. Cruz, and R.P. Hogan. Cardiac output during exercise in sea level residents at sea level and high altitude. *J. Appl. Physiol.* 36:169-172, 1974.
25. Weil, J.V., G. Jamieson, D.W. Brown, and R.F. Grover. The red cell mass-arterial oxygen relationship in normal man. *J. Clin. Invest.* 47:1627-1639, 1968.
26. Weinstein, S.A. General discussion. In: *Biomedicine problems of high terrestrial elevations.* Edited by A. H. Hegnauer. Natick, Mass.: U.S. Army Res. Inst. Environ. Med., 1969, p. 322.

COMPARATIVE ALTITUDE ADAPTABILITY
OF YOUNG MEN AND WOMEN [a]

John P. Hannon

Department of Comparative Medicine
Letterman Army Institute of Research
Presidio of San Francisco, California

It has been said that women adapt more readily to high altitude exposure than men (9,22). Is this a subjective impression or is it an experimentally verifiable fact? At the present time I do not feel that we can answer this question to the complete satisfaction of everyone. Substantive evidence showing enhanced adaptability of women is far from conclusive, largely because female adaptive characteristics are so infrequently studied. Even less attention has been given to direct comparison of male and female response characteristics in studies containing statistically adequate numbers of subjects from both sexes; such studies are virtually non-existent. For the most part, therefore, comparisons must be made on the basis of data acquired in independently conducted investigations. Although this is not the best approach, valuable information can be obtained, provided of course that the experimental conditions and data collection procedures are equivalent, or nearly so. In our studies conducted on Pikes Peak (elev. 4,300 m), there were a number of occasions when such equivalency would seem to apply. Here I shall compare male and female data so obtained. These data show many similarities, some imparting a potential adaptational advantage to women. I shall first describe briefly the subject groups involved, then certain of the nutritional, cardiopulmonary, and acid-base characteristics displayed by these subjects during the course of early high altitude acclimatization.

---

[a] The opinions or assertions contained herein are the private views of the author and are not to be construed as official or as reflecting the views of the Department of the Army or the Department of the Defense.

SUBJECTS

Unless otherwise indicated, the male volunteers were young soldiers stationed at Brooke Army Medical Center in San Antonio, Texas (elev. 200 m). Most were students who had completed a medical technician's course immediately before participating in our studies. Two groups of female volunteers were studied, one from the student population at the University of Missouri (elev. 200 m) and the other from the student population at the University of Oregon (elev. 140 m). Low altitude measurements were made at these three locations after a period of preconditioning to minimize the effects of extraneous variables. The male subjects in Texas, for example, were housed continuously for one week in air conditioned barracks to reduce or eliminate potential confounding effects of heat exposure. When nutritional responses were being investigated, the subjects (both male and female) subsisted on the test diets for seven to ten days prior to any measurements. Both female groups were supplied with supplementary dietary $FeSO_4$ at low altitude to assure adequacy of their body iron stores in view of an anticipated increase in erythropoietic activity at high altitude. Transport to Pikes Peak was made by commercial aircraft to Denver and by automobile to our summit laboratory. The total elapsed time ranged from about five to eight hours. The male subjects remained at altitude for periods ranging from a few days to three weeks, depending on the nature of the experiment. The females from Missouri remained nine weeks, while those from Oregon remained 11 weeks. While at altitude, all of the male subjects lived in our laboratory. The females from Missouri lived in and worked at the Pikes Peak Summit House, a tourist facility, and returned to the laboratory, which was located nearby, for most of the experimental measurements. The females from Oregon resided in our laboratory for the first two weeks and thereafter in the Summit House. In all of these studies, a conscientious effort was made to obtain comparable experimental and environmental conditions, except for elevation, at both low and high altitude.

NUTRITION

All mammalian species so far studied, including humans, display a loss of appetite and concomitant hypophagia during the early stages of exposure to elevations equivalent to those of Pikes Peak or higher (12,13). Comparison of this response in men and women provides us with our first substantive evidence that females may indeed adapt to altitude more readily than males. Accordingly, if the daily energy intake of subjects receiving a normal American diet is plotted so as to remove the influence of body size (Fig. 1), we find that both sexes show similar intake decrements during the first few days of altitude exposure. Thereafter, however, women recover much more rapidly than men; after one week the female values approximate those observed earlier at low altitude, whereas the male values remain depressed even at the end of two weeks. Loss of appetite at high altitude is one of the symptoms

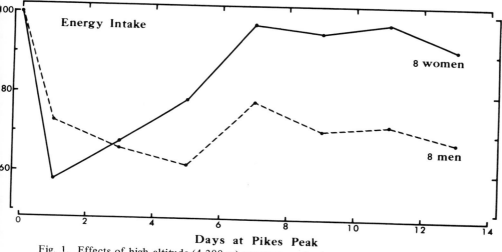

Fig. 1. Effects of high altitude (4,300 m) exposure on food energy intake. Indicated levels are expressed as percent of the low altitude control value: 2,980 kcal/24 hr for men and 1,980 kcal/24 hr in women.

of acute mountain sickness, and in women at least the degree of appetite suppression appears to be directly related to total symptom severity (12). The more transient nature of the female response may be attributable to their rapid recovery from acute mountain sickness (unpublished data). But this offers only a partial explanation, since the male values remain depressed after an exposure interval of two weeks, a time when overt symptoms are essentially absent.

Although energy intake is reduced at high altitude, body energy expenditure remains constant and may even be elevated during the first few days of exposure (12). Energy balance, therefore, becomes negative and the deficit must be met by the utilization of endogenous energy stores. This leads to a loss of body weight that is directly proportional to the energy deficit. The men shown in Fig. 1, for example, lost 3.42 kg, or 4.86% of their initial body weight during the two-week sojourn, while the women lost 0.88 kg, or 1.49% of their initial body weight. Comparable losses for both men and women are reported elsewhere (15). Carbohydrate in the form of glycogen is perhaps the first energy source to be utilized, but the amount available is only sufficient to meet a short-term deficit, i.e. a few hundred kilocalories. Other endogenous energy stores, therefore, must be oxidized. Of these, body protein seems to be the most readily available; thus, negative nitrogen balances are observed even on the first day of altitude exposure (12). Body fat may also serve as an endogenous source of energy, but initially at least, it seems to be less readily available than protein. Depending on the duration and degree of body energy deficit, fat may supplant protein as a primary endogenous energy source. Such fat losses have been recorded in men by means of densitometry, creatine excretion, and total body ^{40}K techniques (24), in women by anthropometry (15), and in laboratory animals by carcass analysis (13). Direct comparison of the two sexes, using the same technique, has not been reported, but on the basis of available evidence, e.g. long-term weight loss, it would appear likely that men utilize body fat stores to a greater extent than women (12). In both sexes, however, extensive fat catabolism probably requires adaptive changes, particularly in terms of capacity for fatty acid oxidation. As summarized in Fig. 2, the period of inadequate food intake is associated with an increase in plasma free fatty acid levels. In women this response is transient, while in men it becomes progressively more pronounced as the exposure period is extended to two weeks. Both sexes show a transient reduction in plasma glucose levels, the initial decrease being somewhat greater in women than in men. These same women showed, in addition, transient increases in plasma ketone and glycerol concentrations; the response pattern in both instances was parallel to

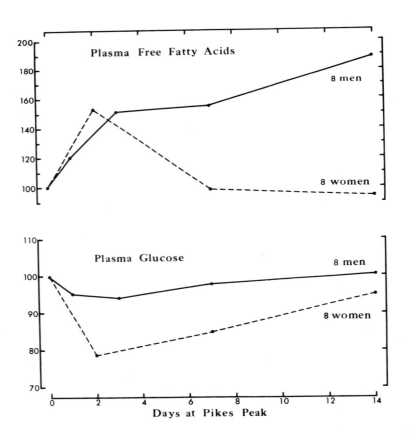

Fig. 2. Effects of high altitude (4,300 m) exposure on plasma free fatty acid and glucose levels. Indicated concentrations are expressed as percent of the low altitude control value: free fatty acids, 420 μ mol/ℓ for males and 473 μ mol/ℓ for women and glucose, 88 mg/100 ml for men, 104 mg/100 ml for women.

changes in free fatty acid concentration (unpublished data). The same men showed sustained decreases in plasma cholesterol and total lipid concentrations and a sustained increase in plasma phospholipid concentration (19).

CARDIOPULMONARY FUNCTION

Altitude acclimatization is associated with a multitude of pulmonary and cardio-vascular changes, which for the most part lead to improved oxygen transport from the ambient environment to the metabolizing cells of the body. These changes have been studied extensively in male subjects, but have received relatively little attention in women. Here I shall limit consideration to two general questions. First, do women exhibit a distinct physiologic advantage relative to men, particularly during the early stages of altitude exposure? And second, if they do, what are the key features that are responsible for this advantage?

Some 75 years ago, a young British woman, Mabel Purefoy FitzGerald, arrived in Colorado as a member of the Anglo-American Expedition to Pikes Peak. While the male expedition members were conducting their now famous studies on the mountain summit (6), Miss FitzGerald traveled from town to town in the mountain-ous regions of the state, collecting data on the pulmonary and blood characteristics of local residents (7). Her techniques were rather simple by modern standards, and certain of her calculations contained inadequate assumptions. Nevertheless, she obtained some of the first evidence that women possess at least one physiologic characteristic that would impart greater altitude adaptability than that seen in men. At various elevations up to 12,500 ft (3,800 m), she showed that hyperventilation led to a decrease in alveolar P_{CO_2}. This effect progressed with elevation; for both men and women the decrement averaged about 4.2 mmHg for each 100 mmHg decrease in barometric pressure. She thus showed that hyperventilation diminished to an appreciable extent the fall in alveolar P_{O_2} associated with altitude exposure. Perhaps more significantly, she showed that women at all elevations had alveolar P_{CO_2} values that were some 3 mmHg lower than those observed in men, and because of this difference, she calculated that at any elevation women had an alveolar P_{O_2} value averaging about 3 mmHg greater than that observed in men. Although Miss FitzGerald did not speculate on the point, the sex difference in alveolar P_{O_2} should lead to an equivalent difference in arterial P_{O_2}. If this were true, then women should have a higher oxygen gradient between blood and metabolizing tissue than men.

Our studies conducted on Pikes Peak have confirmed Miss FitzGerald's observations and the predictions concerning arterial P_{CO_2} and P_{O_2}. We have shown, in addition, that sex differences in gas transport persist over the course of the early acclimatization process. Increases in resting ventilation shortly after the onset of exposure were recorded in both sexes. Initially the response may be more pronounced in women than in men; women showed an 18% increase in ventilation after

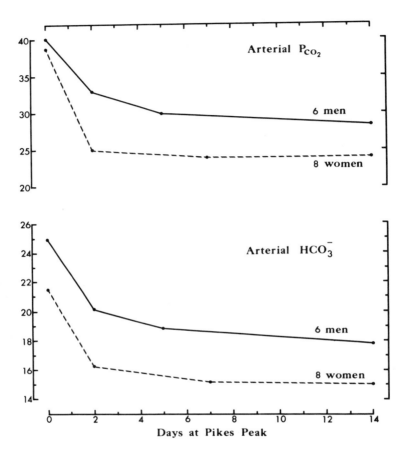

Fig. 3. Effects of high altitude (4,300 m) exposure on arterial P_{CO_2} and HCO_3^- concentration. Ordinate values expressed as mmHg for P_{CO_2} and meq/1 for HCO_3^-

12 hours on Pikes Peak, while an equivalent response was observed in men after two days. Over the first few days of exposure, the increment is largely achieved by an increase in respiratory frequency. However, this effect is transient in both sexes and as acclimatization progresses, the ventilatory increase is maintained by an elevated tidal volume. As discussed elsewhere (17), the latter provides a more efficient process of pulmonary gas exchange. In women full acclimatization of resting ventilation requires about four weeks on Pikes Peak (16). The time period for acclimatization in men at the same site is not known with certainty, since studies of adequate duration have not been conducted. At the end of two weeks, their resting ventilation appears to be increasing at an appreciable rate (17). Other aspects of ventilatory

function at altitude, including male-female comparisons of the increases in maximum breathing capacity and the decreases in vital capacity, are contained in an earlier report (14).

Blood gas measurements reveal (Fig. 3) a marked decrease in arterial P_{CO_2} shortly after the onset of altitude exposure. The response in both men and women is almost a mirror image of increase in ventilation. But in women we find, as predicted earlier, significantly lower values than those observed in men. This effect persists for exposure intervals of at least 14 days. These measurements also revealed marked decrements in arterial bicarbonate concentration which we shall consider in more detail later.

Examination of arterial P_{O_2} values (Fig. 4) also verifies our earlier prediction that because of more pronounced hyperventilation, women do indeed have higher arterial oxygen tensions than men. The differences between men and women ranged from 2 to 6 mmHg, depending on the period of exposure. Men and women are similar in terms of the gradual rise in arterial P_{O_2} as exposure is extended to 14 days. Beyond this point comparative data are not available.

Since the sex differences and acclimatization changes in arterial P_{O_2} amount to only a few mmHg, one might question their physiologic significance. One approach to this question is to calculate, for various elevations, theoretical values for arterial or alveolar P_{O_2} assuming no hyperventilation (i.e. no change in R.Q. or P_{CO_2}), and then compare actually measured values at various exposure periods to these theoretical values. The results of such an endeavor are shown in Fig. 5. The measured

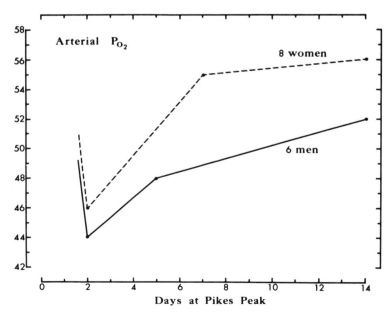

Fig. 4. Effects of high altitude (4,300 m) exposure on arterial P_{O_2}. Ordinate values expressed as mmHg.

alveolar PO_2 values were obtained under basal conditions from women during the course of a 78-day sojourn on Pikes Peak. The theoretical value of 40 adjacent to 4,300 m (labeled O*) was calculated from the Boothby equation (3), assuming PCO_2 to be 38 mmHg and R.Q. to be 0.82 (the average measured values at low altitude). After 12 hours on the mountain summit, the actual measured value for alveolar PO_2 was 46.6 mmHg. Thus, hyperventilation in this context was responsible for a theoretical increase of 6.6 mmHg or a decrease in equivalent altitude of about 2,000 ft (610 m). Thereafter, alveolar PO_2 continued to increase but at a progressively diminishing rate. At 78 days it averaged 54.6 mmHg, an altitude equivalent of 10,300 ft (3,141 m). Over this time period, therefore, hyperventilation was responsible for a theoretical altitude reduction of 4,000 ft (1,220 m). Those who

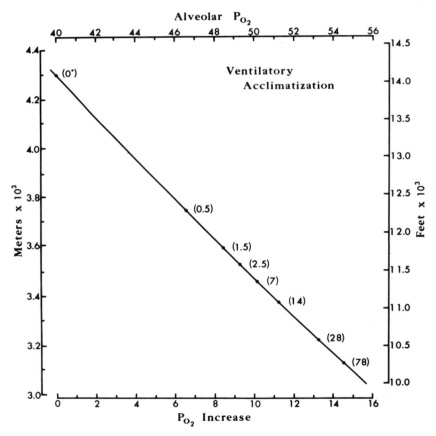

Fig. 5. Effects of hyperventilation on alveolar PO_2 during the course of high altitude (4,300 m) acclimatization. Values in parentheses indicate the exposure interval in days and the altitude at which the measured PO_2 values would be seen if hyperventilation did not occur. (O*) represents an initial value for exposure on Pikes Peak as calculated by the Boothby equation (3), assuming alveolar PCO_2 of 38 and an R.Q. of 0.82.

have sojourned in the high mountains can readily appreciate the physiologic signifi-cance of elevation changes of this magnitude. Although we do not have comparable data for male subjeccts, such data would probably show comparable changes as a function of exposure duration. The acclimatization response in men, however, would differ in one important respect from that of women, namely, male alveolar or arterial PO_2 values would lag a few mmHg behind those of women. A sex differ-ence of 2 mmHg would correspond to about 550 ft (152 m), and 5 mmHg to about 1,400 ft (427 m). At elevations comparable to Pikes Peak, these seemingly small differences could be of physiologic importance.

In addition to hyperventilation, oxygen transport to the tissues is also facilitated by an increase in the oxygen-carrying capacity of the blood, an effect first recog-nized and demonstrated by Bert (2). It is attributable to an increase in circulating red cell and consequently hemoglobin concentrations. The increase is proportion-ate to the altitude or residence. This was elegantly shown by Miss FitzGerald in her studies of men and women living at various elevations in the United States (7,8). In both sexes she found hemoglobin concentrations to be increased by about 10% for every 100 mmHg decrease in barometric pressure.

Increased hemoglobin concentration is attributable to two factors, one found characteristically in the altitude native, the other in the short-term sojourner. In the native, high hemoglobin concentrations are achieved almost entirely by an in-crease in total circulating red cell volume, while in the sojourner, increments of a somewhat smaller magnitude are achieved by hemoconcentration (14). During the course of acclimatization, hemoconcentration serves as a more readily available interim means for increasing the oxygen-carrying capacity of the blood. It is gradu-ally supplanted by the more slowly evolving increase in total circulating red cell volume, a response requiring exposure intervals of a year or more to become fully developed (14,23). In part, therefore, the success of early acclimatization depends upon the degree of hemoconcentration, hence the degree of improvement in blood oxygen-carrying capacity achieved by the individual sojourner. If the response is more pronounced in women, it could contribute to their apparent adaptability advantage.

The first evidence that hemoconcentration may occur is found in a report pub-lished in 1901 by Campbell and Hoagland (4), two Colorado Springs physicians who investigated various conditions leading to hemoconcentration, especially alti-tude exposure. They showed in three rabbits and ten humans that substantial hemo-concentration, as evidenced by red blood cell count, occurred during a 90-min cog train ride from the base to the summit of Pikes Peak. But more importantly, their human data show that the average red blood cell count increased by 5.3% in men and by 11.4% in women. Application of a *t*-test to their data reveals a statistically significant sex difference.

In our studies of men and women we have never attempted to replicate the mea-surements made by Campbell and Hoagland, but we have obtained data that are consistent with their observations as well as with the premise that hemoconcentra-tion is more pronounced in women than in men. At low altitude, as well as at any time point in altitude acclimatization, blood hemoglobin levels in men are always higher than those of women (14). However, if we compare the percentage increase in hemoglobin concentrations in the two sexes, we find (Fig. 6) a more pronounced effect in women than in men. Because of more pronounced hemoconcentration, more pronounced hyperventilation, higher arterial PO_2 and higher O_2 saturation, the women in our studies also showed relatively higher arterial O_2 content values during altitude acclimatization compared to those observed in men (Fig. 6). Again, the absolute values recorded for men at any point in the acclimatization process were always higher than those recorded for women. It is of interest to note that as altitude acclimatization progressed, the arterial oxygen content of women began to exceed sea level values after exposure for four weeks. The time to reach asymptotic

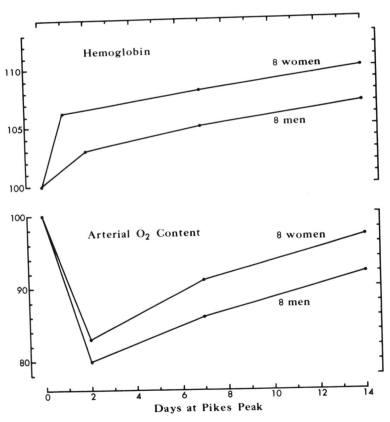

Fig. 6. Effects of high altitude (4,300 m) exposure on venous hemoglobin and arterial O_2 content. Indicated levels are expressed as percent of the low altitude control value: hemoglobin, 14.9 g/100 ml for men, 13.7 g/100 ml for women; and O_2 content, 19.4 ml O_2/100 ml blood for men, 17.3 ml/100 ml blood for women.

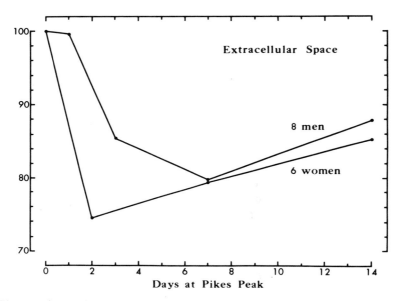

Fig. 7. Effects of high altitude (4,300 m) exposure on extracellular volume, estimated by thiocyanate dilution in men and sucrose dilution in women. Indicated levels are expressed as percent of the low altitude control value: 17.3 ℓ for men, 10.2 ℓ for women.

levels appears to exceed our longest exposure intervals, 78 days. The time course of this change in men remains to be determined.

Hemoconcentration results in large measure, if not entirely, from a reduction in extracellular fluid volume (11). One would expect, therefore, parallel changes in the two variables as a function of altitude exposure. This would seem to occur even though we used different indicators to estimate extracellular fluid volume in men and women. Thiocyanate space was measured in men, while sucrose space was measured in women. In both instances, the volume was estimated by conventional dilution-extrapolation procedures. If it is assumed that altitude exposure affects only the distribution volume of the two indicators, not their tissue distribution characteristics, and if distribution volumes are plotted as percentages of control values obtained at low altitude to allow for differences attributable to body size, then the results shown in Fig. 7 are obtained. Extracellular fluid volume seems to undergo a much more abrupt decrease in women than in men during the first few days of altitude exposure. This would be consistent with the enhanced hemoconcentration seen in women during the same period. Obviously, this interpretation must be accepted with reservation until the results are verified with equivalent technical procedures.

ACID-BASE REGULATION

Since the time of Bert's pioneering investigations (1) or before, it has been known that altitude exposure is associated with a rather marked reduction in the total carbon dioxide content of the blood (20). The earliest workers attributed this loss to hyperventilation and a resultant increase in the rate of dissolved CO_2 removal from the blood. However, it soon became evident that dissolved CO_2 represented only a small fraction (about 5%) of the total amount present in the blood. Most was present as bicarbonate. Thus, shortly after the turn of the century, efforts were being made to determine the mechanisms responsible for the loss of bicarbonate at high altitude. Some postulated hypoxic-acidosis as the cause. That is, hypoxia led to an increased lactic acid production; this caused bicarbonate to be converted to CO_2, which in turn stimulated the respiratory center, leading to an enhanced removal of the CO_2 so formed (10). The failure to find appreciable changes in blood lactic acid or hydrogen ion concentration in altitude sojourners led to the demise of the acidosis theory. In its place, other early workers such as Douglas *et al.* (6) proposed an enhanced rate of renal bicarbonate excretion as the regulatory mechanism. Experiments involving acute, simulated altitude exposure (20) seemed to verify this proposal, as did studies of acute alkalosis induced by normoxic hyperventilation (5). Enhanced renal bicarbonate excretion is now commonly accepted as the primary mechanism for lowering the blood bicarbonate concentration of altitude sojourners (18). Such acceptance, however, does not constitute proof, and in this regard it is noteworthy that very little effort has been made to actually measure changes in bicarbonate excretion during an altitude sojourn or to assess the contributions of other potential mechanisms for regulating acid-base balance. Several of our studies have indirectly or directly addressed this problem. They show that enhanced bicarbonate excretion plays only a minor role, if any, in compensating for the hyperventilatory alkalosis associated with exposure to an elevation of 4,300 m.

On the basis of the data presented in Fig. 3, it is apparent that plasma bicarbonate levels undergo a progressive and rather pronounced reduction during the first week or so of altitude exposure. Appropriate calculations also show that a large fraction of the bicarbonate contained in the body fluids is lost during the course of acclimatization. Thus, total body bicarbonate can be estimated by the following equation:

$$(HCO_3^-)_{TB} = 2/3 \ (HCO_3^-)_p \ x \ T.B.W.$$

where: $(HCO_3^-)_p$ is the measured bicarbonate concentration in plasma water,

the 2/3 factor corrects for intra- and extracellular volume and bicarbonate concentration differences, and

T.B.W is the total body water (estimated by body densitometry measurements.)

In men, calculated losses of 98, 148, and 213 meq were obtained after exposure intervals of 2, 5, and 14 days, respectively. In women, the losses were 114, 135, and 132 meq after 2, 7, and 14 days, respectively. In relative terms total bicarbonate in body water is reduced in both sexes to about 70% of the low altitude value (Fig. 8). The early decrements appear to be more pronounced in women than in men.

In either sex the magnitude of the calculated bicarbonate losses, if excreted by the kidney, should lead to an alkaline urine, especially during the first day or two of altitude exposure. But when urine acidity was actually measured, this was not found to be the case. Accordingly, over a four-day interval at low altitude, over the first week of exposure on Pikes Peak, and periodically thereafter, 24-hr urine collections were obtained from the women shown in Fig. 8. Similar collections were obtained from a group of seven soldiers stationed in Denver (1,610 m) initially, but taken to Pikes Peak for a two-week sojourn. At low altitude, average pH values of about 5.75 were recorded for both sexes. During the first 24-hr period on Pikes Peak the average value for men was 5.84 and for women, 5.92. In subsequent 24-hr collections, average pH values ranged from 5.30 to 5.60. The highest value observed in any subject was 7.06; this occurred in a soldier who, because of severe mountain sickness, lost a considerable amount of acid by vomiting. In view of the distinctly acid urine values observed on Pikes Peak, one would expect very little bicarbonate excretion. This was verified in the soldiers, but not in the women because of technical difficulties. Thus, van Slyke measurements of total urine carbon dioxide and appropriate calculations of urine bicarbonate levels in men showed an average bicarbonate excretion of 1.10 mM/24 hr in Denver and 1.91 mM/24 hr over the first seven days of Pikes Peak; the highest individual value, 14.74 mM/24 hr, was observed on the first day of exposure in the aforementioned soldier who had vomited. Clearly,

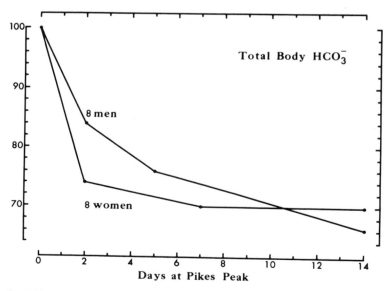

Fig. 8. Effects of high altitude (4,300 m) on total bicarbonate contained in body fluids. Indicated levels expressed as percent of the low altitude control values: 622 meq for men, 445 meq for women.

urinary loss of bicarbonate will not account for the large calculated decrements of total body bicarbonate noted earlier. What alternative mechanisms, therefore, might be responsible for these decrements?

At least a partial answer to this question was obtained from measurements of titratable acid and ammonia excretion. These measurements showed that sizable reductions in total body bicarbonate could be attributed to a diminished rate of bicarbonate regeneration by the renal tubular cells. In the case of women, a marked reduction in both titratable acid and ammonia excretion was observed on the first day of altitude exposure (Fig. 9). Thereafter, ammonia excretion remained low, while titratable acid excretion recovered to low altitude levels after two weeks or so on Pikes Peak. Over the first two week of exposure, comparable data were obtained

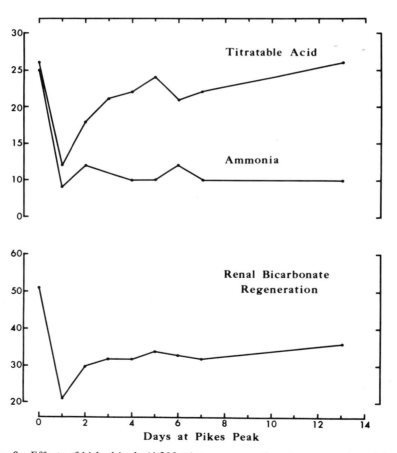

Fig. 9. Effects of high altitude (4,300 m) exposure on the urinary excretion of titratable acid and ammonia, and on the renal regeneration of bicarbonate by women. Ordinate values are expressed meq/24 hr.

from the Denver soldiers. On the basis of these data, we calculated the rate of renal bicarbonate regeneration, since it is well established (21) that for every mole of ammonia or titratable acid excreted, one mole of bicarbonate is formed by the tubular cells and deposited in the tubular venous blood. At low altitude, renal bicarbonate regeneration averaged 51 mM/day, and over the 78-day sojourn at high altitude, average daily values ranged from 21 mM to 43 mM (Fig. 8). Subtraction of the high altitude from the low altitude values and summation of the differences so obtained provides an estimate of the total body deficit that can be attributed to reduced renal regeneration. When this was done, the accumulated deficit was 45 mM after 1.5 days on Pikes Peak, 140 mM after seven days, and 245 mM after 14 days. If these results are compared to those reported earlier for total bicarbonate in the body fluids, it is apparent that except for the first day or so on Pikes Peak, reduced renal bicarbonate regeneration can more than account for the loss of bicarbonate from the body fluids. On the first day or so, regeneration can account for about 40% of the loss; hence, other mechanisms must be operating What they are remains to be determined, but we can at least exclude enhanced urinary excretion of bicarbonate as a major factor.

SUMMARY

Selected data from a series of nutritional, cardiopulmonary, and acid-base studies support the concept that women adapt more readily to altitude than men. Nutritionally, women show a more transient period of hypophagia, and because of this their body energy deficit during the early states of exposure is less pronounced than that observed in men. Both sexes show body weight losses during the early stages of exposure. All of these effects, however, are much greater in males than in females. During the first two weeks of altitude exposure, women hyperventilate to a greater degree than men. At any time point within this period, the female alveolar and arterial values for P_{CO_2} were lower and for P_{O_2} were higher than those recorded in males. In addition, females show a more rapid early reduction in plasma and extracellular fluid volume; hence, they have superior hemoconcentration characteristics. These cardiopulmonary responses lead to enhanced adaptation of arterial O_2 delivery in women as compared to men. Finally, both sexes respond to high altitude respiratory alkalosis by reducing ammonia and titratable acid excretion; the latter is transient, the former is sustained over the course of acclimatization. Most of the compensatory reduction in dissolved body bicarbonate is achieved by a lowered rate of renal bicarbonate regeneration. These changes in renal function, particularly the reduction in titratable acid excretion, seem to proceed more rapidly in women than in men, but this remains to be verified.

REFERENCES

1. Bert, P.   La pression barométrique; recherches de physiologie expérimentale.   Paris: Masson, 1878.
2. Bert, P.   Sur la richesse en hémoglobin du sang des animaux vivant sur les hauts lieux. *Comp. Rend. Acad. Sci.* 94: 805-807, 1882.
3. Boothby, W.M., W.R. Lovelace, II, O.O. Bensen, Jr., and A.F. Strehler. Volume and partial pressures of respiratory gases at altitude. In: *Respiratory physiology in aviation.* Randolph Field, Texas: USAF School of Aviation Medicine, 1954, pp. 39-49.
4. Campbell, W.A. and H.W. Hoagland. The blood count at high altitudes. *Am. J. Med. Sci.* 122: 654-664, 1901.
5. Davies, H.W., J.B.S. Haldane, and E.L. Kenneway. Experiments on the regulation of the blood's alkalinity. *J. Physiol. (London)* 54: 32-45, 1920.
6. Douglas, C.G., J.S. Haldane, Y. Henderson, and E.C. Schneider. Physiological observations made on Pikes Peak, Colorado, with special reference to adaptation to low barometric pressures. *Phil. Trans. Roy. Soc. London, Series B.* 203: 185-318, 1913.
7. FitzGerald, M.P. The changes in the breathing and the blood at various high altitudes. *Phil. Trans. Roy. Soc. London, Series B.* 203: 351-371, 1913.
8. FitzGerald, M.P. Further observations on the changes in the breathing and the blood at various high altitudes. *Proc. Roy. Soc., Series B.* 88: 248-257, 1914-15.
9. Fitzmaurice, F.E. Mountain sickness in the Andes. *J. Roy. Naval Med. Serv.* 6: 403-407, 1920.
10. Haldane, J.S. and E.P. Poulton. The effects of want of oxygen on respiration. *J. Physiol. (London)* 37: 390-407, 1908.
11. Hannon, J.P., K.S.K. Chinn, and J.L. Shields. Alterations in serum and extracellular electrolytes during high-altitude exposure. *J. Appl. Physiol.* 31: 266-273, 1971.
12. Hannon, J.P., G.J. Klain, D.M. Sudman, and F.J. Sullivan. Nutritional aspects of high-altitude exposure in women. *Am. J. Clin. Nutr.* 29: 604-613, 1976.
13. Hannon, J.P. and G.B. Rogers. Body composition of mice following exposure to 4300 and 6100 meters. *Aviat. Space Environ. Med.* 46: 1232-1235, 1975.
14. Hannon, J.P., J. L. Shields, and C.W. Harris. A comparative review of certain responses of men and women to high altitude. In: *Proceedings symposia on arctic biology and medicine, VI. The physiology of work in cold and altitude.* Edited by C. Helfferich. Ft. Wainwright, Alaska: Arctic Aeromedical Laboratory, 1966, pp. 113-245.
15. Hannon, J.P., J.L. Shields, and C.W. Harris. Anthropometric changes associated with high altitude acclimatization in females. *Am. J. Phys. Anthrop.* 31: 77-83, 1969.
16. Hannon, J.P. and D.M. Sudman. Altitude acclimatization and basal ventilatory function of women. *Fed. Proc.* 34:410, 1975.
17. Hannon, J.P. and J.A. Vogel. Oxygen transport during early altitude acclimatization: a perspective study. *Europ. J. Appl. Physiol.* 36: 285-297, 1977.
18. Kellogg, R.H. Altitude acclimatization, a historical introduction emphasizing the regulation of breathing. *Physiologist* 11: 37-57, 1968.
19. Klain, G.J. and J.P. Hannon. Effects of high altitude on lipid components of human serum. *Proc. Soc. Exper. Biol. Med.* 129: 646-649, 1968.
20. Ochwadt, B.   Über Bicarbonatausscheidung und Kohlensauersystem im Harn wahrend akuter Hypoxie. *Pflügers Arch.* 249: 452-469, 1947.
21. Pitts, R.F. Mechanisms for stabilizing the alkaline reserves of the body. *The Harvey Lectures, Ser. XLVIII.* New York: Academic Press, 1954, pp. 172-209.
22. Ravenhill, T.H. Some experiences of mountain sickness in the Andes. *J. Trop. Med. Hyg.* 6: 313-320, 1913.
23. Reynafarje, C. The influence of high altitude on erythropoietic activity. *Brookhaven Symp. Biol.* 10: 132-146, 1957.
24. Surks, M.I., K.S.K. Chinn, and L.O. Matoush. Alterations in body composition in man after acute exposure to high altitude. *J. Appl. Physiol.* 21: 1741-1746, 1966.

OXYGEN UPTAKE AND BLOOD LACTATE
IN MAN DURING MILD EXERCISE AT ALTITUDE

Clark M. Blatteis

Department of Physiology and Biophysics
University of Tennessee Center for the Health Sciences
Memphis, Tennessee

Although the effects of acute and chronic exposures to altitude on the oxygen demands of submaximal work loads have been reported in numerous studies, most have been conducted at exercise intensities greater than 30% of the subjects' sea level aerobic capacity. Most tasks that can be sustained and are performed daily, however, consist of light to moderate work activities (11). Despite evidence that the energy cost of submaximal work loads is unchanged at any altitude (47), it is a common experience that tolerance to ordinary exertion is impaired upon arrival at altitude, then gradually improves thereafter. In view of these subjective observations, we decided to investigate the adaptive responses of lowlanders to four levels of physical performance between 25% and 35% of their $\dot{V}O_2$ max (i.e., within the range of exertion of many normal activities) on arrival and during six weeks at 4,540 m, and to compare these responses with those of highlanders from this altitude. The responses of both groups were also measured after descent to sea level. Indeed, relatively few studies exist on the responses to physical activity of highlanders taken to sea level (or, for that matter, of lowlanders as compared with highlanders under identical conditions). The results are described in this report.

METHODS

These experiments were conducted in the Central Andes of Peru. The subjects included four Peruvian soldiers born and reared at sea level (lowlanders, LOL) and four Andean administrative workers indigenous to altitude (highlanders, HIL).

Their physical characteristics are summarized in Table 1. None of the subjects had any recent (less than one year) or long-term (more than three days) exposure to any other than his own resident altitude. Histories were taken and physical examinations performed on all the subjects to ensure that they were healthy. None of the subjects was highly physically conditioned or athletically trained, although some (especially the soldiers) were more fit than others.

The LOL were tested initially at sea level (SLi, Lima: 150 m, $\overline{P}_B$ = 748 Torr). One week later they were transported to altitude (Morococha: 4,540 m, $\overline{P}_B$ = 444 Torr) by enclosed vehicle, while recumbent on air mattresses and provided with blankets. They were tested within two to 12 hours (from the first to the last subject) upon arrival (Alt Ac) and after one, two, four, and six weeks of residence (Accl 1, 2, 4, and 6, respectively). During the seventh week at altitude, they were returned to Lima and tested after overnight sleep following their descent (SLf) and again one week later (SLf'). Since no significant differences were uncovered between SLf and SLf' , the latter results are not included in this report. Between tests at altitude, the LOL were at their leisure and enjoyed the freedom of Morococha; but at sea level they were quartered and restricted to a military hospital. At both locations they dressed for thermal comfort and consumed *ad libitum* their habitual diet. They followed no formal training regimen, but continued their normal pre-study activities, which included participation in daily recreational soccer games, except during the first two to three weeks at altitude, when their participation was occasional and limited. The HIL, in turn, were tested initially in their own habitat (Alt, Morococha) and one week later, after overnight sleep, following their arrival at sea level (SL, Lima). During this interval, they lived in their own homes and engaged in their regular work and recreational activities; the latter also included nearly daily soccer games. All the subjects were familiarized with the experimental procedure, including walking on the treadmill, one week prior to their participation in this study. They were postabsorptive (overnight fast) for each experiment. The ambient temperature during the tests in Lima was 21.5 $\pm 0.8^{\circ}$C, and in Morococha, 23.8 $\pm 2.5^{\circ}$C.

Table 1. Initial Physical Characteristics of Subjects

	Lowlanders (LOL) (Lima, 150 m)	Highlanders (HIL) (Morococha, 4,540 m)
N	4	4
Age, years	20 (17–21)[a]	20 (17–21)
Weight, kg	62.6 (60.7–64.4)	54.3 (45.2–62.1)
Height, cm	160.9 (158.5–164.0)	154.4 (148.5–162.0)
Surface area, m^2	1.66 (1.63–1.70)	1.51 (1.38–1.64)
Maximal O$_2$ uptake, ml$\cdot$kg$^{-1}\cdot$min^{-1}	55.7 (40.6–57.3)	44.2 (39.3–47.8)
Hematocrit, %	48.4 (43.0–52.0)	65.2 (58.0–71.0)

[a] Mean (range)

For a test, appropriate measuring devices were attached to each subject, clad only in shorts, socks, and boots. The subject was requested to recline on a cot for 1.5 hr and to cover himself for thermal comfort. He then sat for 0.5 hr more on a laboratory stool on the treadmill belt before any measurements were begun. Each experiment consisted of: (1) three consecutive 10-min control periods, the measurements being made while the subjects were sitting in this position; and (2) four 10-min walks on the treadmill at 4.3 kph at a grade of 0.3, 3, 5, and 8%, respectively. These work loads induced initially in the LOL at sea level and the HIL at altitude a two- to three-fold increase in resting O_2 uptake (Table 2) and therefore were "light" to "moderate" by the classification of Wells *et al.* (60). Each walk was followed by a recovery period until baseline values were attained with the subjects sitting as before. The treadmill characteristics were identical at the two altitudes.

Respiratory metabolism was measured by the open-circuit technique. Pulmonary ventilation ($\dot{V}_I$) was measured by reading from the continuously cumulating dial of a low-resistance mass flowmeter incorporated into the inspiratory side of a Collins triple J valve; this meter has been described previously (6). Readings were taken over the entire 10 min of each control period and over 2-min intervals during and after each exercise; the values were corrected for meter calibration differences at altitude. Expired air was collected through a short length of large-bore, low-resistance tube in Douglas bags during the 6-10-min interval of each 10-min control period, the 6-8 and 8-10-min intervals of each exercise period, and during the 0-2, 4-6, and 8-10-min intervals of each post-exercise recovery period. The O_2 in the entire volume collected was analyzed immediately, using a paramagnetic oxygen analyzer (Beckman E2), frequently calibrated. The $\dot{V}_I$ and F_EO_2 measured during the successive 6-8 and 8-10-min intervals of each exercise period were used to calculate O_2 uptake ($\dot{V}O_2$); the average of these two $\dot{V}O_2$'s was submitted to analysis. Preliminary studies had indicated that steady-state was achieved after six min of exercise at the present submaximal work loads, both at sea level and altitude. Measurements during all of the periods were calculated individually. Blood was drawn from the antecubital vein at the end of the control period and within one to two min after termination of each work load; hematocrit (Hct) was determined by the microcapillary technique only on the control sample. All the samples were deproteinated immediately, centrifuged, filtered, and stored at 4^oC until analyzed two days later; preliminary studies had indicated that these handling procedures were without *post factum* effect on lactate and pyruvate concentrations. Lactate was analyzed by Reynafarje's modification of the method of Barker and Summerson (4), and pyruvate by the method of Friedemann and Haugen (21). During the 9-10-min interval of each exercise and recovery period, respiratory (f_R) and cardiac (HR) frequencies were counted by observation and auscultation, respectively. Rectal (T_{re}) and ambient (T_a) temperatures were monitored continuously by means of Yellow Springs thermistor probes and a telethermometer, but recorded only during the 9-10-min intervals of every period. Remarks by the subjects and observations by the investigators were recorded at any time. Aerobic capacity ($\dot{V}_{O_2}$ max) was estimated from the steady-state pulse rate and O_2 uptake during the walk on the 8% grade, using the nomogram of Åstrand and Rhyming (2).

Table 2. Description of Work Loads

Work Load Code	Treadmill Speed (kph)	Treadmill Grades (%)	Work Rates (kgm/min)		% $\dot{V}O_2$ max		Mean Increase in $\dot{V}O_2$ (ml·kg⁻¹·min⁻¹)[a]		HR (beats/min)	
			LOL	HIL	LOL	HIL	LOL	HIL	LOL	HIL
0	4.3	0.3	14	12	25	30	2.1x	2.3x	13	17
3	4.3	3	138	119	28	35	2.4x	2.7x	16	23
5	4.3	5	232	201	30	36	2.6x	2.8x	25	28
8	4.3	8	363	315	35	42	3.0x	3.2x	34	39

[a] Increase times resting level in ml·kg⁻¹·min⁻¹

Table 3. Mean Changes in Body Weight, Hematocrit and Maximal O_2 Uptake

	Weight (kg)	Hct (%)	$\dot{V}O_2$ max (ml·kg⁻¹·min⁻¹)	Max[a]
LOL				
SLi	62.6	48.4	55.7	100
Alt Ac	61.5	50.7	34.6	→ 38
Accl 1	61.5	51.8	35.1	→ 37
Accl 2	60.5	54.8	36.0	→ 35
Accl 4	60.3	57.8	37.3	→ 33
Accl 6	59.6	59.4	46.1	→ 18
SLf	60.7	57.8	70.8	↑ 21
HIL				
Alt	54.3	65.2	44.2	100
SL	54.3	63.6	70.0	↑ 37

[a] Data are elicited steady-state oxygen uptakes, expressed as percentates of $\dot{V}O_2$ max estimated in LOL and HIL initially in their respective habitats.

The control data were analyzed separately from the experimental data. Whenever the subjects were the same at both altitudes, a one-way classification (environment) analysis of variance in randomized blocks (with each subject constituting a block) was used. When different subjects were used in the two environments, a two-way classification (environment and subjects) analysis of variance was employed. In the case of the experimental data, when the subjects were the same in both environments, a two-way classification (environment and work load) analysis of variance in randomized blocks was used, and all possible interactions were tested. Both linear and non-linear regression in the work factor were also tested. When different subjects were used in the two environments, a two-way classification analysis of variance in incomplete blocks (i.e., environment was not blocking, whereas work was blocking) with regression in the work factor and testing of interactions were performed. Missing data were calculated and included. The principal reference used for these procedures was Cochran and Cox's *Experimental Designs* (10). The null hypothesis was rejected at the 5% level of confidence.

RESULTS

Body Weight and Hematocrit

During their six weeks' stay at altitude, the LOL lost a mean 3.0 kg in body weight and raised their Hct a mean 11%, relative to the initial sea level values (Table 3). However, their hematocrit did not attain the high value of the HIL.

Oxygen Uptake

Aerobic capacity. The $\dot{V}O_2$ max of the LOL, as estimated at SLi, was depressed greatly on their arrival at 4,540 m (Table 3, Alt Ac), but gradually recovered (Accl 1-6), especially during the interval between the fourth and sixth weeks of acclimatization. On return to sea level, $\dot{V}O_2$ max was greatly increased, both by comparison with SLi and that at altitude after six weeks of acclimatization. Furthermore, the $\dot{V}O_2$ max of the LOL both on week six of altitude acclimatization and on return to sea level was similar to that of the HIL at Alt and SL, respectively.

Submaximal O_2 uptake. The increases in the steady-state rates of oxygen consumption ($\dot{V}O_2$) produced in the LOL by the present four submaximal work loads were significantly greater on arrival at altitude (Fig. 1, curve b) than at SLi (a). However, one week later (c), they were no longer different from SLi and remained unchanged during the following five weeks at altitude (d-f); they also were not significantly different from those of the HIL at altitude (h). On return to sea level (g), these increases were significantly reduced as compared with all the previous weeks, including that observed at SLi (a). The exercise-induced increase in O_2 uptake of the HIL similarly was significantly reduced at sea level (i) as compared with

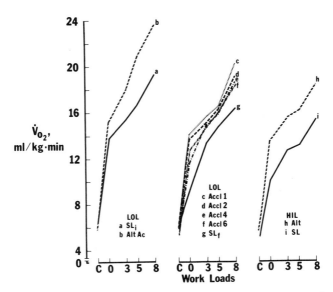

Fig. 1. Steady-state O_2 uptakes ($\dot{V}O_2$) at rest (C) and during various submaximal work loads (4.3 kph, 0%-8% grade) of: *left panel*, lowlanders (LOL) at sea level (SLi) and on arrival at 4,540 m (Alt Ac); *middle panel*, LOL during six weeks at 4,540 m (Accl 1-6) and on return to sea level (SLf); *right panel*, highlanders (HIL) at 4,540 m (Alt) and sea level (SL).

altitude (h), but was not different from those of the LOL returned to sea level (g). The resting $\dot{V}O_2$ of both LOL and HIL, by contrast, were not affected under any of the above conditions; however, the control $\dot{V}O_2$ of the HIL at sea level (i) was significantly smaller than the SLi value of LOL (a).

O_2 extracted from the inspired air. The percentage of O_2 extracted from the inspired air ($F_{I-E}O_2$, Fig. 2) consistently was higher in all the subjects at altitude than at sea level. In the LOL, $F_{I-E}O_2$ was greatest on arrival at 4,540 m (curve b), then gradually diminished during the subsequent six weeks there (c-f). During weeks 2-6, the $F_{I-E}O_2$ of LOL was not different from that of the HIL at altitude (h). On descent to sea level (g), $F_{I-E}O_2$ decreased in the LOL to values significantly lower than during SLi (a); in the HIL, it fell still lower (i). In all cases, exercise induced a rise in $F_{I-E}O_2$, but it was smaller at sea level than at altitude; it was particularly small in the HIL at sea level (i). However, work intensity *per se* had no effect on $F_{I-E}O_2$, except in the HIL at sea level (i) in whom the amount of O_2 extracted increased linearly with work load.

$\dot{V}O_2$ /% $\dot{V}O_2$ max relations. Upon arrival at 4,540 m, the LOL performed submaximal work loads at significantly higher percentages of their $\dot{V}O_2$ max estimated at SLi (Fig. 3, curves a and b). However, the relative intensities of these exercises gradually returned toward their initial values by the sixth week at altitude (c-f), at which time (f) they were similar to those of the HIL at altitude (h). On

returning to sea level (g), the LOL performed these work loads at a significantly lower percentage of their $\dot{V}O_2$ max than initially (a), but not differently from those of the HIL at sea level (i).

It is apparent from Fig. 4 that the higher absolute steady-state $\dot{V}O_2$ of the LOL on arrival at altitude (curve b) was related to the increased relative intensities of these submaximal work loads as compared with SLi (a). By contrast, during the subsequent six weeks (c-f), although the subjects performed these exercises at ever lower percentages of their sea level aerobic capacity, their absolute O_2 uptake decreased only in the first week (c), then stabilized (d-f). Similarly, on return to sea level (g), although these work loads encompassed half the range of their relative

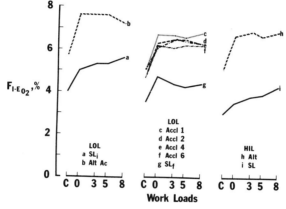

Fig. 2. Steady-state fractions of oxygen extracted from the inspired air ($F_{I-E}O_2$) by LOL and HIL at rest and during work at sea level and 4,540 m. Abbreviations as in Fig. 1.

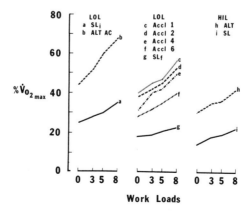

Fig. 3. Steady-state O_2 uptakes, expressed as percentages of the initial, estimated aerobic capacity (% $\dot{V}O_2$ max) of the LOL and HIL in their respective habitats. Abbreviations as in Fig. 1.

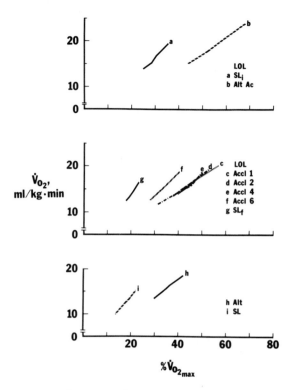

Fig. 4. Relation between $\dot{V}O_2$ and % $\dot{V}O_2$ max of LOL and HIL during work at sea level and 4,540 m. Abbreviations as in Fig. 1.

$\dot{V}O_2$ max at SLi (a), they induced the same increments in absolute $\dot{V}O_2$ under both conditions. The same relationship was apparent in the HIL at the two altitudes (h and i).

Ventilatory Responses

Pulmonary ventilation. On arrival at altitude (Fig. 5, upper, curve b), the LOL displayed a significant increase in resting $\dot{V}_I$ as compared with that observed at SLi (a). The hyperventilation increased during the next two weeks, then stabilized for the remainder of their stay at 4,540 m (c-f). Control $\dot{V}_I$ returned to the SLi value (a) upon return to sea level (g). The HIL similarly exhibited a decrease of resting $\dot{V}_I$ at sea level (i) as compared with altitude (h). The increases in steady-state $\dot{V}_I$ induced in the LOL by the four submaximal work loads were significantly greater at altitude (curve b) than at SLi (a). Moreover, the rate of these increases was significantly greater at altitude than at sea level. However, in contrast to the response of

$\dot{V}O_2$, exercise hyperventilation and its greater increments did not subside with continued stay at 4,540 m (c-f). On return to sea level, however, both the absolute level of $\dot{V}_I$ and the rate of increase with increased work load were reduced; but in contrast to $\dot{V}O_2$, they were not significantly below SLi values (a). Similarly, the $\dot{V}_I$ and the rate of increase of $\dot{V}_I$ in the HIL during exercise were significantly smaller at sea level (i) than at 4,540 m (h). The HIL and LOL had similar levels of $\dot{V}_I$ at both sea level and 4,540 m.

Respiratory frequency. The steady-state respiratory frequencies of both groups of subjects (Fig. 5, lower panels) were increased significantly with exercise, but not with altitude. Resting f_R was not different in the two environments.

Ventilatory equivalent. Ventilatory equivalent (VE) in relation to work load is illustrated in Fig. 6. The control VE of the LOL increased significantly upon arrival at 4,540 m and increased further during their six-week sojourn (c-f). On return to sea level (g), VE returned to its initial value. The control VE of the HIL, by contrast, was not different at altitude and sea level (h and i). Exercise caused the VE of both the LOL and HIL to fall from their respective control value, both at sea level and at altitude, but increasing work load had no further effect. The steady-state

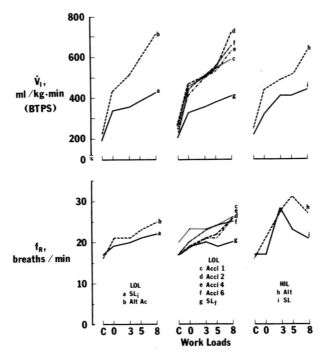

Fig. 5. Steady-state pulmonary ventilation ($\dot{V}_I$, upper panels) and respiratory frequency (f_R, lower panels) of LOL and HIL at rest and during work at sea level and 4,540 m. Abbreviations as in Fig. 1.

exercise VE of the LOL was significantly higher on arrival at 4,540 m (curve b) than at SLi (a), and increased further during the subsequent six weeks (c-f). On descent to sea level (g), they returned toward, but remained significantly higher than, SLi values (a). The exercise VE values of the HIL, by contrast, were not different at altitude (h) and at sea level (i), and were not different from those of the LOL after the first week at altitude (c-f).

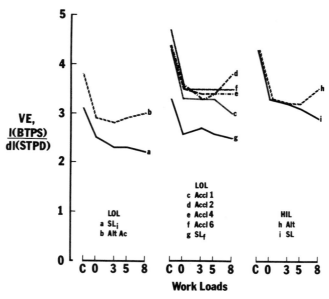

Fig. 6. Steady-state ventilatory equivalents (VE) of LOL and HIL at rest and during work at sea level and 4,540 m. Abbreviations as in Fig. 1.

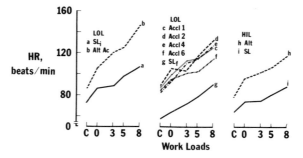

Fig. 7. Steady-state heart rates (HR) of LOL and HIL at rest and during work at sea level and 4,540 m. Abbreviations as in Fig. 1.

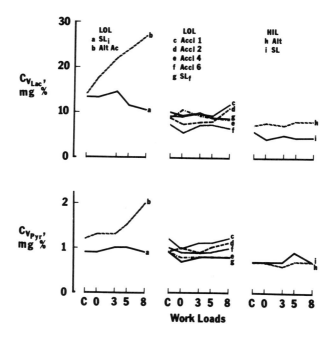

Fig. 8. Venous lactate (C_Vlac, upper panels) and pyruvate (C_Vpyr, lower panels) of LOL and HIL at rest and during work at sea level and 4,540 m. Abbreviations as in Fig. 1.

Heart Rate

The increase in steady-state heart rate (HR) induced by work was, like oxygen uptake, significantly higher on arrival at altitude (Fig. 7, curve b) than at SLi (a). The exercise HR after one week at altitude (c) was less than on initial exposure to altitude (b), but, in contrast to $\dot{V}O_2$, remained above the SLi value until the sixth week at altitude (a). By the sixth week (f), the exercise HR was similar to the SLi value and to that of the HIL at altitude (h). On descent to sea level, the exercise HR of LOL (g) and HIL (i) was significantly lower than at any time at altitude; in the LOL, they also were lower than initially at sea level (a). In contrast to resting $\dot{V}O_2$, the resting HR of both LOL and HIL were at all times significantly higher at altitude than at sea level. Moreover, on return to sea level (g), the resting HR of LOL was significantly lower than initially (a).

Blood Lactate and Pyruvate

Although the present work loads did not alter the blood lactate concentrations of LOL at sea level initially (Fig. 8, upper, curve a), each higher load caused a significant and progressively larger rise of lactate on their arrival at altitude (b). However, during the subsequent six weeks at altitude (c-f), blood lactate during exercise was significantly lower than at SLi and remained so after returning to sea level (g). The blood lactates of the HIL similarly were unchanged during exercise at either altitude, although they were significantly lower at sea level than at altitude (h and i). At altitude the blood lactates of the HIL and LOL were similar, except on the latter's arrival at 4,540 m. At sea level, the blood lactates of HIL were significantly lower than those of LOL. The resting lactate levels of the LOL were not altered on their arrival at 4,540 m (b), but were significantly reduced from their second week at altitude onward (d-f) and were then equal to those of the HIL (h). The control lactates of the altitude-acclimatized LOL and those of the HIL were unchanged on their descent to sea level (g, i).

As with blood lactate, the blood pyruvate levels of LOL were not affected by submaximal exercise at sea level initially (Fig. 8, lower, a), while they were significantly and progressively increased by each incremental work load on their arrival at altitude (b). After one week, however, blood pyruvate levels did not increase with exercise (c-f). On return to sea level (g), blood pyruvate levels were lower than initially (a). In contrast to blood lactate, there were no significant differences among the blood pyruvate levels of the HIL at altitude and at sea level (h and i), nor were there any differences between the control values of the LOL and HIL under any condition.

Rectal Temperature

Both HIL and LOL had similar rectal temperatures at rest of $37.0^{\circ}C \pm 0.1^{\circ}C$. During exercise, T_{re} increased significantly ($0.2^{\circ}C$) in both LOL and HIL only during the walk on the 8% grade.

Subjective Responses

None of the LOL displayed the more distressful symptoms of acute mountain sickness, such as nausea and vomiting, during exercise after arrival at 4,540 m; however, all complained of lightheadedness, greater fatigability, breathlessness, and tachycardia. No change was apparent in their walking gait on the treadmill. One subject developed a headache after his work performance, which persisted for several hours; he and two others reported insomnia on the following day. All three, however, were recovered from these symptoms by the third day at altitude. Nonetheless, during the following three days, all four LOL volitionally engaged in only limited physical activity, complaining of breathlessness and fatigue on exertion. They gradually became more active during their second week at altitude, and by the fourth week, they were participating regularly in afternoon soccer games with local HIL. After descent to sea level, both LOL and HIL reported that the present work loads seemed less strenuous than at any time previously.

Recovery

In the intervals between work loads, under all environmental conditions, $\dot{V}O_2$ and $\dot{V}_I$ returned in both the LOL and HIL to their respective control values within four to six minutes after the end of each exercise. The only exception was in LOL on arrival at altitude, when eight to ten minutes were required for these parameters to return to control levels. HR, f_R, and T_{re} were not different from control when measured at eight to ten minutes post-exercise.

DISCUSSION

The present estimated $\dot{V}O_2$ max values are in good agreement with previously published values for lowland Peruvian military personnel (3,56) in good physical condition and nutritional status, and for Andean HIL, specifically at Morococha (3, 19,56). The calculated reduction in the aerobic capacity of the LOL on arrival at 4,540 m also is with the range of previously published results (3,8,12,14,15,20,22, 24,32,46). The gradual improvement observed in aerobic capacity during the subsequent six weeks, however, is more equivocal, since some workers (8,12,20,22,24, 35,40,47,57) have found no change with altitude acclimatization, while others (3, 14,15,32,46,48,56) have reported variable amounts of recovery. The quantitative reliability of the present observed amelioration is admittedly questionable, since $\dot{V}O_2$ max was assessed indirectly. The large increase in the $\dot{V}O_2$ max of both the altitude-acclimatized LOL on their return to sea level and of the HIL descended to sea level probably is excessive for the same reasons, although this result would agree with other reports (14,15,32,50) that altitude acclimatization may improve maximal work performance on return to sea level; however, these observations (8,12, 20,22,24) are not universal.

The present finding that the O_2 cost of these mild-to-moderate exercises was higher on arrival at 4,540 m than at sea level initially accords with the subjective sensation of greater exertion during physical performance at altitude and is supported by the observation that the same work loads were performed at higher percentages of these subjects' sea level aerobic capacity at altitude than at sea level. The latter result is in agreement with most data in the literature. However, some workers (1,3,9,12,24,29,34,39,43,46,51,52,53) have found $\dot{V}O_2$ values during various intensities of submaximal work to be unchanged at altitudes ranging from 1,600 m to 4,300 m as compared with sea level, while others (13,44,53,54) have observed 11% to 20% decreases in the $\dot{V}O_2$ of subjects exercising at various rates between 250 and 900 kgm/min at altitudes from 2,309 m to 4,300 m as compared with sea level. On the other hand, Klausen et al. (32) noted a 10% increase in the O_2 cost of bicycling at 50% of $\dot{V}O_2$ max 1.5 days after arrival at 3,800 m, but attributed the rise to a higher work load at altitude due to a presumed difference in the calibrations of the ergometers used. Reeves and Daoud (49) found that $\dot{V}O_2$ increased 17% above sea level values during walking on a treadmill for two hours at

30% of $\dot{V}_{O_2}$ max 20 to 30 min after reaching 4,572 m simulated altitude. Billings et al. (5) reported that the O_2 uptake during bicycling at 50% of sea level aerobic capacity was significantly elevated, especially on the second and fifth days after arrival at 3,800 m. Jones et al. (30) observed a rise in $\dot{V}_{O_2}$ during bicycling at 300 and 600 kpm/min when their subjects breathed 10% to 13% O_2 acutely at sea level. Unfortunately, none of the previous studies is strictly comparable with the present one, since no other study used an equivalent series of low work loads. Indeed, it is difficult to reconcile the differences among them, because all employed different intensities, durations, and types of exercise, as well as manners of induction and types of hypoxic stimulus, in subjects of different levels of physical condition.

It cannot be ascertained from the present data precisely why the exercise $\dot{V}_{O_2}$ of these LOL was elevated above sea level values upon their arrival at 4,540 m. It is probable that the difference was not due to work load disparities at the two altitudes, because the treadmill belt characteristics were identical at the two locations, and their velocity and angle of inclination were recalibrated for each new setting. That this rise in $\dot{V}_{O_2}$ may have been due to the calorigenic action of catecholamines released on induction to altitude (9,27) would be contradicted by our earlier findings (7) that acute moderate hypoxia depresses this calorigenic effect; in any case, there was no significant increase in the resting $\dot{V}_{O_2}$ upon arrival at 4,540 m. There can be little doubt that the increased $\dot{V}_{O_2}$ during exercise reflected a greater strain at altitude, as attested also by the accompanying higher heart rates, the lactacidemia, the longer recovery times, and the higher relative intensities (% $\dot{V}_{O_2}$ max) of these work loads. It is likely that the increased work of the heart and of breathing under these conditions, as well as the added energy needed to reconvert lactic acid to glucose during the steady-state phase of the exercises, all contributed to the observed augmentation of the O_2 cost at altitude. However, these factors are relatively small and could not have accounted for the entire increase. Therefore, the balance could have been due to differences in the mechanisms of walking of these subjects upon their arrival at 4,540 m. Thus, it might be speculated that the mechanical power spent by these LOL in gyrating around the center of gravity of their bodies in order to gain and maintain their balance at each step while lifting their limbs relative to the trunk was greater during the first hours at altitude than at sea level. Although these effects were not obvious on simple observation, muscle action and balance conflicts are compatible with the general sense of malaise and weakness commonly experienced by the LOL even during slight exertion and in the absence of overt symptoms of mountain sickness during acute exposure to altitude.

The finding that exercise $\dot{V}_{O_2}$ in LOL was not different from sea level one week after arrival at altitude is in conformity with the results of Douglas et al. (17), and others (1,3,11,24,29,34,39,43,46,51,52,55), indicating that the O_2 cost of a given submaximal exercise is not affected by altitude. This result also accords with the improved sense of well-being and apparent increase in vigor and endurance reported by these subjects after a few days at altitude. Despite the recovery of this objective measure, these exercises continued to be performed at significantly higher heart rates and percentages of these LOL's sea level $\dot{V}O_2$ max during the following three weeks, even though these functions declined gradually over this period. It is interesting that on the sixth week at altitude, HR and % $\dot{V}O_2$ max were no longer different from those at sea level initially, and also were equal to those of the HIL at altitude. This latter finding is in general agreement with the conclusions of Balke (3), Grover and Reeves (22), and Lahiri et al. (36,37), but not with those of Pugh et al. (47), Velasquez (56), and Elsner et al. (19), who found O_2 uptakes at given work loads to be lower, and of Kollias et al. (35), who found them to be higher, in HIL than in either sea level controls or altitude-acclimatized LOL. However, these earlier studies are not readily comparable because of the diverse work intensities, altitudes, and training regimens employed.

The greatly reduced exercise $\dot{V}O_2$ and HR of the altitude-acclimatized LOL and of the HIL upon descent to sea level would suggest that residence at altitude may potentiate physical performance at sea level, at least in the range of the present low work rates. This would accord with the subjective sensation of improved work tolerance under these conditions. As stated earlier, reports on the effects of altitude acclimatization on physical working capacity upon return to sea level are conflicting. It would appear that the potentiating effect, when present, has been attributed less to the altitude stress than to an effect of greater physical activity, whether due to a controlled training program or more leisure time devoted to outdoor activities. In the absence of a control group at sea level engaged in identical activities, a significant training effect during the course of the present experiments cannot be excluded, although the normal daily routine of these LOL was not demonstrably different in Lima and in Morococha. However, the Morococha terrain is rugged and steep, forcing inhabitants to undergo physical exertion even while carrying out normal daily activities. If this were a significant factor, it could have accounted for the gradual decrease of the estimated $\dot{V}O_2$ max of these subjects during their six weeks' stay at altitude. Nevertheless, a significant and direct contribution of altitude *per se* may be inferred from the fact that the HIL similarly performed these mild work loads at a greatly reduced $\dot{V}O_2$ and HR within hours after descent to sea level.

The ventilatory changes that accompanied the present $\dot{V}O_2$ responses are in general agreement with many previous results (5,8,22,24,26,29,30,31,34,42,43,45,46, 55). In the present study, the increased $\dot{V}_I$ of the LOL and HIL at altitude was accomplished, both at rest and during exercise, more by means of a greater depth rather than a higher frequency of breathing. This is partly in contrast to the findings of Hansen et al. (24) and McManus et al. (43), who noted an increased respiratory frequency and unchanged tidal volume in resting LOL acclimatizing to 4,300 m and 3,058 m, respectively, as compared with sea level. It is noteworthy that the increased $\dot{V}_I$ of the LOL during these submaximal exercises was, after the first week at altitude, of the same magnitude as that of the HIL. Grover et al. (22) have reported similar findings, but Lahiri et al. (36,37) and Kollias et al. (35) found that LOL consistently ventilated more at altitude than HIL at equivalent submaximal work loads. At sea level there were no differences between the $\dot{V}_I$ of altitude-acclimatized LOL and those of HIL. This is contrary to the findings of Lahiri et al. (36,37), who observed that the respiration of HIL breathing high oxygen at altitude was depressed less than that of LOL. Hansen *et al.* (24) found the $\dot{V}_I$ of lowlanders was greater on return to sea level, whereas the $\dot{V}_I$ of the present LOL was unchanged at sea level after altitude acclimatization. These various discrepancies probably were due to the small sample size and special experimental conditions employed in each different study.

The finding that the ventilatory equivalents of these LOL were higher upon arrival at 4,540 m than at sea level accords with previous reports (42,49). Their further rise with continued residence at altitude also has been reported before (8,12,56). In the present study, this rise was due to the maintenance of high $\dot{V}_I$ in spite of the decreased $\dot{V}_{O_2}$. The higher VE at sea level after, as contrasted to before, their altitude exposure similarly was due to an unchanged $\dot{V}_I$ despite a significantly lower $\dot{V}_{O_2}$; the latter was achieved by their lesser O_2 extraction from the inspired air. Although the $\dot{V}_I$ and $\dot{V}_{O_2}$ of the HIL and altitude-acclimatized LOL were not significantly different at sea level, the $\dot{V}_I$ of the HIL tended to be higher and their $\dot{V}_{O_2}$ lower, thereby accounting for their still higher VE. The HIL extracted significantly less O_2 from the inspired air at sea level than the returned LOL; they thus ventilated relatively more at sea level at the same $\dot{V}_{O_2}$ than did the LOL. Since the % $\dot{V}_{O_2}$ max at which the LOL performed these exercises decreased with duration at altitude, while their $\dot{V}_I$ remained unchanged, their $\dot{V}_I$ for a given relative work intensity in fact gradually increased as they acclimatized to altitude.

The increase in both resting and exercise HR seen in the present LOL on arrival at 4,540 m as compared with sea level is in agreement with many previous reports (1,3,22,26,29,31,34,36,37,43,45,53,57,59). The progressively lower steady-state heart rates at equivalent work loads during the subsequent six weeks at altitude also agrees with previous findings (45,46,48,51,57,59). Finally, the present observed decrease of the LOL's HR, to below initial values both at rest and during exercise on return to sea level, also accords with previous results (1,31,52,57,59). On the other hand, Saltin *et al.* (51) and Lahiri *et al.* (36) have found that the HR response to submaximal exercise of altitude-acclimatized LOL was not appreciably decreased by breathing high oxygen at altitude. The similarity between the control and exercise HR of the altitude-acclimatized LOL and the HIL, both at 4,540 m and at sea level, also agrees with previous findings (22,25,35,58). However, Lahiri *et al.* (36) reported lower resting and higher exercise HR in HIL than in acclimatized LOL at 2,745 m and 4,880 m. The decrease in exercise HR observed in the present study is consistent with the reduced oxygen uptake at each work load of both the altitude-acclimatized LOL and HIL on descent to sea level. Indeed, the HR responses generally were commensurate with the $\dot{V}_{O_2}$ responses, although the former were slower in returning to their original sea level values than the latter.

Although much data exist regarding blood lactate in work at altitude, precise comparisons of the present results with others is difficult because relatively few studies are based on similar work loads and environmental and other conditions. Thus, venous blood lactate and pyruvate did not accumulate during the present low work rates at sea level, in conformity with the findings of Dill and Sacktor (16) and

Margaria *et al.* (41); on the other hand, Huckabee (28), Knuttgen (33), and Hansen *et al.* (23) found rises in arterial lactate and pyruvate. Differences in methodologies, therefore, presumably account for these discrepancies. Similarly, the lack of a rise in the present lactate and pyruvate levels at rest on the LOL's arrival at 4,540 m conforms to some (40,43), but contradicts other reports (18,23,32,37,44,48) and may be ascribed to the same differences. There is better agreement that blood lactate and pyruvate levels for a given submaximal work load are significantly higher on acute exposure to altitude than at sea level (18, 23,26,29,30,39,44,48,52,55), although some authors (34,42,43) have found no such increase. The observed decline in the resting levels of lactate and pyruvate after two weeks' altitude acclimatization and the persistence of these low values on return to sea level have been reported previously (23,31), but are as yet unexplained. By the same token, declines in exercise blood lactates and pyruvates in LOL with altitude acclimatization and on return to sea level, and lower values in HIL as compared with LOL at altitude, also have been found (18,23,31,36,40). The pyruvic and lactic acidemias in LOL produced by these work loads, on their arrival at 4,540 m, and their subsequent abatement, were consistent with the corresponding oxygen cost of these exercises. In the absence of other correlative measures, any explanation of these results can be only speculative. The initial accumulation of lactate and pyruvate might have occurred because the cardiovascular and respiratory adaptations of LOL to altitude were not complete, as manifested by their concurrent $\dot{V}_I$ and HR responses, thus leading to a fall in muscle tissue PO_2. The subsequent abolition of these responses with altitude acclimatization might be due, in part, to the associated improvement of oxygen transport and to physical training. The latter, however, could not have been an important factor during the first two weeks at altitude when these LOL remained largely inactive. Other additional, more slowly developing, hypoxia-induced acclimatory processes (38) (e.g., increased capillary density, hemoglobin, myoglobin, mitochondrial density, lactic dehydrogenase-M, and other glycolytic and respiratory enzymes, etc.) might therefore be collectively implicated.

SUMMARY

These results substantiate that the subjective reduction in tolerance to ordinary work activities experienced by most LOL on arrival at altitude is related to a significantly increased energy cost of such tasks relative to sea level, and that a portion of the work at altitude under these conditions is supported by anaerobic processes. Discomforting effects may be reflections of subclinical, asymptomatic acute mountain sickness, itself a reflection of the insufficiency of immediate, adaptive oxygen transport mechanisms. However, the excessive metabolic response of LOL to physical work is abated after one week at altitude as the limitations to adequate tissue oxygenation are compensated; this is reflected in an improved sense of well-being and apparent increase in vigor and endurance. The compensatory changes occurring in all steps of the O_2 transport system with continued residence at altitude result in gradual, further benefits to the economy of the body until the response to ordinary activity of LOL does not differ from that of HIL. These adaptive effects of altitude, moreover, are not immediately reversible on descent to sea level; indeed, they appear to potentiate the capacity for submaximal exercise at sea level of both HIL and altitude-acclimatized LOL.

ACKNOWLEDGMENTS

The skilled technical assistance of Dr. H. R. Scholnick, Dr. B. Reynafarje, Ms. L. Oyola H., Mr. W. Lafferty, and Mr. L. Sirvio is greatly acknowledged.

I am also greatly indebted to the Peruvian Army Medical Service for generously detailing to this study the sea level subjects and various support personnel, as well as providing facilities at the Hospital Militar Central in Lima; and to Dr. Tulio Velasquez, then director of the Institute of Andean Biology, who graciously allowed me the use of both the Lima and Morococha laboratories and went, together with his staff, considerably out of his way to assist me in many ways.

A portion of this work was supported by Contract DADA 17-68-C-8136 from the U. S. Army Medical Research and Development Command to the University of Tennessee Center for the Health Sciences Biometric Computer Center.

The investigator has complied with the recommendations from the Declaration of Helsinki.

REFERENCES

1. Åstrand, P.-O., and I. Åstrand. Heart rate during muscular work in man exposed to prolonged hypoxia. *J. Appl. Physiol.* 13: 75-80, 1958.
2. Åstrand, P.-O., and I. Ryhming. A nomogram for calculation of aerobic capacity (physical fitness) from pulse rate during submaximal work. *J. Appl. Physiol.* 7: 218-221, 1954.
3. Balke, B. Work capacity and its limiting factors at high altitude. In: *The physiological effects of high altitude.* Edited by W.H. Weihe. Oxford: Pergamon, 1964, pp. 233-247.
4. Barker, S.B., and W. H. Summerson. The colorimetric determination of lactic acid in biological material. *J. Biol. Chem.* 138: 538-554, 1941.
5. Billings, C.E., R. Bason, D.K. Matthews, and E.L. Fox. Cost of submaximal and maximal work during chronic exposure at 3,800 m. *J. Appl. Physiol.* 30: 406-408, 1971.
6. Blatteis, C.M. and L.O. Lutherer. Effect of altitude exposure on thermoregulatory response of man to cold. *J. Appl. Physiol.* 41: 848-858, 1976.
7. Blatteis, C.M. and L.O. Lutherer. Reduction by moderate hypoxia of the calorigenic action of catecholamines in dogs. *J. Appl. Physiol.* 36: 337-339, 1974.
8. Buskirk, E.R., J. Kollias, R.F. Akers, R.K. Prokop, and E.P. Reategui. Maximal performance at altitude and on return from altitude in conditioned runners. *J. Appl. Physiol.* 23: 259-266, 1967.
9. Clancy, L.J., J.A.J.H. Critchley, A.G. Leitch, B.J. Kirby, A. Ungar, and D.C. Flenley. Arterial catecholamines in hypoxic exercise in man. *Clin. Sci. Molec. Med.* 49: 503-506, 1975.
10. Cochran, W.G. and G.M. Cox. *Experimental designs,* 2nd ed. New York: Wiley, 1957.
11. Consolazio, C.F., R.E. Johnson, and L.J. Pecora. *Physiological measurements of metabolic functions in man.* New York: McGraw-Hill, 1963, pp. 328-333.

12. Consolazio, C.F., L.O. Matoush, and R.A. Nelson. Energy metabolism in maximum and submaximum performance at high altitudes. *Fed. Proc.* 25:1380-1287, 1966.
13. Cronin, R.F.P. and D.J. MacIntosh. The effect of induced hypoxia on oxygen uptake during muscular exercise in normal subjects. *Can. J. Biochem. Physiol.* 40: 717-726, 1962.
14. Dill, D.B. and W.C. Adams. Maximal oxygen uptake at sea level and at 3,090 m altitude in high school champion runners. *J. Appl. Physiol.* 30: 854-859, 1971.
15. Dill, D.B., L.G. Myhre, D.K. Brown, K. Burrus, and G. Gehlsen. Work capacity in chronic exposures to altitude. *J. Appl. Physiol.* 23: 555-560, 1967.
16. Dill, D.B. and B. Sacktor. Exercise and the oxygen debt. *J. Sport Med.* 2: 66-72, 1962.
17. Douglas, C.G., J.S. Haldane, Y. Henderson, and E.C. Schneider. Physiological observations made on Pike's Peak, Colorado, with special reference to adaptation to low barometric pressures. *Phil. Trans. Roy. Soc. London, Ser. B* 203: 185-318, 1913.
18. Edwards, H.T. Lactic acid in rest and work at high altitudes. *Am. J. Physiol.* 116: 367-375, 1936.
19. Elsner, R.W., A. Bolstad, and C. Forno. Maximum oxygen consumption of Peruvian Indians native to high altitude. In: *The physiological effects of high altitude.* Edited by W.H. Weihe. Oxford: Pergamon, 1964, pp. 217-223.
20. Faulkner, J.A., J. Kollias, C.B. Favour, E.R. Buskirk, and B. Balke. Maximum aerobic capacity and running performance at altitude. *J. Appl. Physiol.* 24: 685-691, 1968.
21. Friedemann, T.E. and G.E. Haugen. Pyruvic acid. II: The determination of keto acids in blood and urine. *J. Biol. Chem.* 147: 415-442, 1943.
22. Grover, R.F., J.T. Reeves, E.B. Grover, and J.E. Leathers. Exercise performance of athletes at sea level and 3100 meters altitude. *Med. Thorac.* 23: 129-143, 1966.
23. Hansen, J.E., G.P. Stelter, and J.A. Vogel. Arterial pyruvate, lactate, pH, and P_{CO_2} during work at sea level and high altitude. *J. Appl. Physiol.* 23: 523-530, 1967.
24. Hansen, J.E., J.A. Vogel, G.P. Stelter, and C.F. Consolazio. Oxygen uptake in man during exhaustive work at sea level and high altitude. *J. Appl. Physiol.* 23: 511-522, 1967.
25. Hartley, L.H., J.K. Alexander, M. Modelski, and R.F. Grover. Subnormal cardiac output at rest and during exercise in residents at 3,100 m altitude. *J. Appl. Physiol.* 23: 839-848, 1967.
26. Hermansen, L. and B. Saltin. Blood lactate concentration during exercise at acute exposure to altitude. In: *Exercise at altitude.* Edited by R. Margaria. Amsterdam: Excerpta Medica, 1967, pp. 48-53.
27. Hoon, R.S., S.C. Sharma, Y. Balasubramanian, and K.S. Chadha. Urinary catecholamine excretion on induction to high altitude (3,658 m) by air and road. *J. Appl. Physiol.: Respirat. Environ. Exercise Physiol.* 42: 728-730, 1977.
28. Huckabee, W.B. Relationships of pyruvate and lactate during anaerobic metabolism. II. Exercise and formation of O_2 debt. *J. Clin. Invest.* 37: 255-263, 1958.
29. Hughes, R.L., M. Clode, R.H.T. Edwards, T.J. Goodwin, and N.L. Jones. Effect of inspired O_2 on cardiopulmonary and metabolic responses to exercise in man. *J. Appl. Physiol.* 24: 336-347, 1968.
30. Jones, N.L., D.G. Robertson, J.W. Kane, and R.A. Hart. Effect of hypoxia on free fatty acid metabolism during exercise. *J. Appl. Physiol.* 33: 733-738, 1972.
31. Klausen, K., D.B. Dill, and S. M. Horvath. Exercise at ambient and high oxygen pressure at high altitude and at sea level. *J. Appl. Physiol.* 29: 456-463, 1970.
32. Klausen, K., S. Robinson, E.D. Micahel, and L.G. Myhre. Effect of high altitude on maximal working capacity. *J. Appl. Physiol.* 21: 1191-1194, 1966.
33. Knuttgen, H.G. Oxygen debt, lactate, pyruvate, and excess lactate after muscular work. *J. Appl. Physiol.* 17: 639-644, 1962.
34. Knuttgen, H.G. and B. Saltin. Oxygen uptake, muscle high-energy phosphates, and lactate in exercise under acute hypoxic conditions in man. *Acta Physiol. Scand.* 87: 368-376, 1973.

35. Kollias, J., E.R. Buskirk, R.F. Akers, E.K. Prokop, P.T. Baker, and E. Picón-Reategui. Work capacity of long-time residents and newcomers to altitude. *J. Appl. Physiol.* 24: 792-799, 1968.

36. Lahiri, S., J.S. Milledge, H.P. Chattopadhyay, A.K. Bhattacharyya, and A.K. Sinha. Respiration and heart rate of Sherpa highlanders during exercise. *J. Appl. Physiol.* 23: 545-554, 1967.

37. Lahiri, S., J.S. Milledge, and S.C. Sørensen. Ventilation in man during exercise at high altitude. *J. Appl. Physiol.* 32: 766-769, 1972.

38. Lenfant, C. and K. Sullivan. Adaptation to high altitude. *New Eng. J. Med.* 284: 1298-1308, 1971.

39. Linnarsson, D., J. Karlsson, L. Fagraeus, and B. Saltin. Muscle metabolites and oxygen deficit with exercise in hypoxia and hyperoxia. *J. Appl. Physiol.* 36: 399-402, 1974.

40. Maher, J.T., L.G. Jones, and L.H. Hartley. Effects of high-altitude exposure on submaximal endurance capacity of men. *J. Appl. Physiol.* 37: 895-898, 1974.

41. Margaria, R., P. Cerretelli, P.E. Di Prampero, C. Massari, and G. Torelli. Kinetics and mechanism of oxygen debt contraction in man. *J. Appl. Physiol.* 18: 371-377, 1963.

42. Masson, R.G. and S. Lahiri. Chemical control of ventilation during hypoxic exercise. *Resp. Physiol.* 22: 241-262, 1974.

43. McManus, B.M., S.M. Horvath, N. Bolduan, and J.C. Miller. Metabolic and cardiorespiratory responses to long-term work under hypoxic conditions. *J. Appl. Physiol.* 36: 177-182, 1974.

44. Naimark, A., N.L. Jones, and S. Lal. The effect of hypoxia on gas exchange and arterial lactate and pyruvate concentration during moderate exercise in man. *Clin. Sci.* 28:1-13, 1965.

45. Pugh, L.G.C.E. Cardiac output in muscular exercise at 5,800 m (19,000 ft). *J. Appl. Physiol.* 19: 441-447, 1964.

46. Pugh, L.G.C.E. Athletes at altitude. *J. Physiol. Lond.* 192: 619-646, 1967.

47. Pugh, L.G.C.E., M.B. Gill, S. Lahiri, J.S. Milledge, M.P. Ward, and J.B. West. Muscular exercise at great altitudes. *J. Appl. Physiol.* 19: 431-440, 1964.

48. Raynaud, J., J.P. Martineaud, J. Bordachar, M.C. Tillous, and J. Durand. Oxygen deficit and debt in submaximal exercise at sea level and high altitude. *J. Appl. Physiol.* 37: 43-48, 1974.

49. Reeves, J.T. and F. Daoud. Increased alveolar-arterial oxygen gradients during treadmill walking at simulated high altitude. *Adv. Cardiol.* 5: 41-48, 1970.

50. Roskamm, H., F. Landry, L. Samek, M. Schlager, H. Weidemann, and H. Reindell. Effects of a standardized ergometer training program at three different altitudes. *J. Appl. Physiol.* 27: 840-847, 1969.

51. Saltin, B., R.F. Grover, G.C. Blomquist, L.H. Hartley, and R.L. Johnson, Jr. Maximal oxygen uptake and cardiac output after 2 weeks at 4,300 m. *J. Appl. Physiol.* 25: 400-409, 1968.

52. Scheen, A., F. Pirnay, J. Juchmès, and A. Cession-Fossion. Influence du blocage des recepteurs B-adrénergiques sur l'hyperlactacidémie d'exercice in hypoxie. *Arch. Internat. Physiol. Biochim.* 82: 737-761, 1974.

53. Sime, F., D. Peñaloza, L. Ruiz, N. Gonzales, E. Corvarrubias, and R. Postigo. Hypoxemia, pulmonary hypertension, and low cardiac output in newcomers at low altitude. *J. Appl. Physiol.* 36: 561-565, 1974.

54. Soni, J., I. Chavez Rivera, and F. Mendoza. The effects of altitude on non-acclimatized athletes during effort. In: *The international symposium on the effects of altitude on physical performance.* Edited by R.F. Goddard. Albuquerque, N.M.: The Athletic Institute, 1966, pp. 76-79.

55. Stenberg, J., B. Ekblom, and R. Messin. Hemodynamic response to work at simulated altitude, 4,000 m. *J. Appl. Physiol.* 21: 1589-1594, 1966.

56. Velasquez, T. Response to physical activity during adaptation to altitude. In: *The physiological effects of high altitude.* Edited by W.H. Weihe. Oxford: Pergamon, 1964, pp. 289-300.

57. Vogel, J.A., J.E. Hansen, and C.W. Harris. Cardiovascular responses in man during exhaustive work at sea level and high altitude. *J. App. Physiol.* 23: 531-539, 1967.

58. Vogel, J.A., L.H. Hartley, and J.C. Cruz. Cardiac output during exercise in altitude natives at sea level and high altitude. *J. Appl. Physiol.* 36: 173-176, 1974.

59. Vogel, J.A., L.H. Hartley, J.C. Cruz, and R.P. Hogan. Cardiac output during exercise in sea-level residents at sea level and high altitude. *J. Appl. Physiol.* 36: 169-172, 1974.

60. Wells, J.G., B. Balke, and D.D. Van Fossen. Lactic acid accumulation during work. A suggested standardization of work classification. *J. Appl. Physiol.* 10: 51-55, 1957.

SLEEP HYPOXEMIA AT ALTITUDE: ITS RELATIONSHIP TO ACUTE MOUNTAIN SICKNESS AND VENTILATORY RESPONSIVENESS TO HYPOXIA AND HYPERCAPNIA

A. C. P. Powles
J. R. Sutton

McMaster University
Hamilton, Ontario, Canada

G. W. Gray

Defence and Civil Institute of Environmental Medicine
Toronto, Ontario, Canada

A. L. Mansell
M. McFadden

University of Toronto
Toronto, Ontario, Canada

C. S. Houston

The University of Vermont
Burlington, Vermont

Work performed at McMaster University
and Arctic Institute of North America
Yukon, Canada

INTRODUCTION

During sleep, ventilatory responsiveness to hypoxia is maintained but ventilatory responsiveness to hypercapnia is decreased (4,7). At sea level, the resulting sleep hypoventilation causes small increases in $PaCO_2$ and decreases in PaO_2, but no arterial desaturation (4,18), except very transiently during the initial phase of eye movement in REM sleep (1). At high altitude, such hypoventilation might be expected to produce considerable arterial desaturation, as above 4,000 m, subjects will usually be on the steep part of the oxyhemoglobin dissociation curve. We have previously shown an association between the severity of acute mountain sickness (AMS) and impairment of pulmonary gas exchange (20), which would aggravate the degree of arterial desaturation at any given level of alveolar ventilation. It was also observed that the slope of Phase III of the nitrogen washout curve was increased most in those subjects with the worst AMS and most deterioration in gas exchange (21). We postulated that sleep hypoxemia would be more marked in those persons with more severe symptoms of acute mountain sickness. Further, we considered that the degree of sleep hypoxemia would be related to the ventilatory response of hypoxia, since this has been shown to be related to the severity of AMS (9).

The mechanism by which hypoxemia causes altitude illness is unclear; cerebral edema has been found at autopsy in patients dying of high altitude illness (8), and vasogenic and cytotoxic theories have been put forward (5,22). In milder cases of AMS, alteration of mental function may be produced by interference with neurotransmission, particularly catecholamine release (3). Such alteration in catecholamine release may perpetuate the hypoxemia, as studies using α-blockade have demonstrated a much attenuated discharge from the carotid body in response to hypoxemia (10). A reduced ventilatory response would then occur, as observed in those persons with AMS symptomatology (16). During the 1976 and 1977 High Altitude Physiology Studies Programme (HAPS) on Mount Logan, we took advantage of the advances in ear oximetry (19) to conduct studies of sleep hypoxemia at altitude.

METHODS

The 20 subjects were fit young men and women, aged 20 to 35, free of cardiorespiratory disease, except for one subject (B) who had mild bronchial asthma controlled with an aerosol bronchodilator. Ventilatory responses to hypoxia and hypercapnia were measured at 100 m above sea level prior to ascent to altitude. Hypercapnic ventilatory responsiveness (V_E/CO_2) was measured using the method of Read (13). The isocapnic mixed venous PCO_2 method (V_E/SaO_2) of Rebuck and Campbell (14) was used to measure hypoxic ventilatory responsiveness.

The logistics of ascent to altitude on Mount Logan allowed study of "acclimatized" and "non-acclimatized" groups. The "acclimatized" group was studied after four weeks' residence at 5,360 m and the "non-acclimatized" group within two days of arrival at 5,360 m, having been flown in to 3,280 m for four days' staging and then climbing to 5,360 m in eight days.

During the sleep studies, arterial oxygen saturation (SaO_2) was recorded from an ear oximeter (HP 47201A). The awake SaO_2 was determined as the average value with the subject supine prior to sleep. Sleep SaO_2 values were monitored continuously and the highest and lowest values recorded each five minutes. Subjects were observed during sleep for respiratory irregularity, snoring, changes in position, and for arousals (15). Extremely low values of SaO_2 due to breath-holding while shifting position or high values associated with coughing were excluded. In 1977, magnetometers were used to follow breathing pattern. In four subjects, indwelling radial artery catheters were used to obtain arterial blood gas measurements during sleep; PO_2, PCO_2, and pH were measured on an Instrumentation Laboratories Blood Gas Analyser, and arterial saturation was measured using an OSM1 cuvette oximeter.

Each subject showed considerable fluctuation in SaO_2 during sleep. For statistical purposes, a 'mean' sleep SaO_2 was derived for each subject by averaging the highest and lowest SaO_2 values recorded during each five-min period of sleep. Significance of changes in SaO_2 was sought using paired and non-paired t tests for individual and group data as appropriate.

Each day, symptoms of AMS were evaluated using a previously described questionnaire (20).

RESULTS

A considerable inter-individual range of awake and asleep SaO_2 values were observed in both the non-acclimatized and acclimatized groups (Tables 1 and 2). All subjects showed decreased arterial saturation during sleep, but there was no significant difference in the severity of sleep hypoxemia between the acclimatized and non-acclimatized groups (Table 3).

The severity of sleep hypoxemia was not related to the sea level ventilatory responses to either hypercapnia or hypoxia (Figs. 1 and 2). Marked fluctuation in SaO_2 occurred during sleep, and the magnetometer tracings showed that the fluctuations were associated with either periodic (Fig. 3) or irregular ventilation (Fig. 4). The mean SaO_2 during periodic or irregular breathing was not necessarily lower than that observed during regular breathing (Figs. 3 and 4). Periodicity of respiration, although not quantified, appeared more marked in those subjects with the highest ventilatory responses to CO_2 and was accentuated during snoring.

In four subjects studied with indwelling arterial lines, saturation values determined by analysis of arterial blood with an OSM1 cuvette oximeter were 1% to 5% higher than the SaO_2 reading of the ear oximeter at the time of drawing the arterial blood (Fig. 5). Similar differences were observed during exercise (23).

Symptoms of AMS were absent in all subjects studied after one month at 5,360 m.

Table 1. Asleep and Awake Arterial Oxygen Saturations
in Acclimatized Subjects at 5,360 m

Subject	Awake $SaO_2 \%$	Sleep[a] $SaO_2 \%$	Δ $SaO_2 \%$
A	87	73	14
B	79	57	22
C	79	64	15
D	62	42	20
E	81	63	18
F	74	72	2
G	74	60	14
H	75	74	1
I	69	67	2
J	72	71	1

[a] Mean of values recorded at five-min intervals.

Table 2. Awake and Asleep Arterial Oxygen Saturations
in Non-Acclimatized Subjects at 5,360 m

Subject	Awake $SaO_2 \%$	Sleep[a] $SaO_2 \%$	Δ $SaO_2 \%$
K	63	50	13
L	76	71	5
M	66	60	6
N	75	70	5
O	76	64	12
P	83	64	19
Q	83	82	1
R	78	63	15
S	76	57	19
T	78	63	25

[a] Mean of values recorded at five-min intervals.

Table 3. Awake and Asleep Arterial Oxygen Saturations in
Acclimatized and Non-Acclimatized Subjects at 5,360 m

	Awake $SaO_2 \%$	Sleep $SaO_2 \%$	P
Acclimatized	Mean = 75.2 ± S.D. = 6.9	Mean = 64.3 ± S.D. = 9.7	< 0.01
Non-Acclimatized	Mean = 75.4 ± S.D. = 6.4	Mean = 64.4 ± S.D. = 8.6	< 0.01

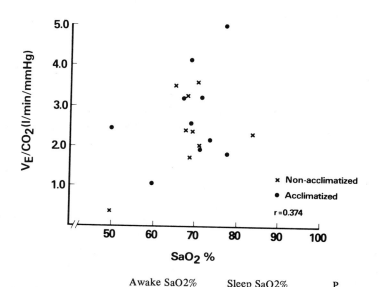

Awake SaO2%   Sleep SaO2%   P

Fig. 1. The lack of relationship between ventilatory responsiveness to hypercapnia at sea level (V_E/CO_2) and severity of sleep hypoxemia (SaO_2) at 5,360 m, whether acclimatized (•) or not (X).

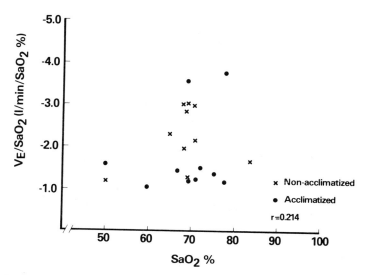

Fig. 2. There was no relationship between ventilatory responsiveness to isocapnic hypoxia (V_E/SaO_2) at sea level and sleep hypoxemia (SaO_2) at altitude (5,360 m).

In the non-acclimatized group, headache and insomnia were more frequently reported by those subjects who demonstrated the most severe sleep hypoxemia.

DISCUSSION

The results indicate that during sleep at 5,360 m, arterial desaturation is increased to a similar degree in subjects recently arrived and after one month at that altitude. Marked fluctuations in SaO_2 were associated with periodic or irregular respiration. No relationship was found between the ventilatory response to hypercapnia or hypoxemia measured at sea level and the severity of sleep hypoxemia at altitude; thus, these procedures appear to be of no predictive value.

Ear oximetry was used to monitor arterial O_2 saturation while awake and asleep. We have previously documented the accuracy of the ear oximetry used (19), but not in the range encountered during sleep at high altitudes. Comparison of the results obtained by direct measurement of arterial saturation using a cuvette oximeter with the reading of the ear oximeter at the time of drawing the arterial sample suggests that the ear oximeter reads 1% to 5% low in the range of saturations

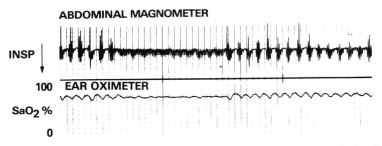

Fig. 3. An example of fluctuations in SaO_2 recorded by the ear oximeter during sleep associated with periodic ventilation, as indicated by the abdominal magnetometer trace.

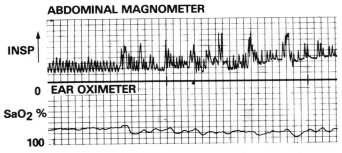

Fig. 4. Fluctuations in SaO_2 during sleep associated with irregular ventilation indicated by the abdominal magnetometer trace.

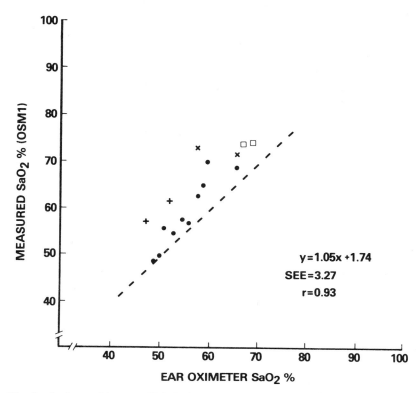

Fig. 5. In four subjects studied during sleep at 5,360 m, arterial saturation determined using an OSM1 cuvette oximeter on arterial blood was 1% to 5% higher than the reading of the ear oximeter at the time the arterial blood sample was drawn. (Shown by line of identity - - - - - which passes through the origin.)

encountered during sleep. This suggests that we may be overestimating the degree of sleep hypoxemia, although any such systematic error would not affect our interpretation of qualitative changes in SaO_2 or in inter-individual differences.

Periodicity of ventilation occurred during sleep in all our subjects, as has been observed previously (17). We found that periodicity of ventilation was more marked in those subjects with high ventilatory responses to CO_2. This would be in keeping with oscillation within the ventilatory feedback system, possibly from increased gain of the CO_2 sensor, as has been observed in patients with disease of the central nervous system (12) and in normal individuals at altitude (6). Our own observations that periodicity was increased when subjects snored, when coupled with those of Reite *et al.* (17), that oxygen administration terminated periodic respiration, suggest that a high CO_2 responsiveness may be permissive to periodic ventilation rather than causative. The degree of respiratory alkalosis may be important in that the exaggerated Bohr effect will allow considerable periods of apnea (2).

The association of more profound sleep hypoxemia with increased morning head-ache does not necessarily imply a causal role for the hypoxemia. Although the more profound hypoxemia may have been due to impairment of pulmonary gas exchange, with maintenance of arterial PCO_2 at awake levels, it may also have been due to diminished alveolar ventilation with increase in arterial PCO_2. We did not routinely monitor arterial PCO_2, but in those four subjects in whom arterial blood samples were obtained during sleep, decrease in SaO_2 was usually accompanied by increase in $PaCO_2$. We therefore cannot exclude an association of more severe hypoxemia with relative hypercapnia, which has been shown to exacerbate AMS (10). The occurrence of profound sleep hypoxemia after four weeks at altitude in two subjects who were asymptomatic suggests that alleviation of sleep hypoxemia is not essential for successful acclimatization to high altitude and implies additional adaptive mechanisms.

We did not confirm that relationship between the ventilatory response to hypoxia and the severity of AMS found by King and Robinson (9). The reason for this is unclear, but may be related to differences in ascent profile and techniques for measuring hypoxic ventilatory response. Their chamber study was of acute exposure, whereas the partial acclimatization afforded our subjects may have allowed adaptation that negated the effects of a low V_E/SaO_2.

SUMMARY

Arterial oxygen saturation was followed during sleep by ear oximetry in 20 subjects at 5,360 m. Hypoxemia was more pronounced during sleep than awake in all subjects ($P < 0.01$) and was not significantly different whether subjects were recently arrived or had been four weeks at that altitude. There was no relation between the degree of sleep hypoxemia observed at altitude and the ventilatory responsiveness to hypercapnia or hypoxia measured at sea level prior to ascent. Morning head-ache was more frequent and severe in those with more marked sleep hypoxemia on acute exposure to 5,360 m. However, severe sleep hypoxemia occurred in two individuals who were asymptomatic after four weeks at 5,360 m, suggesting that acclimatization occurs independently of sleep hypoxemia, which further suggests additional adaptive mechanisms for successful acclimatization.

ACKNOWLEDGMENTS

We wish to thank the following for their particular contributions to these studies: Mrs. E. Inman, Mrs. B. Weatherston, Miss S. Coons, P. Hackett, J. Kane, A. Menkis, J. Rigg, M. Robertson, P. Rondi, and W. Woodley; and the subjects for their willing cooperation and assistance.

The HAPS Programme is supported by NIH grant HL 14102-05, under the auspices of the Arctic Institute of North America, with logistic support from the Canadian Armed Forces.

The HP 47201A ear oximeter, which made these studies feasible, was lent by Hewlett-Packard.

REFERENCES

1. Aserinsky, E. Periodic respiratory pattern in conjunction with eye movement during sleep. *Science* 150:763-766, 1965.

2. Brown, H.W. and F. Plum. The neurologic basis of Cheyne-Stokes respiration. *Am. J. Med.* 30: 849-860, 1961.

3. Brown, R.M., W. Kehr, and A. Carlsson. Functional and biochemical aspects of catecholamine metabolism in brain under hypoxia. *Brain Res.* 85: 491-509, 1975.

4. Bulow, K. Respiration and wakefulness in man. *Acta Physiol. Scand.* Suppl. 209, 1963.

5. Fishman, R.A. Brain edema. *New Eng. J. Med.* 293: 706-711, 1975.

6. Forster, H.V., J.A. Dempsey, M.L. Birnbaum, W.G. Reddan, J. Thoden, R.F. Grover, and J. Rankin. Effect of chronic exposure to hypoxia on ventilatory response to CO_2 and hypoxia. *J. Appl. Physiol.* 31: 586-592, 1971.

7. Honda, Y. and T. Natsui. Effect of sleep on ventilatory response to CO_2 in severe hypoxia. *Resp. Physiol.* 3: 220-228, 1967.

8. Houston, C.S. and J. Dickinson. Cerebral forms of high altitude illness. *Lancet* 2:758-761, 1975.

9. King, A.B. and S. M. Robinson. Ventilation response to hypoxia and acute mountain sickness. *Aerospace Med.* 43: 419-421, 1972.

10. Maher, J.T., A. Cymerman, J. Reeves, J. Cruz, J. Denniston, and R. Grover. Acute mountain sickness: increased severity in eucapnic hypoxia. *Aviat. Space Environ. Med.* 46: 826-829, 1975.

11. Meyer, J.R., J.V. Weil, and J.T. Reeves. Sympathetic control of carotid body function. *Fed. Proc.* 32: 386, 1973 (Abstr.).

12. Plum, F. Neural mechanisms of abnormal respiration in humans. *Arch. Neurol.* 3: 484-487, 1960.

13. Read, D.J.C. A clinical method for assessing the ventilatory response to carbon dioxide. *Aust. Ann. Med.* 16: 20-32, 1967.

14. Rebuck, A.S. and E.J.M. Campbell. A clinical method for assessing the ventilatory response to hypoxia. *Am. Rev. Resp. Dis.* 109: 345-350, 1974.

15. Rechtschaffen, A. and A. Kales, eds. *A manual of standardised terminology, techniques and scoring system for sleep stages of human subjects.* U.S. Dept. of Health, Education and Welfare, Bethesda, Md. National Institute of Health Publication 204, 1968.

16. Reed, D.J. and R. H. Kellogg. Changes in respiratory response to CO_2 during natural sleep at sea level and at altitude. *J. Appl. Physiol.* 13: 325-330, 1958.

17. Reite, M., D. Jackson, R.L. Cahoon, and J.V. Weil. Sleep physiology at high altitude. *Electroenceph. Clin. Neurophysiol.* 38: 463-471, 1975.

18. Robin, E.D., R.D. Whaley, C.H. Crump, and D. M. Travis. Alveolar gas tensions, pulmonary ventilation and blood pH during physiologic sleep in normal subjects. *J. Clin. Invest.* 37: 981-989, 1958.

19. Saunders, N.A., A.C.P. Powles, and A.S. Rebuck. Ear oximetry: accuracy and practicality in the assessment of arterial oxygenation. *Am. Rev. Resp. Dis.* 113: 745-749, 1976.

20. Sutton, J.R., A.C. Bryan, G.W. Gray, E.S. Horton, A.S. Rebuck, W. Woodley, D.I. Rennie, and C.S. Houston. Pulmonary gas exchange in acute mountain sickness. *Aviat. Space Environ. Med.* 47: 1032-1037, 1976.

21. Sutton, J.R., G. Gray, M. McFadden, A.C. Bryan, E.S. Horton, and C.S. Houston. Nitrogen washout studies in acute mountain sickness. *Aviat. Space Environ. Med.* 48: 108-110, 1977.

22. Sutton, J.R. and N. Lassen. Pathophysiology of acute mountain sickness and high altitude pulmonary oedema. (Unpublished, 1977.)

23. Sutton, J.R., A.C.P. Powles, G.W. Gray, J. Kane, A. Mansell, M. McFadden, M. Robertson, P. Rondi, and C.S. Houston. Arterial hypoxemia during maximum exercise at altitude. *Clin. Res.* 25: 673 (Abstr.).

VI
Summary

SUMMARY

Steven M. Horvath

Institute of Environmental Stress
University of California, Santa Barbara
Santa Barbara, California

The final session of the conference was designed to serve as an open forum — to provide an opportunity for attendees and speakers to air their views and to derive satisfaction for their intense concern for interdisciplinary research and its significance in clarifying man's responses and adaptation to his environment. Although each questioner and respondent was identifiable on the tapes, it was decided that providing an overview of the session would be more effective than to present the usual stilted question-and-answer format. The session was divided into two panels, one including only the keynote speakers and the other the various session chairmen. Despite attempts of the chair to maintain some degree of accord as to time and tenor of the remarks being made, it was evident that in interdisciplinary circles such restrictions could not be maintained, and free play resulted. In view of all these factors, the following represents the bias of the chairman of this session and does not necessarily reflect upon the audience or the panel members, whose views may be slighted by this summary.

Thermal stress was considered to be a most important and valuable tool to unravel the complexities of both immediate physiological responses and adaptative processes. It was apparent that much more research was required to clarify differences in response with respect to age, sex, race, and state of physical condition. The importance of considering individual differences in evaluating responses was discussed by a number of participants who inevitably had noted that one or more

of the individual subjects in their studies reacted differently from the mean alterations reported. Interestingly, similar comments were made by those investigators
studying the reaction of man to air pollutants. The noise in these data deserves
careful consideration. An interesting aspect, especially to those studying hot environments, was the impact of the circadian rhythm on the observed responses. It
was pointed out by several panelists that when a subject was tested in the afternoon
instead of the usual morning session, the responses were more intense, i.e. if the
study was an acclimatization one, the subjects appeared to have regressed in their
acclimatization. Since man is exposed to thermal stress at various times of the day
(not just at the convenience of the investigators), more attention should be given to
evaluation of circadian rhythm changes. Seasonal effects should also be considered.
Japanese studies have emphasized the need to make additional long-term experiments. There was also some concern expressed regarding the tendency to lump all
heat stresses into one compartment. This has resulted in considerable confusion in
interpretation, since physiological responses differ widely when man is exposed to
hot-dry vs. hot-wet ambients. A strong plea was made urging investigators to utilize
multiple stressors, since it is obvious that man is never being exposed to a single situation but to a complex series of stressors. We do not live in a single stress world;
potentiation, synergism, etc. must and could occur in certain situations.

The main goal of environmental research should not be the elimination of all
stresses but the discovery of the proper combination for the ultimate in individual
adaptation. In this sense, concern was expressed by some that environmental research may be hampered by the increased restrictions being placed on the use of
human subjects, even at levels of stressors found in our environment. However, it
was believed by most investigators that carefully designed studies with attention to
explaining the experimental procedures fully to subjects in order to obtain their informed consent would alleviate these concerns. The outlook for future research in
all areas of environmental physiology, utilizing stressors that occur naturally or are
man-induced (such as exercise or air and water pollutants) was considered to be
excellent. Not only do the basic problems remain unresolved even after many
decades of intense research activity, but new ones continue to be generated as investigators become more active in utilizing more complex biochemical and physiological
tools.

Some specific unanswered questions relating to environmental stress were raised
and discussed during these summary sessions. The descriptions of mechanisms for
active vasodilation in hot environments were considered and found wanting. The
nature of the interaction between vasodilation and vasoconstriction systems is a
matter of primary research concern. It was noted that clarification of the role of
various hormones in responding to thermal stressors offers an exciting opportunity
for investigators. Plasma volume shifts, during both initial heat stress and the process of acclimatization, need to be more completely evaluated; dilation of the skin
vessels during heat exposure is associated with large volume shifts. Reduction in
skin blood flow (and decreased venous compliance) may be most important for the

acclimatization process. Are circulatory changes more important than body temperature changes in providing for effective acclimatization? Is improvement in central venous pressure mainly responsible for the changes in stroke volume observed during acclimation? Or are other factors involved? Thermal regulatory models still hold fascination, but the difficulties of obtaining adequate data to fill the gaps presently occurring in such models suggest the necessity for investigators to consider the requirements of our model builders in their research activities.

Do we adapt or acclimatize to altitude? Do we need to impose a second stress before we can determine whether or not we are considering a restoration of function (partial or complete) as the primary effect on whether or not adaptation has really occurred? Can man adapt to altitude while asleep? Perhaps our attitude toward altitude needs modification; sea level is just another altitude. During man's evolutionary development, he had to adapt to a hyperoxic environment. Are we misconstruing altitude problems? Sojourners to altitude alter their performance but rarely attain complete compensation for altitude hypoxia; yet Andean natives perform as well at altitude as most people do at sea level.

The need to study aging individuals, both sexes, and ethnic groups was emphasized. It was noted that older individuals reponded differently to acute altitude exposure, and it was suggested that future studies not only consider this variable, but also take into account the problems of individual variabilities in response. It was also suggested that studies on prenatal organisms could provide valuable insight into our understanding of altitude physiology.

Variability also raised its head during discussions on air pollution research, only it now became sensitivity. The observation that approximately 10% to 15% of subjects (animal or human) were exquisitely sensitive to various air pollutants raised the question as to what distinguishes these individuals from the remaining population. No solution was offered as to methods for distinguishing between these two groups. An important question was raised by several groups in that we need to understand the physiological/medical consequences of exposing young children to polluted environments. Present studies, mostly on young male adults, provide only a crude index of what is happening. The past history of exposure may be most important in evaluating the results of later exposure. The lung continues to develop until the individual is approximately nine years old. Does pollution exposure during these early years determine the magnitude and direction of the response? What else should we be looking at in order to evaluate pollutant effects? The interplay among pollutants, drugs (like ethanol), and normally non-pathogenic organisms may markedly alter the sensitivity of exposed individuals.

Exercise! A man-induced stress. Man apparently prefers to be active at a relatively small percentage of his maximum capacity. Nonetheless, he is eager to learn of his potential and induces himself to be stressed to his maximum level of attainment. In this respect, as in all other stresses discussed by those attending the symposium, the need to normalize data with respect to percentage of capacity of the individual was considered a major problem. This is true with relation to sex, age, obesity, and general state of physical condition. The genetic contribution to successful performance and adaptative capability was discussed at length. Self-selection for various stressful conditions was deemed a factor of considerable importance in determining man's response to his environment. The usefulness of scaling responses to man's ability was recognized by all participants; therefore, utilizing the energy output/capacity ratio to determine the significance of change consequent to a stressor becomes a most important element in evaluating strain-stress responses.

What are the consequences of hard physical training in early life? Does this result in beneficial or deleterious effects on the individual's physiology in later life? Have we ignored motivational factors in assessing performance? It was indicated that inadequate attention has been given to this factor. The role of interspersing extraneous factors such as involvement of central nervous system functions into stress situations was suggested to be a matter of considerable importance. As an example, simple muscular fatigue resulting from finger ergographic tests could be overcome by inserting a period of simple activity such as mental arithmetic before repeating the test, resulting in improved performance. Another example is to perform a work bout with eyes closed or open. It's your guess as to the change in performance. Other indications that interventions through the central nervous system can modify physiological responses were provided, suggesting another modifier of stressor responses. As in all discussions of exercise physiology, fiber typing was discussed. Can genetic predisposition be modified by extensive training regimes? More attention needs to be given not just to fiber type distribution but also to how, in these differing fiber types, the functional capability of the bridges between actin and myosin determine the ability to perform. More experimental studies need to be conducted on long-term work. We can only approximate man's potential if we limit our studies to performance requiring minutes. Repeated long-term work periods of eight hours would be especially valuable. The influence of the circadian rhythm and hormonal interactions requires further clarification. As was true in other discussions, the most important factors in evaluating exercise stress were considered to be age, sex, ethnic origin, nutritional status, state of training, etc. As one member of the panel stated, "We may not have solved the questions regarding exercise physiology posed by investigators some 40 years ago." We have much to learn and more to study.

A number of additional major points were raised in this last morning of the conference. The first was concerned with the relationship of scientists to the general public. Political decision-making has become a prime responsibility of environmental scientists, because their observations have been and will continue to be utilized in setting standards for human performance under a variety of environmental stressors, both physiological and psychological. The second is concerned with adaptative potential — Have we reached our limits or do we have untapped resources that need identifying? It was further noted that the prevailing tendency to devote primary attention to the immediate and direct effects of a stressor may lead to misleading conclusions. We have not evaluated recovery adequately. It was noted that several recent investigations have shown that long-term residual effects occur. Some have persisted for up to one week. Are they present for longer periods and if so, how do they influence physiological responses to repeated stressful situations? A fourth concern expressed is the need to perform studies on an international scale. Equivalent studies should be conducted in different countries with cooperating scientists involved. Cross tolerance to stressors was minimally explored at this conference. Nonetheless, it was evident that states of physical conditioning and the conditions under which such was produced resulted in better, more rapid adaptation to heat or cold stress. Similar comments were made as to pollutant stressors in that exposure to sublethal levels of one contaminant resulted in tolerance to lethal doses of another pollutant. Cross adaptation has been only minimally evaluated. The final suggestion was to the effect that in many countries there is a need to conduct studies over several generations (e.g. the Japanese-American). Gradual changes in adaptative qualities apparently occur, and these have received only minimal attention.

An interesting sidelight to the scientific discussions at the conference was the expression of concern regarding training and developing of additional personnel in enviromental medicine and physiology. It was noted that the need for replacements for the diminishing group of environmentalists is not being met. There are only a few laboratories devoted exclusively to environmental research. It was pointed out that generally only one or at the most two individuals in standard academic departments are involved in such research, and frequently the interest of these is only tangential to the main thrust. The opportunity for training in the broad environmental field is consequently somewhat minimal. Apparently there is considerable graduate student interest in environmental research. It may be of some importance to readers to note that there was a strong recommendation by the Second Task Force on Research Planning in Environmental Science to the National Institute of Environmental Health Sciences to support multidisciplinary environmental health research training programs. There are many questions yet unanswered, and new problems are continually arising as man changes his environment. Of even greater significance, the increasing interest in environmental problems will bring new recruits to this exciting field of inquiry. The future for environmental medicine and environmental physiology seems assured.

Index